Still Only One Earth

Progress in the 40 Years Since the First UN Conference on the Environment

ISSUES IN ENVIRONMENTAL SCIENCE AND TECHNOLOGY

ISSUES IN ENVIRONMENTAL SCIENCE AND TECHNOLOGY

EDITORS: R.E. HESTER AND R.M. HARRISON

40
Still Only One Earth
Progress in the 40 Years Since the First UN Conference on the Environment

Issues in Environmental Science and Technology No. 40

Print ISBN: 978-1-78262-076-1
PDF eISBN: 978-1-78262-217-8
ISSN 1350-7583

A catalogue record for this book is available from the British Library

Published by The Royal Society of Chemistry,
Thomas Graham House, Science Park, Milton Road,
Cambridge CB4 0WF, UK

Registered Charity Number 207890

For further information see our web site at www.rsc.org

Printed and bound by CPI Group (UK) Ltd, Croydon, CR0 4YY

Preface

The first United Nations Conference on the Human Environment took place in Stockholm from 5–16 June 1972. It received huge attention internationally and produced a final declaration which "having considered the need for a common outlook and for common principles to inspire and guide the peoples of the world in the preservation and enhancement of the human environment" proclaimed 26 principles. These were carefully considered high-level objectives which have undoubtedly influenced policy on the environment internationally, but – in some aspects and by some UN members – have been substantially ignored. Noting that this is the 40th volume of *Issues in Environmental Science and Technology*, we considered it appropriate to look back over the past 40 years to evaluate progress in environmental management in the context of the bold vision set out by the proclamation from the 1972 conference.

In more recent years, high-level United Nations activity on the environment has focussed very much on the atmosphere. Topics affecting the global atmosphere, and therefore requiring major international agreements, have been depletion of stratospheric ozone (the ozone hole) and global warming. In the first chapter, Martyn Chipperfield of the University of Leeds outlines progress in relation to the former topic, explaining the evidence for depletion of ozone in the Antarctic and at other latitudes, reviewing the international actions on regulation and control, and considering the future outlook for stratospheric ozone. Regarding greenhouse gases, John Sottong, Mark Broomfield, Joanna MacCarthy, Anne Misra, Glen Thistlethwaite and John Watterson of Ricardo-AEA in the second chapter provide a perspective on climate change, emissions and atmospheric concentrations of major greenhouse gases, and the related international policy actions and challenges. These two chapters provide a distinct contrast in that on the one hand the Montreal Protocol and subsequent international agreements have led to major action which is already showing benefits for the stratosphere, while on the other hand progress has been much slower in relation to the

Issues in Environmental Science and Technology No. 40
Still Only One Earth: Progress in the 40 Years Since the First UN Conference on the Environment
Edited by R.E. Hester and R.M. Harrison
© The Royal Society of Chemistry 2015
Published by the Royal Society of Chemistry, www.rsc.org

mitigation of global warming through reductions in greenhouse gas emissions.

Local air quality has been a major issue throughout the past 40 years. In developed countries, huge progress had been made in improving air quality prior to 1972 but subsequent improvements in epidemiological methods showed clearly that there were still major adverse effects on human health. In the third chapter, on Trends in Local Air Quality, Roy Harrison, Francis Pope and Zongbo Shi of the University of Birmingham review progress since 1970, both in developed countries and in less-developed parts of the world, such as India and China, where air quality remains very poor and impacts on human health are substantial. One of the very controversial pollution issues in 1972 was that of the use of lead as a motor fuel additive which led to substantial local air pollution issues, which have now been resolved in most parts of the world through the cessation of use of lead additives. In the fourth chapter, Robert Mason of the University of Connecticut examines the environmental presence and behaviour of two toxic metals: lead and mercury. Mercury was a major source of concern as early as the 1950s through contamination of the marine environment and the toxicity of contaminated fish. Mercury pollution remains a concern, although now more in relation to emissions to the atmosphere from sources such as coal combustion. The chapter makes it clear how regulation has tightened substantially over the period considered.

Some of the most recognised environmental problems at the time of the 1972 Stockholm Conference related to the presence of persistent organic pollutants, such as the organochlorine pesticides, as residues in the environment. The phenomenon of eggshell thinning and chick mortality in raptors and oceanic birds was brought to public attention by Rachel Carson's classic book *Silent Spring* published in 1962. In the fifth chapter, Mohamed Abdallah of the Universities of Birmingham and Assiut describes the most important persistent organic pollutants, their behaviour in the environment, temporal trends and future scenarios. This is clearly an area in which international action has been quite effective, but because of the very long environmental lifetimes of the compounds, many problems still remain. In the sixth chapter, Shane Snyder and Tarun Anumol of the University of Arizona consider emerging chemical contaminants. This classification includes both new industrial chemicals and substances which have been in use for many years but for which problems have only recently been recognised. The chapter considers, in particular, pharmaceutical and personal care products, perfluorinated compounds and endocrine disrupting chemicals. The importance of ever-improving analytical techniques is highlighted and consideration given to the implications for water sustainability.

One of the other major changes over the past 40 years has related to the management of solid waste. From a situation in which solid waste was seen as material to dispose of (and forget), the emphasis has now changed to re-use, recycling and waste-derived products. In the seventh chapter, Ian

Williams of the University of Southampton explains the trends in waste management and examines some of the current practices designed to make optimal use of materials previously considered simply as waste for disposal. In the final chapter by David Taylor of WCA, progress in the management of effluent discharges to the aquatic environment is considered. In the 1970s, the quantities of effluents discharged to water courses frequently far exceeded the natural purification capacity of the waters; this chapter explores some of the associated problems and the solutions that have been arrived at through improved effluent management.

In its totality, this volume tracks much of the progress in management and enhancement of the human environment over the past 40 years. It is a very mixed story, with many successes but also some notable failures. There are undoubtedly lessons for the future. We are grateful to our distinguished group of authors for their contributions and commend the volume to both students and practitioners in environmental science, engineering and policy as a valuable record of the progress made in many of the most important areas of environmental pollution, and as a key reference on environmental management.

Ronald E. Hester
Roy M. Harrison

Contents

Issues in Environmental Science and Technology No. 40
Still Only One Earth: Progress in the 40 Years Since the First UN Conference on the Environment
Edited by R.E. Hester and R.M. Harrison
© The Royal Society of Chemistry 2015
Published by the Royal Society of Chemistry, www.rsc.org

Global Atmosphere – Greenhouse Gases   **34**
John Sottong, Mark Broomfield, Joanna MacCarthy, Anne Misra,
Glen Thistlethwaite and John Watterson

Trends in Local Air Quality 1970–2014   **58**
Roy M. Harrison, Francis D. Pope and Zongbo Shi

Mercury and Lead   107

Robert P. Mason

Persistent Organic Pollutants   150

Mohamed Abou-Elwafa Abdallah

Emerging Chemical Contaminants: How Chemical Development   187
Outpaces Impact Assessment

Shane A. Snyder and Tarun Anumol

A Change of Emphasis: Waste to Resource Management   207

I. D. Williams

Editors

Ronald E. Hester, BSc, DSc (London), PhD (Cornell), FRSC, CChem

Ronald E. Hester is now Emeritus Professor of Chemistry in the University of York. He was for short periods a research fellow in Cambridge and an assistant professor at Cornell before being appointed to a lectureship in chemistry in York in 1965. He was a full professor in York from 1983 to 2001. His more than 300 publications are mainly in the area of vibrational spectroscopy, latterly focusing on time-resolved studies of photoreaction intermediates and on biomolecular systems in solution. He is active in environmental chemistry and is a founder member and former chairman of the Environment Group of the Royal Society of Chemistry and editor of 'Industry and the Environment in Perspective' (RSC, 1983) and 'Understanding Our Environment' (RSC, 1986). As a member of the Council of the UK Science and Engineering Research Council and several of its sub-committees, panels and boards, he has been heavily involved in national science policy and administration. He was, from 1991 to 1993, a member of the UK Department of the Environment Advisory Committee on Hazardous Substances and from 1995 to 2000 was a member of the Publications and Information Board of the Royal Society of Chemistry.

Roy M. Harrison, BSc, PhD, DSc (Birmingham), FRSC, CChem, FRMetS, Hon MFPH, Hon FFOM, Hon MCIEH

Roy M. Harrison is Queen Elizabeth II Birmingham Centenary Professor of Environmental Health in the University of Birmingham. He was previously Lecturer in Environmental Sciences at the University of Lancaster and Reader and Director of the Institute of Aerosol Science at the University of Essex. His more than 400 publications are mainly in the field of environmental chemistry, although his current work includes studies of human health impacts of atmospheric pollutants as well as research into the chemistry of pollution phenomena. He is a past Chairman of the Environment Group of the Royal Society of Chemistry for whom he edited 'Pollution: Causes, Effects and Control' (RSC, 1983;

Fifth Edition 2014). He has also edited "An Introduction to Pollution Science", RSC, 2006 and "Principles of Environmental Chemistry", RSC, 2007. He has a close interest in scientific and policy aspects of air pollution, having been Chairman of the Department of Environment Quality of Urban Air Review Group and the DETR Atmospheric Particles Expert Group. He is currently a member of the DEFRA Air Quality Expert Group, the Department of Health Committee on the Medical Effects of Air Pollutants, and Committee on Toxicity.

List of Contributors

Mohamed Abou-Elwafa Abdallah, Division of Environmental Health & Risk Management, University of Birmingham, Edgbaston, Birmingham, B15 2TT, UK and Department of Analytical Chemistry, Faculty of Pharmacy, Assiut University, 71526 Assiut, Egypt. Email: M.Abdallah@bham.ac.uk

Tarun Anumol, Chemical and Environmental Engineering, University of Arizona, Tucson, Arizona 85721, USA.

Mark Broomfield, 1 Ricardo-AEA, Gemini Building, Fermi Avenue, Harwell, OX11 0QR, UK.

Martyn P. Chipperfield, School of Earth and Environment, University of Leeds, Leeds, UK. Email: M.Chipperfield@leeds.ac.uk

Joanna MacCarthy, 1 Ricardo-AEA, Gemini Building, Fermi Avenue, Harwell, OX11 0QR, UK.

Roy M. Harrison, Division of Environmental Health and Risk Management, School of Geography, Earth and Environmental Sciences, University of Birmingham, Edgbaston, Birmingham, B15 2TT, UK and Department of Environmental Sciences/Center of Excellence in Environmental Studies, King Abdulaziz University, PO Box 80203, Jeddah, 21589, Saudi Arabia.

Robert P. Mason, Departments of Marine Sciences and Chemistry, University of Connecticut, Groton, CT 06340, USA. Email: robert.mason@uconn.edu

Anne Misran, 1 Ricardo-AEA, Gemini Building, Fermi Avenue, Harwell, OX11 0QR, UK.

Francis D. Pope, Division of Environmental Health and Risk Management, School of Geography, Earth and Environmental Sciences, University of Birmingham, Edgbaston, Birmingham, B15 2TT, UK.

Zongbo Shi, Division of Environmental Health and Risk Management, School of Geography, Earth and Environmental Sciences, University of Birmingham, Edgbaston, Birmingham, B15 2TT, UK.

Shane A. Snyder, Chemical and Environmental Engineering, University of Arizona, Tucson, Arizona 85721, USA. Email: snyders2@email.arizona.edu

John Sottong, 1 Ricardo-AEA, Gemini Building, Fermi Avenue, Harwell, OX11 0QR, UK. Email: john.sottong@ricardo-aea.com

David Taylor, WCA, Brunel House, Volunteer Way, Faringdon, Oxon, SN7 7YR, UK. Email: david.taylor@wca-consulting.com

Glen Thistlethwaite, 1 Ricardo-AEA, Gemini Building, Fermi Avenue, Harwell, OX11 0QR, UK.

John Watterson, 1 Ricardo-AEA, Gemini Building, Fermi Avenue, Harwell, OX11 0QR, UK.

Ian D. Williams, Centre for Environmental Sciences, Faculty of Engineering and the Environment, University of Southampton, Highfield, Southampton, Hampshire, SO17 1BJ, UK. Email: idw@soton.ac.uk

Global Atmosphere – The Antarctic Ozone Hole

MARTYN P. CHIPPERFIELD

ABSTRACT

The Antarctic ozone hole has been one of the major environmental issues of the past few decades. Its discovery in the mid 1980s was completely unexpected and prompted intense research activities to determine its cause. That cause was, ultimately, the emission of chlorine and bromine-containing compounds by human activity. In parallel with the discovery of the Antarctic ozone hole, an international agreement to limit the emission of ozone depleting substances, the Montreal Protocol, was signed. This chapter reviews the discovery and scientific cause of the Antarctic ozone hole. The current situation is described and the timescale for the recovery of the ozone layer is discussed.

1 Introduction

Ozone (O_3) is a vitally important component of the Earth's atmosphere. About 90% of atmospheric ozone resides in the stratosphere where it filters out harmful ultraviolet radiation and stops wavelengths shorter than about 300 nm from reaching the surface.[1] The remaining 10% resides in the troposphere where it is a photochemical pollutant.[2] Concern over the depletion of the stratospheric ozone layer first focussed on the detrimental effects of increased surface UV on humans, animals and plants. As ozone also absorbs

Issues in Environmental Science and Technology No. 40
Still Only One Earth: Progress in the 40 Years Since the First UN Conference on the Environment
Edited by R.E. Hester and R.M. Harrison
© The Royal Society of Chemistry 2015
Published by the Royal Society of Chemistry, www.rsc.org

in the infrared (IR), the stratospheric ozone layer also plays a key role in the Earth's climate balance. This role of ozone in climate has become increasingly apparent over the past decade or so and is a second principal motivation for ongoing research into stratospheric ozone.

Ozone in the stratosphere is produced naturally by the photolysis of O_2 molecules by short wavelength UV radiation ($\lambda < 242$ nm).[3]

$$O_2 + h\nu \rightarrow 2O \tag{1}$$

Atomic oxygen rapidly recombines with an O_2 molecule to form O_3, and $O + O_3$ are treated together in stratospheric chemistry as the 'odd oxygen' family. While there is only one source of ozone, there are many reactions which remove it *via* catalytic cycles (see Section 3.2). These cycles can comprise species which are present naturally in the stratosphere (*e.g.* hydrogen and nitrogen compounds) or species which are largely present through human activities. Overall there is a dynamic equilibrium which maintains the balance of stratospheric ozone through competing loss and destruction cycles. Production by sunlight is favoured at lower latitudes while loss can occur at higher latitudes following transport of ozone *via* the stratospheric Brewer–Dobson circulation. Given that the natural production of ozone varies only slightly (*e.g.* over the 11 year solar cycle)[4,5] human perturbations to the ozone layer have occurred by changes to the rate of catalytic loss. As described below, the ultimate cause of the Antarctic ozone hole has been the large increase in stratospheric bromine and chlorine in the period from 1960 onwards due to increased use of gases such as chlorofluorocarbons (CFCs).

Many thousands of research papers have been published on the topic of stratospheric ozone depletion and the Antarctic hole over the past three decades. A number of comprehensive reviews have already been published. Solomon[6] reviewed stratospheric ozone depletion in general, with a concise summary of the science of the Antarctic ozone hole as of 1999. More recently Müller and coauthors[7] have provided a comprehensive summary of stratospheric ozone in a Royal Society of Chemistry publication. That book has separate detailed chapters dealing with varied aspects relevant to the chemistry and dynamics of the Antarctic ozone hole. The state of knowledge in stratospheric ozone research is assessed every 4 years in the World Meteorological Organisation/United Nations Environment Programme (WMO/UNEP) assessments.[8] These publications provide a detailed summary of the most recent literature in the field. An extensive account of the activities of policy makers involved in the signing of the Montreal Protocol is given in Anderson and Sarma.[9]

The aim of this review is to provide a concise summary of the Antarctic ozone hole from its discovery, the determination of its cause, the policy action taken to solve the problem and the outlook for the future. The aim is to provide the reader with an up-to-date overview of the subject and he or she is referred to the more detailed reviews cited above for further information. Section 2 discusses observations of Antarctic ozone during the period of the springtime depletion. The processes which lead to this depletion are described in Section 3. Although not the focus of this review, Section 4 briefly gives a summary of ozone depletion which has occurred at other latitudes, and in particular the Arctic. Section 5 summarises the Montreal Protocol, the international agreement to limit the emission of ozone-depleting substances (ODSs). Finally, Section 6 describes some areas of current research in stratospheric ozone and discusses what may happen to the ozone hole in the future.

2 Observations of Antarctic Ozone

The Antarctic ozone hole is one of the best known geophysical phenomena among both scientists and the general public. False-colour satellite images of total column ozone at the end of Austral spring are a classic and ubiquitous image in environmental science. An example is given in Figure 1 for October 9, 2006. By this day in the season the ozone column over the Antarctic had been reduced to <90 Dobson Units (DU) with the area below 220 DU, which is commonly taken to indicate the size of the hole, extending over the entire Antarctic continent.

The Antarctic ozone hole was discovered in the mid 1980s when a sharp downward trend in total ozone column over Halley was observed by Farman, Gardiner and Shanklin.[11] Figure 2 shows the updated time series of Halley column ozone which dates back to the 1950s. Between 1960 and the mid-1990s the October monthly mean decreased by over 50%. Similar behaviour is seen in the satellite data of the October minimum column ozone and the ozone hole area. Since the mid 1990s the size and depth of the ozone hole, as determined by various metrics, has not increased. In recent years the observations in Figure 2 indicate a slight decrease in size with larger variability. This is consistent with recent decreases in ODSs (see Section 5) but it is not yet possible to determine with certainty that Antarctic ozone has started to recover[12] (see Section 6).

The ozone hole develops after the end of Austral polar night and the rapid ozone loss occurs when sunlight returns to the polar region. Figure 3a shows the time series of daily column ozone values at Halley for 1957–1973 and 1990–2009. This shows that in contrast to the earlier period, in 1990–2009 column ozone decreased strongly from late August through to early October. In late winter at Halley, temperatures at 100 hPa (~ 15 km) are cold (see Figure 3b) but increase through to December as sunlight warms the polar stratosphere. The delay in the warming in 1990–2009 compared to 1957–1973 is due to the role that ozone plays in absorbing sunlight and warming the stratosphere. The decreased ozone in the latter period delays the warming of the stratosphere – an example of ozone chemistry-climate feedbacks (see Section 6).

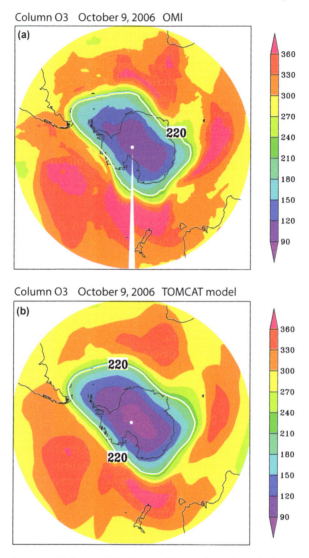

Figure 1 Total ozone (Dobson Units) over the southern hemisphere on October 9, 2006 as: (a) observed by the Ozone Monitoring Instrument (OMI); and (b) simulated by the TOMCAT 3-D chemical transport model[10] at moderate resolution ($5° \times 5°$). The 220 DU contour is commonly used as the definition for the edge of the ozone hole and is indicated in white.
(Plot courtesy of Sandip Dhomse, University of Leeds).

The depletion of ozone is not spread uniformly throughout the column but occurs in the lower stratosphere. Indeed, Figure 4 shows that in the Antarctic ozone hole by early October there has been complete ozone loss between about 13 and 21 km. Ozone at higher altitudes is not significantly depleted. The restriction of the rapid ozone loss to the lower stratosphere

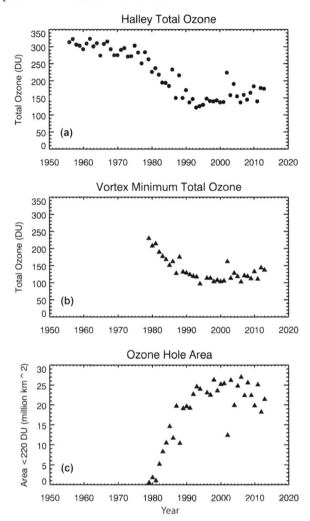

Figure 2 Long-term evolution of the Antarctic ozone hole through 2013: (a) the October monthly mean total at Halley (75.6°S, 26.6°W); (b) the minimum total ozone over Antarctica between September 21 and October 16 of each year; and (c) the mean area with total ozone less than 220 DU south of 45°S between September 7 and October 13 of each year.
(Data courtesy of J. D. Shanklin, British Antarctic Survey and P. Newman, NASA).

provides a natural limit to the depth the ozone hole and explains the levelling off in the observations of column ozone in Figure 2.

3 Causes of Antarctic Ozone Depletion

In the mid 1980s the appearance of the Antarctic ozone hole came as a complete surprise to the atmospheric science community. Based on the early

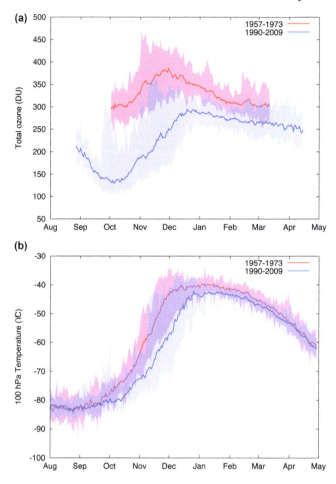

Figure 3 Measurements of: (a) total ozone; and (b) temperature at Halley for 1957–1973 and 1990–2009. The central line in each dataset is the mean for the period and the shaded area shows the complete range of observations for that day. The years 1990–2009 show lower column ozone and a delayed springtime warming.
(Data courtesy of J. D. Shanklin, British Antarctic Survey; plot courtesy L. Abraham, University of Cambridge). Reproduced with permission from Harris and Rex.[13]

work on CFCs in the 1970s,[14,15] it was expected that any ozone depletion due to these compounds would be modest and would occur in the upper stratosphere at around 40 km. The large loss of ozone in the high latitude lower stratosphere was not predicted by chemical models being used at that time.[16] Because of the surprise occurrence of the ozone hole numerous different theories involving different chemical and dynamical processes were proposed to explain it. A number of measurement campaigns were conducted starting with ground-based observations in 1986 and a larger

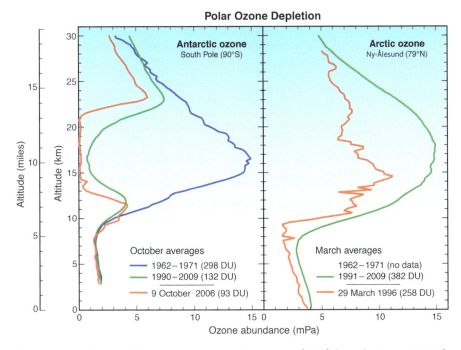

Figure 4 Vertical profiles of ozone partial pressure (mPa) based on ozonesonde
balloon measurements at the South Pole (90°S) and the Arctic station of
Ny-Ålesund (79°N). For the South Pole the figure shows the October means
for the periods 1962–1971 (blue) and 1990–2009 (green). Also shown is the
individual sonde on October 9, 2006 (red; see Figure 1 for column ozone
map). For Ny-Ålesund the figure shows the March average for 1991–2009
and the individual sonde of March 29, 1996.
Reproduced with permission from WMO 2010.[8]

aircraft-based campaign in 1987. These campaigns rapidly established that
the cause of the depletion was chlorine (and bromine)-catalysed ozone
loss.[17] However, the occurrence of high levels of ozone-destroying reactive
chlorine in the Antarctic spring was not expected and new heterogeneous
chemical processes needed to be invoked to explain this.[18] Overall, the
annual occurrence of the ozone hole is due to a remarkable combination of
dynamical and chemical processes which occur in the Antarctic winter/
spring lower stratosphere. These processes are described in this section. It
was very fortunate for society that the first large warning that we had for
global ozone depletion by anthropogenic emissions occurred in this specific
region of the atmosphere, rather than on a global scale.

3.1 Dynamical Preconditioning

In the 24 hour darkness of polar night, the Antarctic lower stratosphere cools
and a strong westerly circulation forms around the pole – the so-called polar
vortex (see Figure 5). This vortex isolates the high latitudes in a 'containment

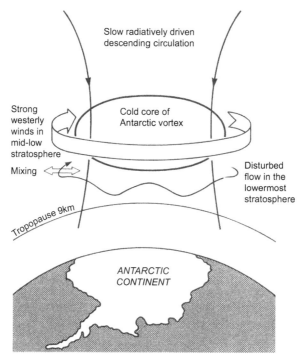

Figure 5   Schematic of the Antarctic polar vortex.
Adapted from SORG 1988.[19]

vessel' for the subsequent chemistry of ozone hole. Temperatures inside the wintertime Antarctic vortex can fall to around 190 K or less, making it the coldest location in the stratosphere outside of the tropical tropopause. The polar vortex persists until late spring, when warming due to solar heating causes the winter westerly circulation to revert to summer easterlies as the vortex breaks down.

3.2   Polar Stratospheric Clouds

The stratosphere is very dry due to the very cold temperatures (~ 190 K) which occur at the tropical tropopause and cause water in air parcels to condense out as they are transported upwards from the troposphere. Therefore, extremely cold temperatures are needed for clouds to form in the stratosphere. For ice clouds these temperatures will only occur in the polar regions or just above the cold tropopause at lower latitudes. Historically, sightings of 'mother of pearl' clouds in the polar regions have been reported since the 1880s[20] but their role in atmospheric chemistry only became apparent after the observation of the Antarctic ozone hole.

We now know that Polar Stratospheric Clouds (PSCs) can be composed of a number of particle types. In the lower stratosphere there exists a ubiquitous layer of liquid binary sulfate aerosols (H_2SO_4–H_2O). At cold temperatures (around 195 K) these aerosols grow and take up HNO_3 to form supercooled liquid ternary solutions (STS, H_2SO_4–H_2O–HNO_3).[21] At temperatures below about 195 K the formation of solid nitric acid trihydrate particles[22,23] becomes thermodynamically possible, although some degree of supersaturation is required. Finally, below the ice point ($\sim$188 K) solid ice particles can form. These liquid and solid particles provide surfaces for heterogeneous reactions,[18] the most important of which are:

$$N_2O_5 + H_2O \rightarrow 2HNO_3 \tag{2}$$

$$HCl + ClONO_2 \rightarrow Cl_2 + HNO_3 \tag{3}$$

$$ClONO_2 + H_2O \rightarrow HOCl + HNO_3 \tag{4}$$

$$HOCl + HCl \rightarrow Cl_2 + H_2O \tag{5}$$

$$BrONO_2 + H_2O \rightarrow HOBr + HNO_3 \tag{6}$$

The discovery that heterogeneous reactions can occur on the surface of PSCs was a key step in explaining the cause of the ozone hole. In particular, reactions (2)–(6) cause the conversion of stable reservoir chlorine species (HCl and $ClONO_2$) into more active forms. Solid PSC particles (NAT, ice) can grow large enough to sediment from the stratosphere and cause permanent removal of HNO_3 (denitrification)[23] and H_2O (dehydration).

PSCs in the atmosphere can be detected by a variety of *in situ* and remote observations.[24] In recent years such observations have been revolutionised by the launch in 2006 of the CALIPSO satellite with the CALIOP lidar onboard.[25,26] This active lidar technique has revealed the full three-dimensional structure of PSCs and can be used to classify the different type and composition of the particles. The instrument is still performing well and has accumulated a long time series of measurements. Figure 6 shows CALIPSO observations of the daily PSC area in the 8 Antarctic winters from 2006–2013. PSCs tend to occur from May to late September between about 12 and 25 km. There is very little interannual variability in the PSC occurrence, reflecting the regular nature of the Antarctic polar vortex. Figure 7 shows the mean distribution of four types of PSC over these 8 winters. Pure Supercooled Ternary Solution (STS) tends to be favoured early in the season with a smaller maximum in late September. The larger 'denitrifying' NAT particles occur preferentially at lower altitudes, but are always present as a mixture with liquid STS particles. Smaller NAT particles are also present in a mixture with STS and maybe solid ice.

The role of heterogenous PSC chemistry is illustrated in Figure 8 with satellite observations of chemical species in the Antarctic vortex of 2005.[27] In the middle of winter, at altitudes below about 25 km, PSCs cause the almost complete conversion of HCl and $ClONO_2$ to active forms. This active chlorine is revealed as ClO when sunlight returns to the polar region in September. The formation of PSCs also removes HNO_3 from the gas phase (which is measured by the satellite) either by condensation into PSCs or through sedimentation of solid particles (denitrification).

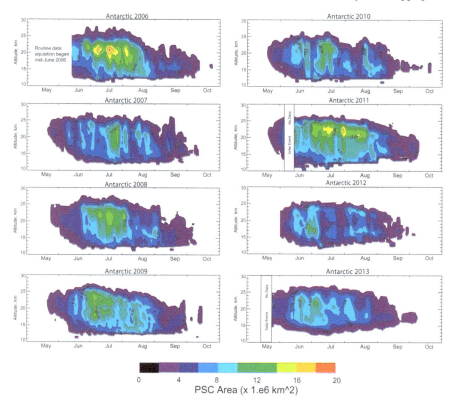

Figure 6   Daily time series of CALIPSO PSC area (km^2) for the eight Antarctic winters
from 2006 until 2013.
(Plot courtesy of Michael Pitts, NASA).

3.3   Catalytic Ozone Loss Cycles

In the early days of research into atmospheric chemistry it was a mystery
how the observed concentrations of stratospheric ozone could be maintained based on production from solar ultraviolet radiation and loss
through the direct chemical reaction of $O + O_3 \rightarrow 2O_2$. Although ozone is
present at levels of 1–10 parts per million by volume (ppmv) in the stratosphere, this is still a factor of 100 or more greater than the concentration of
chemical species which are able to destroy it. The answer to this puzzle lies
in the fact that reactive radical chemical species, present in concentrations
much lower than ozone, can destroy ozone through catalytic cycles. An
example is:

$$ClO + O \rightarrow Cl + O_2 \tag{7}$$

$$Cl + O_3 \rightarrow ClO + O_2 \tag{8}$$

$$\text{Net } O + O_3 \rightarrow 2O_2$$

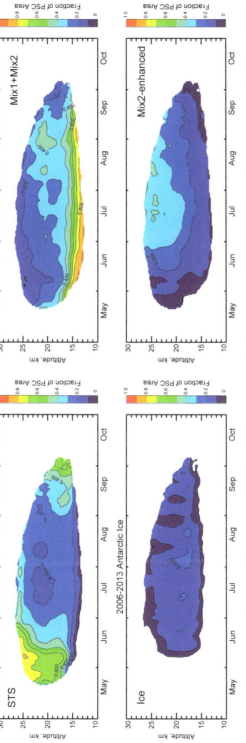

Figure 7 Relative contribution of four types of PSC (or PSC mixture) to the 8-year mean CALIPSO PSC area from data shown in Figure 6. The panels show the relative occurrence of STS, denitrifying NAT (and STS) mixture, ice and other NAT (and STS) mixtures. (Plot courtesy of Michael Pitts, NASA).

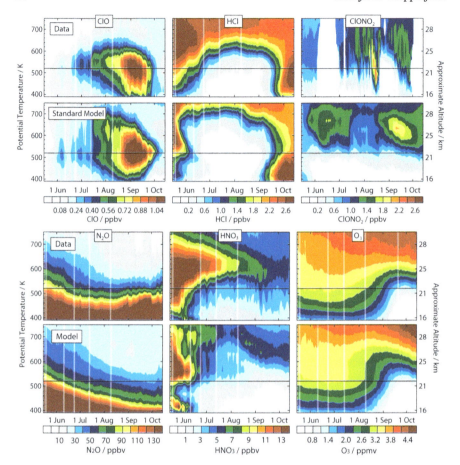

Figure 8 Time series over the 2005 Antarctic winter of vortex-averaged quantities calculated within the 1.6×10^{-4} s^{-1} contour of scaled potential vorticity as a function of potential temperature. (Top row) ClO and HCl data from Aura Microwave Limb Sounder (MLS) and ClONO$_2$ data from Atmospheric Chemistry Experiment Fourier Transform Spectrometer (ACE-FTS). Only daytime data are shown for ClO. ACE-FTS ClONO$_2$ data have been smoothed to a slightly greater degree to enhance the legibility of the plots, but the large gaps arising from the sparse sampling of the ACE-FTS measurements within the polar vortex at the beginning and end of the observation period have not been filled. The black horizontal line in each panel marks the 520 K level. (Second row) Corresponding TOMCAT/ SLIMCAT 3-D model results, sampled at the MLS measurement locations and times. (Third row) N$_2$O, HNO$_3$ and O$_3$ data from Aura MLS. (Fourth row) Corresponding TOMCAT/SLIMCAT 3-D model results. Reproduced with permission from Wiley from Santee *et al.*[27]

Note how the Cl and ClO species are reformed in this cycle and that the overall stoichiometry is that same as the direct reaction between O + O$_3$. In this way, chemical species containing hydrogen, nitrogen, chlorine and bromine are able to destroy significant amounts of stratospheric ozone.

However, during the 1980s our understanding of catalytic stratospheric ozone destruction cycles was far from complete. Cycles of the form given above, which catalyse the reaction $O + O_3 \rightarrow 2O_2$, are very slow in the lower stratosphere because of the low concentration of atomic oxygen there. Observations (see Section 2) showed that the Antarctic ozone depletion occurred in this region on a timescale of just a few weeks. This problem was resolved by the discovery of cycles involving ClO and BrO radicals which catalyse the reaction $2O_3 \rightarrow 3O_2$, and therefore do not depend on atomic oxygen. There are two main cycles which are responsible for the rapid loss in the Antarctic ozone hole. The first is based on the formation of the ClO dimer:[28]

$$ClO + ClO + M \rightarrow Cl_2O_2 + M \tag{9}$$

$$Cl_2O_2 + h\nu \rightarrow Cl + ClO_2 \tag{10}$$

$$ClO_2 + M \rightarrow Cl + O_2 + M \tag{11}$$

$$2(Cl + O_3 \rightarrow ClO + O_2) \tag{8}$$

$$\text{Net } 2O_3 \rightarrow 3O_2$$

The second involves the interaction of chlorine and bromine:[29]

$$ClO + BrO \rightarrow Cl + Br + O_2 \tag{12}$$

$$Cl + O_3 \rightarrow ClO + O_2 \tag{8}$$

$$Br + O_3 \rightarrow BrO + O_2 \tag{13}$$

$$\text{Net } 2O_3 \rightarrow 3O_2$$

A minor variant of this second cycle also occurs with the initial formation of BrCl ($BrO + ClO \rightarrow BrCl + O_2$ followed by BrCl photolysis). The ability of these cycles to quantitatively reproduce observed ozone loss is demonstrated in Section 3.4.

3.4 Modelling of Polar Ozone Depletion

Atmospheric numerical models are based on our best current understanding of the relevant physics and chemistry expressed in a mathematical form. In order to be realistic, models need a representation of all of the important chemical, physical and dynamical processes; a model cannot contain unknown processes. The atmospheric models available in the 1970s and 1980s famously failed to predict the occurrence of the Antarctic ozone hole. The lack of representation of heterogeneous PSC chemistry alone would have caused this failure, but the catalytic ClO dimer cycle was also unknown and not contained in the models. Models do now include a treatment of these and other processes relevant to the Antarctic ozone hole. Nevertheless, we should not forget the fact that our understanding of atmospheric science is incomplete and our best current models can only be as good as our current understanding.

Numerical models which are now used to simulate the Antarctic ozone hole can be summarised as:

- Chemical Box or Trajectory models. These models have a detailed representation of chemistry which occurs in a single air parcel. They contain all of the relevant gas-phase and heterogeneous chemical reactions and they are computationally cheap. They can be used to integrate the chemistry at a single point or along an air mass trajectory.
- Off-Line Three-Dimensional Chemical Transport Models (3-D CTMs). These models simulate the global atmospheric domain in an array of grid-boxes. They also contain a detailed description of stratospheric chemistry. Off-line models are driven by external meteorological fields of winds and temperature[30] which makes them computationally efficient and well suited to interpretation of past observations.
- Coupled Chemistry-Climate Models (CCMs). These 3-D models have similar chemical schemes to CTMs but calculate their own winds and temperatures. They are therefore much more computationally expensive but these are the only models which can be used for future predictions. These models will be able to capture the coupled interaction of atmospheric chemistry and climate: for example they can simulate the impact of changing ozone concentrations on atmospheric temperature (*e.g.* a feedback illustrated in Figure 3).

Chemical box models which employ the observed levels of ClO (1–2 ppbv) and BrO (5–10 pptv) in the Antarctic polar vortex are able to quantitatively reproduce the observed rates of ozone loss, giving us confidence in our understanding of the cycles described above. An example is given in Figure 9 from the work of Frieler *et al.*[31] With updated information on the kinetics of reaction (10) their model was able to reproduce the rates of O_3 loss derived from a Lagrangian analysis of Antarctic (and Arctic) ozonesonde observations.[32]

Figure 8 also includes results from a state-of-the-art 3-D CTM, TOMCAT/ SLIMCAT. This model has a detailed treatment of polar ozone chemistry including heterogeneous chemistry on PSCs and the relevant catalytic loss cycles.[33] The model generally captures the depletion of HCl in midwinter and the production of ClO. The model also captures the decrease in HNO_3 in midwinter below about 21 km (denitrification) and the lower stratosphere ozone loss in September. A comparison of column O_3 from the same CTM for early October 2006 is shown in Figure 1. This clearly shows that the model is able to reproduce the ozone hole. Although there are some differences in detail in the comparisons in Figure 8, for example $ClONO_2$, the comparisons show that we now generally have a good quantitative understanding of polar ozone depletion gained through the intensive research

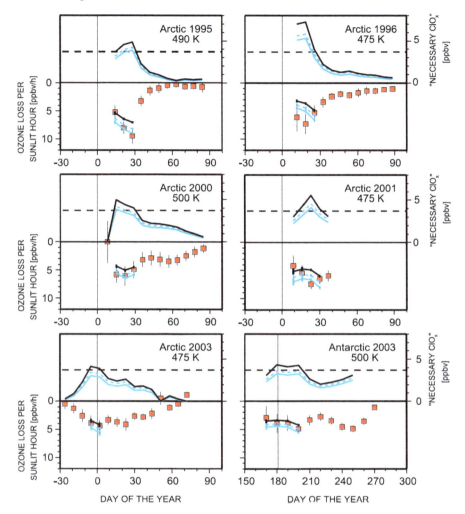

Figure 9 Chemical O$_3$ loss rate in the Arctic and Antarctic polar vortices based on the Match ozonesonde analyses (red boxes; error bars are 1σ uncertainty). The abundance of active chlorine (ClO$_x$) necessary to account for the measured O$_3$ loss ("necessary ClO$_x$") and the modelled O$_3$ loss assuming ClO$_x$ = 3.7 ppbv ("maximum possible ozone loss") are shown in the upper and lower parts of each plot, respectively. The dashed line in the upper part of each plot marks the level of 3.7 ppbv ClO$_x$. Maximum possible ozone loss is shown only for the time periods where nearly complete chlorine activation is likely to occur. Black lines: reference run (standard kinetics + 2-D model bromine); dashed blue lines: "new Cl$_2$O$_2$ kinetics" + 2-D model bromine; solid blue lines: "new kinetics" + bromine derived from atmospheric BrO measurements which include contribution from VSLS. Reproduced with permission from Wiley from Frieler *et al.*[31]

efforts since the mid 1980s. Nonetheless, we should still be careful not to repeat the mistake of the late 1970s of assuming that our models are complete and accurate in all situations.

4   Ozone Depletion at Other Latitudes

The Antarctic ozone hole is the largest manifestation of human impact on the ozone layer, and indeed one of the most dramatic impacts on the whole global environment. Ozone depletion has also been observed at other latitudes, though to a smaller extent.

4.1   Arctic

The dynamical and chemical processes described in Section 3 also occur in the Arctic, but to smaller extent. The difference in the behaviour of the two polar regions in winter and spring is driven by the difference in surface orography and its impact on stratospheric temperatures and dynamics. Northern mid-high latitudes are much more mountainous than the southern mid-latitudes which are largely ocean. This orography causes the Arctic winter polar vortex to be smaller and more disturbed. As a result winter polar temperatures in the Arctic are warmer than the Antarctic and also much more variable. Some 'warm' Arctic winters may see no PSC activity at all and therefore essentially zero ozone loss. In contrast, 'cold' Arctic winters may experience relatively large ozone depletion, but confined to a much smaller vortex than in the south.[34]

One of the largest episodes of Arctic ozone depletion occurred in the recent winter of 2010–2011.[35] Temperatures in this winter were unusually cold which led to extensive PSC activity and activation of chlorine in the Arctic polar vortex. Extensive ozone loss then ensued (see Figure 10). Although the Arctic loss was very large, this was not unexpected given the meteorology. Models which had been developed to model the large loss in the Antarctic (*e.g.* Figure 1) could reproduce the Arctic loss of 2010–2011 using the same chemistry.[36] Therefore, this winter turned out as a corroboration of our understanding. This recent winter did, however, show the ongoing potential for large ozone loss despite the reductions in stratospheric chlorine and bromine that have started to happen.

4.2   Mid-latitudes

Ozone depletion at mid latitudes (30°S–60°S and 30°N–60°N) is much smaller than in the polar regions and occurs all year round. The largest depletion observed was a reduction of around 8% in the north and 10% in the south.[8] This loss is also linked to the increase in atmospheric chlorine and bromine through human activities, though the activation of chlorine occurs on sulfate aerosols rather than PSCs. These aerosols are enhanced following large volcanic eruptions which reach the stratosphere. The most recent such eruption was Mt Pinatubo in June 1991 which was responsible for the lowest northern hemisphere mid-latitude ozone columns observed so far.[37]

4.3   Tropics

The ozone layer is naturally much thinner over the tropics (30°S–30°N) compared to mid and especially high latitudes. The observed trend in column ozone in the tropics is very small. Observations in the stratosphere do show evidence of a small downward trend but this is not apparent in the column, possibly due to compensating increases in tropospheric ozone.[38] This lack of large trend is consistent with our understanding of the tropics being a source region of ozone which is then transported to higher latitudes. However, the low ozone column and low average daytime solar zenith angle means that the Earth's surface in the tropics experiences large UV indices. Even in the absence of large ozone depletion, humans can experience a much larger UV dose in the tropics than in the Antarctic region in late October under the ozone hole.[39]

5 Regulation and Control: The Montreal Protocol

As summarized above, stratospheric ozone depletion, especially in the Antarctic ozone hole, is caused by chlorine and bromine radicals which are released from long-lived reservoirs. This section discusses the compounds responsible for the transport of chlorine and bromine to the stratosphere and the international agreement to limit production and release of the most damaging man-made examples.

5.1 Chlorine and Bromine Source Gases

Chlorine and bromine is carried to the stratosphere in the form of source gases. These are compounds which are emitted at the surface but have long atmospheric lifetimes meaning that they are transported to the stratosphere before they degrade.[40,41] The major natural source gas of stratospheric chlorine is CH_3Cl, with a tropopause mixing ratio of 0.6 ppbv. There are a range of other anthropogenic source gases and the use of these compounds resulted in increases of around a factor 6 in chlorine and a factor 2 in bromine compared to natural levels by the late 1990s.

The major anthropogenic chlorine source gases are the CFCs, CF_2Cl_2 and $CFCl_3$. Their atmospheric abundances grew along with their production at $\sim 3\%$ per year during much of the 1970s and 1980s due to their widespread use as aerosol propellants, refrigerants, foam blowing agents, *etc.* Neither compound reacts in the troposphere, making them particularly useful, but at high enough altitudes they can be broken down by UV radiation and by reaction with $O(^1D)$. For example

$$CF_2Cl_2 + h\nu \rightarrow 2Cl + \text{other products}$$

$$CFCl_3 + h\nu \rightarrow 3Cl + \text{other products}$$

Notice that both molecules carry more than one Cl to the stratosphere. The time to reach the sufficient altitudes for destruction leads to very long lifetimes, between 50 and 100 years for these compounds. In effect there is a wide range of compounds which carry ozone-depleting halogens to the stratosphere but they can be classified into a few groups which are summarized in Table 1 with some specific compounds in Table 2.

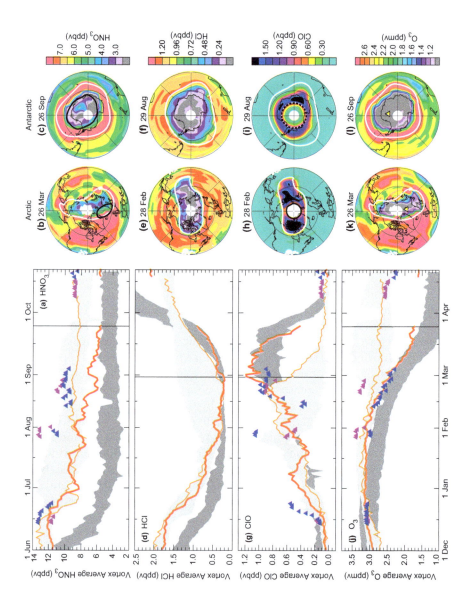

5.2 The Montreal Protocol

The *Montreal Protocol on Substances that Deplete the Ozone Layer* was a seminal international agreement signed in September 1987. In retrospect, it can be viewed as a triumph for the 'precautionary principle'. The main motivation for the Protocol was the belief, in the 1970s and early 1980s, that CFCs *may* damage the ozone layer. Recall that at the time atmospheric models predicted only modest ozone depletion from CFCs. Supporters for the Protocol argued that given the very long residence times in the atmosphere, waiting for definite scientific evidence of ozone depletion (as requested by the opponents) was too dangerous. As it turned out, preparations for the Montreal Protocol were well underway when the dramatic losses in the Antarctic were discovered and attributed to chlorine.

In 1985, a treaty called the Convention for the Protection of the Ozone Layer was signed by twenty nations in Vienna. The signing nations agreed to take appropriate measures to protect the ozone layer from human activities. The Vienna Convention supported research, exchange of information, and

Table 1 Examples of ozone-depleting substances controlled by the Montreal Protocol.

Name	Examples	Typical uses
Chlorofluorocarbons (CFCs)	CFC-11 ($CFCl_3$)	Refrigeration
	CFC-12 (CF_2Cl_2)	Foam blowing
	CFC-113 (CCl_2FCClF_2)	Aerosol propellants
Solvents	CCl_4	Cleaning
	CH_3CCl_3	
Halons	Halon-1301 (CF_3Br)	Fire extinnguisher
	Halon-1211 ($CBrClF_2$)	
Hydrochlorofluorocarbons (HCFCs)	HCFC-22 (CHF_2Cl)	Replacement for CFCs
	HCFC-123 ($CHCl_2CF_3$)	

Figure 10 Chemical composition in the lower stratosphere from AURA Microwave Limb Sounder (MLS) observations. Maps (right) and vortex-averaged time series (left) at 485 K potential temperature ($\sim$20 km, $\sim$50 hPa) for four different gases: HNO_3 (a, b, c), HCl (d, e, f), ClO (g, h, i) and O_3 (j, k, l). Averaging for the time series is done within the white contour shown on the maps. Blue (purple) triangles on time series show 1995–96 (1996–97) values from UARS MLS. Light (dark) grey shading shows range of Arctic (Antarctic) values for 2005–2010. Red and orange lines show the 2010–11 and 2004–05 Arctic winters, respectively. Antarctic dates are shifted by six months (top axis on time series) to show the equivalent season. Vertical lines show dates of maps in 2011 (2010) in the Arctic (Antarctic). Black overlays on HNO_3 maps indicate the temperature for chlorine activation (T_{act}, $\sim$196 K at this level); HNO_3 may be sequestered in PSCs at lower temperatures. Dotted black/white contour on ClO maps show the location of the 92° solar zenith angle, poleward of which measurements were taken in darkness.
Reproduced by permission from Macmillan Publishers Ltd (Nature, copyright 2011) from Manney *et al.*[35]

Table 2 Atmospheric lifetime,[40] Ozone Depletion Potential (ODP)[8] and Global Warming Potential (GWP)[8] of selected gases. Gases which are controlled by the Montreal Protocol are indicated in bold. The ODP values are those used in the Montreal Protocol. The values will change as our understanding of these gases change (*e.g.* better estimates of the atmospheric lifetimes).[40]

Molecule	Lifetime (years)	ODP	GWP (over time horizon)		
			20 years	50 years	500 years
CO_2	–	0	1	1	1
CFC-11	52	1	6730	4750	1620
CFC-12	102	1	11 000	10 900	5200
CFC-113	93	0.8	6540	6130	2690
HCFC-22	13	0.055	5130	1790	545
HCFC-123	1	0.02	273	77	24
Halon 1301	72	10	8480	7140	2760
CCl_4	44	1.1	2700	1400	435
CH_3CCl_3	5	0.1	506	146	45
CH_3Br	1.5	0.6	19	5	2
HFC-23	228	0	11 900	14 200	10 700
SF_6	3200	0	16 300	22 800	32 600

future protocols. In response to growing concern, the Montreal Protocol was signed in 1987 and ratified in 1989. The Protocol established legally binding controls for developed and developing nations on the production and consumption of halogen source gases known to cause ozone depletion. As the scientific basis of ozone depletion became more certain in subsequent years and substitutes and alternatives became available for the principal halogen source gases, the Protocol was strengthened with Amendments and Adjustments. These added new controlled substances, accelerated existing control measures, and scheduled phaseouts of the production of certain gases. The initial Protocol called for only a slowing of CFC and halon production. The 1990 London Amendments to the Protocol called for a phaseout of the production of the most damaging ozone-depleting substances in developed nations by 2000 and in developing nations by 2010. The 1992 Copenhagen Amendments accelerated the date of the phaseout to 1996 in developed nations. Further controls on ozone-depleting substances were agreed upon in later meetings in Vienna (1995), Montreal (1997), Beijing (1999), and Montreal (2007). Figure 11 illustrates the stratospheric chlorine loading that would have occurred without the Montreal Protocol and under different amendments and adjustments. Recall that the Antarctic ozone hole first became detectable in the mid 1980s when the relative equivalent effective stratospheric chlorine (EESC) loading was 1 on the scale in the figure. Without the Montreal Protocol, EESC would have increased to about 4 by 2020 at which point severe ozone depletion would have become widespread.

The 1987 provisions of the Montreal Protocol would have only slowed the approach to the large effective chlorine values by one or more decades in the 21st century. Not until the 1992 Copenhagen Amendments and Adjustments

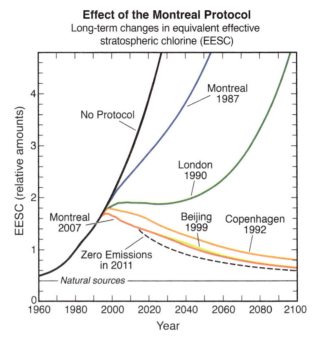

Effect of the Montreal Protocol
Long-term changes in equivalent effective
stratospheric chlorine (EESC)

Figure 11 The impact of the Montreal Protocol and its Amendments and Adjustments on atmospheric halogen loading. The atmospheric halogen loading is expressed as the equivalent effective stratospheric chlorine (EESC) which combines the chlorine and bromine content of ozone-depleting substances. EESC (based on past observations and future projections) is shown for: no protocol provision; the provisions of the 1987 Montreal Protocol and some of its subsequent amendments and adjustments; and zero emissions of ozone-depleting substances starting in 2011. The city names and years indicate where and when changes to the original 1987 Protocol provisions were agreed.
Reproduced with permission from WMO 2010.[8]

did the Protocol show a decrease in future effective stratospheric chlorine values. Now, with full compliance to the Montreal Protocol and its Amendments and Adjustments, the emissions of the major human-produced ozone-depleting gases will ultimately be phased out and the effective stratospheric chlorine value will slowly decay to reach pre-"ozone-hole" values in the mid 21st century. This sets the timescale for the future disappearance of the Antarctic ozone hole, if other atmospheric variables remain the same (see Section 6).

The Montreal Protocol provides for the transitional use of hydrochlorofluorocarbons (HCFCs) as substitute compounds for principal halogen source gases such as CFC-12. HCFCs differ chemically from CFC source gases in that they contain H atoms in addition to the Cl and F. HCFCs are used for refrigeration, blowing foams, and as solvents, which were primary uses of CFCs. HCFCs are 1 to 10% as effective as CFC-12 in depleting stratospheric ozone because they are partially chemically removed in the troposphere by reaction with OH (see Ozone Depletion Potentials or ODPs in Table 2). In contrast, CFCs and other halogen source gases are chemically inert in the troposphere and hence reach the stratosphere without incurring any removal. Because HCFCs still contribute to the halogen abundance in the stratosphere, the Montreal Protocol requires the production and consumption of HCFCs to end in developed and developing nations by 2040.

The observed behaviour and future expected evolution of a range of chlorine and bromine source gases is shown in Figure 12. The atmospheric concentration of the most abundant CFCs has already started to decrease, slightly earlier for CFC-11 as it has an atmospheric lifetime around 50 years compared to 100 years for CFC-12. The concentration of solvents such as methyl chloroform and carbon tetrachloride is also decreasing strongly. Under the provision of the Montreal Protocol the concentration of HCFCs is still increasing, but this too will decrease in the near future. In summary, the

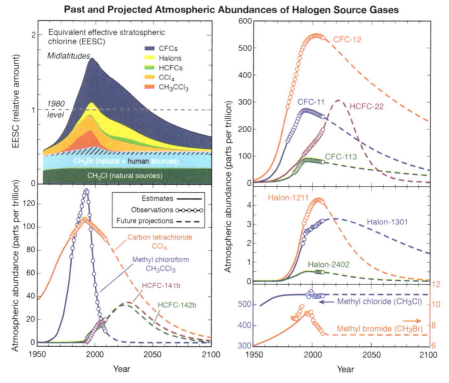

Figure 12 The top left panel summarises the development of atmospheric halogen loading measured as the equivalent effective stratospheric chlorine (EESC). The colour shading indicates the contribution to EESC from different classes of ozone-depleting substances (ODSs). The figure is based on observations up to 2009 and then future projections. The other panels show the surface observations (ppt) and future projections for individual ODSs which contribute to the total EESC.
Reproduced with permission from WMO 2010.[8]

overall atmospheric chlorine and bromine loading peaked around 1998 and has been decreasing since. The behaviour of the gases shown in Figure 12 is generally consistent with our expectations based on the Montreal Protocol and gives us confidence that the agreement is working. At present one gas whose atmospheric behaviour is not fully understood is carbon tetrachloride. It is decreasing in the atmosphere but at a rate slower than that expected based on reported emission reductions.[42]

5.3   Reasons for Success of the Montreal Protocol

The meeting to discuss the final wording of the Montreal Protocol was held in September 1987, coincidentally at the same time as the first major aircraft campaign was in the Antarctic to determine the cause of the newly discovered hole.[17] The agreement would then come into force when it was ratified by at least 11 Parties representing at least two thirds of the 1986 estimated global consumption of regulated species. This happened on January 1, 1989 when 29 counties, representing 83% of global consumption, signed up.

There has been near universal participation in the agreements to protect the ozone layer: 183 countries ratified the treaty and only 11 did not. Key innovations in the Montreal Protocol process which helped ensure this were: (i) having modest initial control measures in the first instance, which allowed countries to sign up, along with the mechanism for strengthening the agreement if later evidence suggests this is necessary, and (ii) the provision for financial/technical assistance for developing countries to help implement the Protocol.

Although the MP was ratified quickly, initially there was not universal support. In particular, major developing countries (especially India and China) were dissatisfied, arguing that they had done little to cause the problem but were being required to impose the same cuts in consumption. Moreover, their economies were less able to cover the costs of switching to alternative substances. For this reason the first meeting to amend the Protocol (London 1990) set up the Multilateral Fund. This is a fund which is paid into by developed countries and then used to support the compliance of developing (so-called "Article 5") countries. An Executive Committee of 14 members (seven each from developed and developing countries) award the money in response to requests to help fund the transition from ozone-depleting substance use to alternatives in developing countries.

5.4   Climate Benefit of the Montreal Protocol

The chlorine and bromine source gases listed in Table 1 are also very efficient greenhouse gases (GHGs). These molecules have strong absorption features in the IR and have fairly long atmospheric lifetimes from a few years to around a century (see Table 2). Indeed, their Global Warming Potential (GWP) is much larger than for CO_2 over the time frame of their atmospheric lifetime, as they have low abundances and atmospheric absorption is not saturated at the different wavelengths where these molecules absorb. As pointed out by Velders *et al.*,[43] by leading to the reduction in these gases, the Montreal Protocol has had a very large benefit in reducing climate change. It is estimated[8] that by 2020 the Montreal Protocol will already have avoided adding the equivalent of 22 Gt of CO_2 to the atmosphere and will have reduced the direct radiative forcing by these compounds by 0.6 Wm^{-2}. This compares with the estimated actual 1750–2010 radiative forcings of ~ 1.68 Wm^{-2} for CO_2 emissions, ~ 0.97 Wm^{-2} for CH_4 emissions, and ~ 0.17 Wm^{-2} for N_2O emissions.

Past global stratospheric ozone depletion has partially offset some of the positive radiative forcing caused by ODS emissions. However, the net effect has been an increase in radiative forcing overall and these effects should not be separated. The overall climate impacts need to include the direct effect of the GHGs and the indirect effect through ozone depletion.[44]

Hydrofluorocarbons (HFCs) are also used as substitute compounds for CFCs and other halogen source gases. HFCs contain no chlorine or bromine so they do not cause ozone depletion. Because of this, HFCs are not controlled by the Montreal Protocol but like all long-lived source gases they are effective GHGs (see Table 2). HFCs therefore fall under classes of gases to be controlled in treaties to prevent climate change.

6   Outlook

While the Montreal Protocol appears to be on track to decrease stratospheric chlorine and bromine levels over the rest of this century, there are other factors which may perturb the stratospheric ozone layer, and other important ways in which a changing stratosphere may interact with the Earth system. This section discusses some of those topics which represent active areas in stratospheric research.

6.1   *Very Short-lived Species*

In recent years it has become apparent that the stratospheric bromine loading is larger than can be explained by the long-lived source gases of CH_3Br and halons alone. Observations of inorganic degradation products in the stratosphere show around 3–8 pptv additional bromine which can be explained through the transport of brominated very short-lived species (VSLS) to the stratosphere.[45,46] The term VSLS is used for species with an atmospheric lifetime of 6 months or less. The transport of these short-lived species to the stratosphere depends on rapid vertical transport to the upper troposphere, for example in tropical convection systems. The two main brominated species are bromoform ($CHBr_3$) and dibromomethane (CH_2Br_2) which are naturally emitted from the ocean. At present there is large uncertainty in these surface emissions. Figure 13 shows profiles of bromine calculated from a 3-D model[47] with four different surface emission datasets

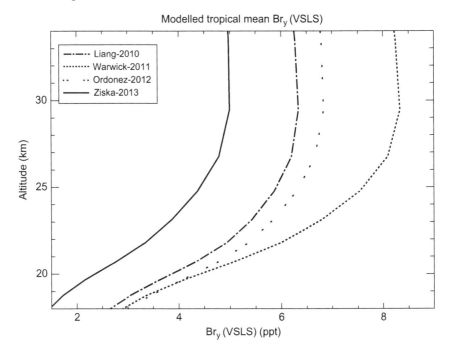

Figure 13 Modelled 2011 tropical ($\pm 30°$ latitude) mean profiles of total inorganic bromine (ppt) derived from the very short-lived species (VSLS): CHBr$_3$, CH$_2$Br$_2$, CHBr$_2$Cl, CH$_2$BrCl and CHBrCl$_2$ in the stratosphere. Profiles are shown for model runs based on four different surface emission datasets (labelled Liang, Warwick, Ordonez and Ziska).
Adapted from Hossaini *et al.*[47]

for CHBr$_3$ and CH$_2$Br$_2$. The model results show that bromine from VSLS can reach the stratosphere with a predicted range of 4–8 pptv.

Current research into halogenated VSLS therefore combines tropospheric and stratospheric chemistry. VSLS contribute about 5 (3–8) pptv bromine out of a total stratospheric loading of $\sim$20 pptv, *i.e.* $\sim$25%. This bromine from VSLS species reaches the polar lower stratosphere where it contributes to polar ozone loss *via* the BrO + ClO cycle (see Section 3). Therefore, there is a direct coupling between transport of brominated VSLS in the tropics and polar ozone loss. This contribution of natural bromine to the stratosphere may change in the future either due to changing emissions from the ocean (*e.g.* through changing temperature) or from changing transport. The latter process was tested in a climate model by Hossaini *et al.*[48] who found that the transport of CHBr$_3$ to the tropical upper troposphere was enhanced under large future increases in GHGs (see Figure 14).

6.2 Recovery of the Ozone Layer

As a consequence of the Montreal Protocol (see Section 5.2), it is predicted that the stratospheric EESC will return to 1980 levels by about the year 2050.

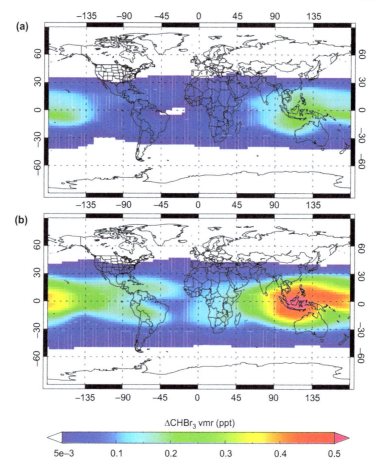

Figure 14   Chemical–Climate Model calculation of the mean December–February (DJF) increase in CHBr$_3$ volume mixing ratio (pptv) at $\sim$17 km between 2000 and 2100 for conditions of: (a) moderate climate change (RCP 4.5); and (b) large climate change (RCP 8.5).
Reproduced with permission from Wiley from Hossaini *et al.*[48]

This is expected to lead to the 'recovery' of the ozone layer from the effect of chlorine and bromine-induced depletion. However, the atmosphere of the latter half of this century will be very different to that of 1980. In particular, increasing greenhouse gases will cool the stratosphere and change its chemical composition. A cooler stratosphere will generally have more ozone, through slower gas-phase loss in catalytic cycles,[49] though a colder polar lower stratosphere could cause more PSCs. Increases in CH$_4$ will change the H$_2$O and HO$_x$ (OH + HO$_2$) budget of the stratosphere. Increases in N$_2$O will lead to more NOy.[50] A prediction of the recovery of the ozone layer therefore requires detailed chemistry–climate models which combine all of the known feedbacks.

Figure 15 shows results from a number of CCMs from groups worldwide which have been used to simulate the ozone layer from 1960–2100.[51,52] The models included changing ODSs and GHGs, and model experiments were used to separate these impacts. In the Antarctic lower stratosphere, the region of the ozone hole (see Figure 15d), the models with both ODS and GHG changes show that ozone depletion peaked between 2000 and 2010 and will slowly recover for the rest of this century. The return to 1980 levels will occur

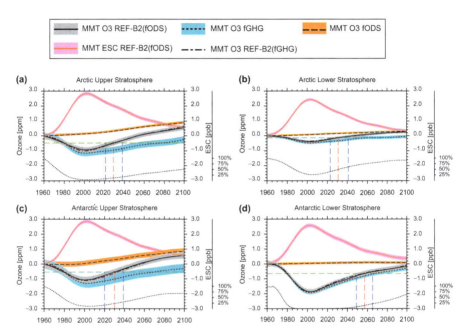

Figure 15 Predicted change in ozone from CCM model simulations for the Arctic (March mean upper row) and Antarctic (October mean lower row). The upper stratosphere is shown for the 5 hPa (∼35 km) level and the lower stratosphere for the 50 hPa (∼20 km) level. The panels show polar mean 1960 baseline-adjusted ozone projections and 95% confidence interval for the multi-model trend (MMT) of CCMVal REF-B2 CCM runs which include both ODS and GHG changes (MMT O3 REF-B2(fODS); black line and grey shaded area). Results are also shown from a different combination of these model runs (MMT O3 REF-B2(fGHG); black dashed-dotted line). Also shown is the multi-model trend plus 95% confidence interval for CCM runs with fixed ODSs (MMT O3 fODS; black dotted line and orange shaded area), CCM runs with fixed GHGs (MMT O3 fGHG; black dashed line and blue shaded area), and Equivalent Stratospheric Chlorine (ESC) (MMT ESC REF-B2(fODS); red solid line and pink shaded area). The red vertical dashed line indicates the year when the multi-model mean of the 9 CCMs in REF-B2 returns to 1980 values (green horizontal dashed line) and the blue vertical dashed lines indicate the uncertainty in these return dates. The thin dotted black line in the bottom of each panel shows the results of a t-test's confidence level that the multi-model means from fODS and REF-B2 are from the same population.
Reproduced with permission from Eyring *et al.*[51]

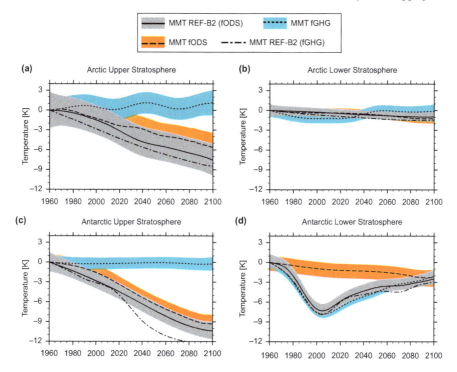

Figure 16   Same as Figure 15, but for temperature.
Reproduced with permission from Eyring *et al.*[51]

by about 2060. Sensitivity experiments with just ODS or just GHG changes reveal how these two forcing terms affect O_3. In the Antarctic lower stratosphere the evolution of ozone is dominated by the change in ODSs and the impact of climate change through increasing GHGs is small. A different behaviour is predicted for the Antarctic upper stratosphere (see Figure 15c). The change in ODS alone would cause a recovery to 1980 levels by 2060. However, stratospheric cooling through increasing GHGs (see Figure 16) is driving an increase in upper stratospheric ozone which accelerates the recovery. Similar results are predicted for the Arctic, but the magnitude of the depletion is smaller.

6.3   Impact of Ozone Depletion on Surface Climate

As ozone is such a radiatively important gas, changes to its distribution in the stratosphere can affect not only the climate of the stratosphere but also the climate at the surface of the Earth. As the Antarctic ozone hole represents the largest perturbation to the ozone layer, with near 100% removal in the lower stratosphere in spring, it would be expected that such impacts would be most strongly seen at southern high latitudes. The cooling of the Antarctic lower stratosphere associated with the ozone hole has caused a delay in the breakdown of the Antarctic polar vortex.[8] The shift in wind patterns extends to the surface during the spring and summer months and a poleward shift in the tropospheric jet has been observed.[53] In fact, the dynamical influence of the ozone hole on surface climate is largely manifested through changes to the structure of the Southern Annular Model (SAM).

As the ozone hole recovers in the future the reverse influences would be expected. Figure 17 shows results from a range of climate models and illustrates the negative impact on the predictions of models which fail to account for stratospheric ozone depletion.[54] The figure shows the change in

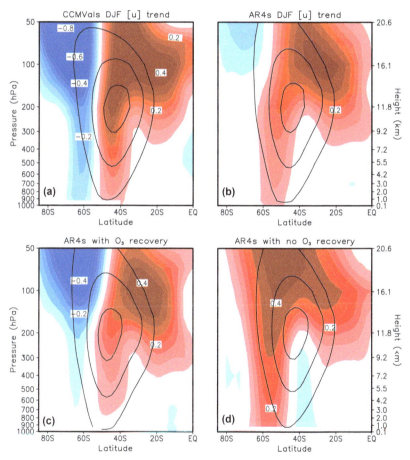

Figure 17 Trends in December-to-February (DJF) zonal-mean zonal wind. The multimodel mean trends between 2001 and 2050 are shown for: (a) the CCMVal models; (b) the IPCC AR4 models; (c) the AR4 models with prescribed ozone recovery; and (d) the AR4 models with no ozone recovery. Shading and contour intervals are 0.05 ms^{-1} decade^{-1}. Deceleration and acceleration are indicated with blue and red colours, respectively, and trends weaker than 0.05 ms^{-1} decade^{-1} are omitted. Superimposed black solid lines are DJF zonal-mean zonal wind averaged from 2001 to 2010, with a contour interval of 10 ms^{-1}, starting at 10 ms^{-1}. Reproduced with permission from AAAS from Son *et al.*[54]

DJF zonal mean zonal wind for 2050 compared to 2000 for CCMVal models, with a fully resolved stratosphere and representation of the ozone hole, and different types of models from the IPCC 4th Assessment Report (AR4). The CCMVal models (see Figure 17a) show a decreasing trend in the SH mid-latitude jet, a reversal of the observed increasing trend during the past period of growing Antarctic depletion. In stark contrast, AR4 models which do not represent stratospheric ozone changes (see Figure 17d) do not show this behaviour and therefore do not produce accurate forecasts of the surface climate. These results show that climate models need to have a realistic treatment of stratospheric ozone. It should be noted that meteorological agencies are also increasing the altitude of their model top boundaries (well above the stratopause) in order to improve surface weather forecasts.[30]

7 Summary

This article has briefly reviewed the science of the Antarctic ozone hole and the policy action taken to prevent further stratospheric ozone loss. The Montreal Protocol is currently acting to reduce stratospheric chlorine and bromine and should lead to the recovery of the ozone layer during this century. Two-way interactions between ozone and climate will likely affect the rate and extent of this recovery, however. Overall, the story of the Antarctic ozone hole and the Montreal Protocol is a seminal example of worldwide cooperative scientific research and successful policy which has allowed the world to avoid catastrophic destruction of the ozone layer.

References

1. R. P. Wayne, *Chemistry of Atmospheres*, Oxford University Press, 2000, p. 806.
2. J. A. Logan, M. J. Prather, C. S. Wofsy and M. B. McElroy, *J. Geophys. Res.*, 1981, **86**, 7210.
3. S. Chapman, *Philos. Mag.*, 1930, **10**, 369.
4. L. L. Hood, in *Solar Variability and its Effect on the Earth's Atmosphere and Climate System*, ed. J. Pap *et al.*, American Geophysical Union, Washington DC, 2004, pp. 283–304.
5. S. S. Dhomse, M. P. Chipperfield, W. Feng, W. T. Ball, Y. C. Unruh, J. D. Haigh, N. A. Krivova, S. K. Solanki and A. K. Smith, *Atmos. Chem. Phys.*, 2013, **13**(10), 113.
6. S. Solomon, *Rev. Geophys.*, 1999, **37**, 275.
7. *Stratospheric Ozone Depletion and Climate Change*, ed. R. Müller, Royal Society of Chemistry, Cambridge, 2012, ISBN 978-1-84973-002-0.
8. WMO (World Meteorological Organisation), *Scientific Assessment of Ozone Depletion: 2010*, Global Ozone Research and Monitoring Project Report No. 52, Geneva, Switzerland, 2011.
9. S. O. Anderson and K. M. Sarma, *Protecting the Ozone Layer*, Earthscan Publications Ltd, London, 2002, ISBN 1-85383-905-1.

10. M. P. Chipperfield, *Q. J. R. Meteorol. Soc.*, 2006, **132**, 1179.
11. J. C. Farman, B. G. Gardiner and J. D. Shanklin, *Nature*, 1985, **315**, 207.
12. A. J. T. de Laat and M. van Weele, *Sci. Rep.*, 2011, **1**, 38.
13. N. R. P. Harris and M. Rex, *Stratospheric Ozone Depletion and Climate Change*, ed. R. Muller, Royal Society of Chemistry, Cambridge, 2012, pp. 145–165.
14. M. J. Molina and F. S. Rowland, *Nature*, 1974, **249**, 810.
15. R. S. Stolarski and R. J. Cicerone, *Can. J. Chem.*, 1974, **52**, 1610.
16. J. A. Pyle, *Pure Appl. Geophys.*, 1980, **118**, 355.
17. J. G. Anderson, W. H. Brune and M. H. Proffitt, *J. Geophys. Res.*, 1989, **94**(11), 465.
18. S. Solomon, R. R. Garcia, F. S. Rowland and D. J. Wuebbles, *Nature*, 1986, **321**, 755.
19. United Kingdom Stratospheric Ozone Review Group, *Stratospheric Ozone 1988*, Her Majesty's Stationery Office, London, ISBN 0 11 752148 5, 1988.
20. C. Piazzi Smyth, *Nature*, 1884, **51**, 148.
21. K. S. Carslaw, B. P. Luo, S. L. Clegg, T. Peter, P. Brimblecombe and P. J. Crutzen, *Geophys. Res. Lett.*, 1994, **21**, 2479.
22. P. J. Crutzen and F. Arnold, *Nature*, 1986, **324**, 651.
23. O. B. Toon, P. Hammill, R. P. Turco and J. Pinto, *Geophys. Res. Lett.*, 1986, **13**, 1284.
24. T. Peter and J.-U. Grooss, *Stratospheric Ozone Depletion and Climate Change*, ed. R. Muller, Royal Society of Chemistry, Cambridge, 2012, pp. 108–139.
25. M. C. Pitts, L. R. Poole and L. W. Thomason, *Atmos. Chem. Phys.*, 2009, **9**, 7577.
26. M. C. Pitts, L. R. Poole, A. Lambert and L. W. Thomason, *Atmos. Chem. Phys.*, 2013, **13**, 2975.
27. M. L. Santee, I. A. MacKenzie, G. L. Manney, M. P. Chipperfield, P. F. Bernath, K. A. Walker, C. D. Boone, L. Froidevaux, N. J. Livesey and J. W. Waters, *J. Geophys. Res.*, 2008, **113**, D12307.
28. L. T. Molina and M. J. Molina, *J. Phys. Chem.*, 1987, **91**, 433.
29. M. B. McElroy, R. J. Salawitch, S. C. Wofsy and J. A. Logan, *Nature*, 1986, **321**, 759.
30. D. P. Dee, S. M. Uppala, A. J. Simmons, P. Berrisford, P. Poli, S. Kobayashi, U. Andrae, M. A. Balmaseda, G. Balsamo, P. Bauer, P. Bechtold, A. C. M. Beljaars, L. van de Berg, J. Bidlot, N. Bormann, C. Delsol, R. Dragani, M. Fuentes, A. J. Geer, L. Haimberger, S. B. Healy, H. Hersbach, E. V. Holm, L. Isaksen, P. Kallberg, M. Kohler, M. Matricardi, A. P. McNally, B. M. Monge-Sanz, J.-J. Morcrette, B.-K. Park, C. Peubey, P. de Rosnay, C. Tavolato, J.-N. Thepaut and F. Vitart, *Q. J. R. Meteorol. Soc*, 2011, **137**, 553.
31. K. Frieler, M. Rex, R. J. Salawitch, T. Canty, M. Streibel, R. M. Stimpfle, K. Pfeilsticker, M. Dorf, D. K. Weisenstein and S. Godin-Beekman, *Geophys. Res. Lett.*, 2006, **33**, L10812.
32. M. Rex, P. von der Gathen, G. O. Braathen, S. J. Reid, N. R. P. Harris, M. Chipperfield, E. Reimer, A. Beck, R. Alfier, R. Krger-Carstensen, H. De

Backer, D. Balis, C. Zerefos, F. O'Connor, H. Dier, V. Dorokhov, H. Fast, A. Gamma, M. Gil, E. Kyrv, M. Rummukainen, Z. Litynska, I. S. Mikkelsen, M. Molyneux and G. Murphy, *J. Atmos. Chem.*, 1999, **32**, 35.

33. W. Feng, M. P. Chipperfield, S. Davies, B. Sen, G. Toon, J. F. Blavier, C. R. Webster, C. M. Volk, A. Ulanovsky, F. Ravegnani, P. von der Gathen, H. Jost, E. C. Richard and H. Claude, *Atmos. Chem. Phys.*, 2005, **5**, 139.

34. M. Streibel, M. Rex, P. von der Gathen, R. Lehmann, N. R. P. Harris, G. O. Braathen, E. Reimer, H. Deckelmann, M. Chipperfield, G. Millard, M. Allaart, S. B. Andersen, H. Claude, J. Davies, H. De Backer, H. Dier, V. Dorokov, H. Fast, M. Gerding, E. Kyro, Z. Litynska, D. Moore, E. Moran, T. Nagai, H. Nakane, C. Parrondo, P. Skrivankova, R. Stubi, G. Vaughan, P. Viatte and V. Yushkov, *Atmos. Chem. Phys.*, 2006, **6**, 2783.

35. G. L. Manney, M. L. Santee, M. Rex, N. J. Livesey, M. C. Pitts, P. Veefkind, E. R. Nash, I. Wohltmann, R. Lehmann, L. Froidevaux, L. R. Poole, M. R. Schoeberl, D. P. Haffner, J. Davies, V. Dorokhov, H. Gernandt, B. Johnson, R. Kivi, E. Kyrö, N. Larsen, P. F. Levelt, A. Makshtas, C. T. McElroy, H. Nakajima, M. Concepcion Parrondo, D. W. Tarasick, P. von der Gathen, K. A. Walker and N. S. Zinoviev, *Nature*, 2011, **478**, 469.

36. B.-M. Sinnhuber, G. Stiller, R. Ruhnke, T. von Clarmann, S. Kellmann and J. Aschmann, *Geophys. Res. Lett.*, 2011, **38**, L24814.

37. X. X. Tie, G. P. Brasseur, B. Briegleb and C. Granier, *J. Geophys. Res.*, 1994, **99**(20), 545.

38. T. G. Shepherd, D. A. Plummer, J. F. Scinocca, M. I. Hegglin, V. E. Fioletov, M. C. Reader, E. Remsberg, T. von Clarmann and H. J. Wang, *Nat. Geosci.*, 2014, **7**, 443.

39. R. L. McKenzie, J. B. Liley and L. O. Bjorn, *Photochem. Photobiol.*, 2009, **85**, 88.

40. SPARC *Report on the Lifetimes of Stratospheric Ozone-Depleting Substances, Their Replacements, and Related Species*, ed. M. Ko, P. Newman, S. Reimann and S. Strahan, SPARC Report No. 6, WCRP-15/2013, 2013.

41. M. P. Chipperfield, Q. Liang, S. E. Strahan, O. Morgenstern, S. S. Dhomse, N. L. Abraham, A. T. Archibald, S. Bekki, P. Braesicke, G. Di Genova, E. L. Fleming, S. C. Hardiman, D. Iachetti, C. H. Jackman, D. E. Kinnison, M. Marchand, G. Pitari, J. A. Pyle, E. Rozanov, A. Stenke and F. Tummon, *J. Geophys. Res.*, 2014, **119**, 2555.

42. Q. Liang, P. A. Newman, J. S. Daniel, S. Reimann, B. D. Hall, G. Dutton and L. J. M. Kuijpers, *Geophys. Res. Lett.*, 2014, **41**, 5307.

43. G. J. M. Velders, S. O. Anderson, J. S. Daniel, D. W. Fahey and M. McFarland, *Proc. Natl. Acad. Sci. U. S. A.*, 2007, **104**, 4814.

44. D. Shindell, G. Faluvegi, L. Nazarenko, K. Bowman, J.-F. Lamarque, A. Voulgarakis, G. A. Schmidt, O. Pechony and R. Ruedy, *Nat. Clim. Change*, 2013, **3**, 567.

45. R. J. Salawitch, D. K. Weisenstein, L. J. Kovalenko, C. E. Sioris, P. O. Wennberg, K. Chance, M. K. W. Ko and C. A. McLinden, *Geophys. Res. Lett.*, 2005, **32**, L05811.

46. M. Dorf, J. H. Butler, A. Butz, C. Camy-Peyret, M. P. Chipperfield, L. Kritten, S. A. Montzka, B. Simmes, F. Weidner and K. Pfeilsticker, *Geophys. Res. Lett.*, 2006, **33**, L24803.

47. R. Hossaini, H. Mantle, M. P. Chipperfield, S. A. Montzka, P. Hamer, F. Ziska, B. Quack, K. Krueger, S. Tegtmeier, E. Atlas, S. Sala, A. Engel, H. Boenisch, T. Keber, D. Oram, G. Mills, C. Ordonez, A. Saiz-Lopez, N. Warwick, Q. Liang, W. Feng, F. Moore, B. R. Miller, V. Marecal, N. A. D. Richards, M. Dorf and K. Pfeilsticker, *Atmos. Chem. Phys.*, 2013, **13**, 11819.

48. R. Hossaini, M. P. Chipperfield, S. Dhomse, C. Ordóñez, A. Saiz-Lopez, N. L. Abraham, A. Archibald, P. Braesicke, P. Telford, N. Warwick, X. Yang and J. Pyle, *Geophys. Res. Lett.*, 2012, **39**, L20813.

49. J. D. Haigh and J. A. Pyle, *Nature*, 1979, **279**, 222.

50. L. K. Randeniya, P. F. Vohralik and I. C. Plumb, *Geophys. Res. Lett.*, 2002, **29**, 10-1–10-4.

51. V. Eyring, I. Cionni, G. E. Bodeker, A. J. Charlton-Perez, D. E. Kinnison, J. F. Scinocca, D. W. Waugh, H. Akiyoshi, S. Bekki, M. P. Chipperfield, M. Dameris, S. Dhomse, S. M. Frith, H. Garny, A. Gettelman, A. Kubin, U. Langematz, E. Mancini, M. Marchand, T. Nakamura, L. D. Oman, S. Pawson, G. Pitari, D. A. Plummer, E. Rozanov, T. G. Shepherd, K. Shibata, W. Tian, P. Braesicke, S. C. Hardiman, J. F. Lamarque, O. Morgenstern, D. Smale, J. A. Pyle and Y. Yamashita, *Atmos. Chem. Phys.*, 2010, **10**, 9451.

52. SPARC CCMVal, *Report on the 715 Evaluation of Chemistry-Climate Models*, ed. V. Eyring, T. G. Shepherd, 716 and D. W. Waugh, SPARC Report No. 5, WCRP-132, WMO/TD-No. 1526, 2010.

53. D. W. J. Thompson and S. Solomon, *Science*, 2002, **296**, 895.

54. S.-W. Son, L. M. Polvani, D. W. Waugh, H. Akiyoshi, R. Garcia, D. Kinnison, S. Pawson, E. Rozanov, T. G. Shepherd and K. Shibata, *Science*, 2008, **320**, 1486.

Global Atmosphere – Greenhouse Gases

JOHN SOTTONG,* MARK BROOMFIELD, JOANNA MacCARTHY,
ANNE MISRA, GLEN THISTLETHWAITE AND JOHN WATTERSON

ABSTRACT

There is compelling evidence that warming of the climate system due to human influence is taking place. The mechanisms for these processes include increasing levels of greenhouse gases (GHGs) in the atmosphere. In May 2013, carbon dioxide levels in the atmosphere exceeded 400 ppm for the first time in several hundred millennia. Levels of carbon dioxide in the remote atmosphere have increased by 0.28 and 2.93 parts per million (ppm) per year since 1960. Ongoing increases in levels of GHGs can be expected to result in a rise in global temperatures, which could trigger a wide range of risk scenarios. A level of 450 ppm is considered to be the atmospheric concentration of carbon dioxide at which temperature change can still be limited to 2 °C. This is a benchmark of climate change specified so as to limit the dangerous effects of climate change. The international community has initiated a series of programmes to address GHG emissions, underpinned by a significant effort to develop robust emission inventories: however, these initiatives have had limited success in limiting the rising trends in GHG emissions. The key issues encountered during these programmes include: population and economic growth; scientific credibility; political priorities; outsourcing of emissions; and difficulties in implementing renewable energy technologies. In this context, the scientific community has an ongoing role to produce data and analysis to link individual, corporate and state actions and policies

---

*Corresponding author.

Issues in Environmental Science and Technology No. 40
Still Only One Earth: Progress in the 40 Years Since the First UN Conference on the Environment
Edited by R.E. Hester and R.M. Harrison
© The Royal Society of Chemistry 2015
Published by the Royal Society of Chemistry, www.rsc.org

to evaluations of GHG emissions, and analysis of evidence for changes in the global climate. The scientific and policy communities should continue to engage with the public, to provide reliable information about the vitally important issues associated with global climate change. Public support for positive action and investment to mitigate climate change is essential to support the decisions which will need to be taken in the coming decades.

1 The Greenhouse Effect

1.1 What is the Greenhouse Effect?

The term "climate change" refers to "a change in the state of the climate that can be identified (*e.g.*, by using statistical tests) by changes in the mean and/or the variability of its properties, and that persists for an extended period, typically decades or longer."[1] The Intergovernmental Panel on Climate Change (IPCC) states that "warming of the climate system is unequivocal," and goes on to highlight warming of the atmosphere and oceans, diminishing amounts of snow and ice, rises in sea levels, and increases in the concentrations of greenhouse gases.[2] The US National Research Council described the case for human influence on climate change as "compelling," and stated that "hypotheses about climate change are supported by multiple lines of evidence and have stood firm in the face of serious debate and careful evaluation of alternative explanations".[3]

1.2 Radiative Forcing

The radiative forcing mechanisms (balance between atmospheric energy inputs and energy losses) behind climate change are now well established. The key physical and chemical processes are as shown in Figure 1, with the key processes as follows:[4]

(a) The presence of increasing levels of well-mixed greenhouse gases (GHGs) in the atmosphere results in an increase in the incident solar radiation absorbed by the earth. This alters the balance of incoming and outgoing energy in the Earth atmosphere system. Levels of most GHGs are increasing, with the primary contributors being carbon dioxide (CO_2), methane (CH_4) and nitrous oxide (N_2O). Halocarbons are also important contributors, although levels of some halocarbons are currently reducing.

(b) Short-lived GHGs in the atmosphere also contribute to radiative forcing, with the primary contributors ozone and stratospheric water vapour.

(c) Changes to the surface characteristics of the earth affect the balance between absorption and reflection of sunlight. Increases in the presence of black carbon on snow tends to increase radiative forcing, whereas there have been net reductions in radiative forcing due to land-use changes.

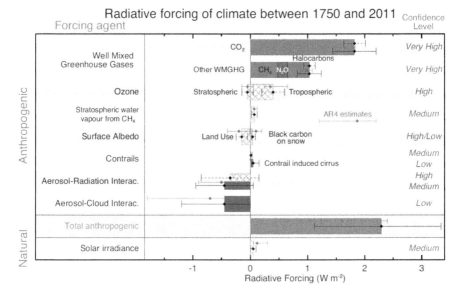

Figure 1   Radiative forcing (RF) and Effective Radiative Forcing (ERF) of climate change during the Industrial Era.[4] (Note: Forcing by concentration change between 1750 and 2011 is indicated with associated uncertainty range (solid bars are ERF, hatched bars are RF, diamonds and associated uncertainties are for RF)).

(d) Reductions in radiative forcing result from aerosol–radiation and aerosol–cloud interactions. The IPCC states, with high confidence, that aerosols have offset a substantial portion of GHG forcing.

Processes (c) and (d) are not directly associated with emissions to the atmosphere of greenhouse gases.

1.3   Uncertainty

At the global scale, calculating GHG emissions and forecasting the environmental consequences of climate forcing is subject to some uncertainty. The IPCC takes a systematic approach to evaluating confidence in the available evidence, and uncertainty in forecast outcomes (see ref. 4, Box TS1). This enables the confidence in data and forecast impacts to be taken into account when interpreting the findings. When creating GHG inventories at national scales, GHG inventory compliers use guidance created by the IPCC[5] and adopted by the UNFCCC to estimate uncertainty in the annual totals and in the trend of emissions over time. If estimates of emissions are being used in emissions trading schemes (such as the European Union Emissions Trading System), the accuracy of annual estimates is particularly important. Typically, the uncertainties at a 95% confidence interval for annual estimates of emissions of CO_2 in the inventories of Annex I countries are approximately 2% to 5%. Estimates of uncertainties associated with the non-CO_2 gases are typically much higher, and those associated with certain individual categories, for example direct N_2O emissions from agricultural soil, can be orders of magnitude higher than for CO_2. If GHG reductions have been set over a time period (for example the obligations for some Annex I countries under the first commitment period of the Kyoto Protocol) then the accuracy (or uncertainty) associated with the trend is particularly important. The National Inventory Reports from Annex I countries reveal that the uncertainties associated with the trends in GHGs are often much smaller than those on the national totals, and when countries are reducing their emissions, this observation provides confidence that reductions in GHG emissions over time are actually occurring.

1.4   Greenhouse Gas Emissions

The United Nations Environment Programme (UNEP) reported that global emissions of well-mixed GHGs amounted to 49 Gigatonnes carbon dioxide equivalent ($GtCO_2e$) in 2010.[6] Emissions have continued to increase since 2010, although at a slower pace than was observed during the preceding decade. Of the 49 $GtCO_2e$ emissions released in 2010, the energy supply sector was responsible for 35%, 24% came from Agriculture, Forestry and Other Land Use (AFOLU), 21% from industry, 14% from the transport sector and 6.4% came from buildings. Since 2000, GHG emissions have been growing in all sectors, except AFOLU.[7] Carbon dioxide accounted for 76% of global GHG emissions in 2010; 16% came from sources of methane, 6% from nitrous oxide and 2% from fluorinated gases.[7]

Broken down by economies, North America accounted for 14.5%, OECD Europe 11.0% and non-OECD G20 countries (*i.e.* Argentina, China, Brazil, India, Indonesia, the Russian Federation, Saudi Arabia and South Africa) 42.5% of GHG emissions in 2010. Other developing and least developed countries accounted for almost all of the balance of 32.0% of GHG emissions.[6]

There are uncertainties in GHG inventories due to differences in methodology and completeness of inventories. Trends in GHG emissions can typically be evaluated with greater confidence than absolute values. The long-term rising trends in GHG emissions are well established. Emissions have continued to increase since the start of the Industrial Revolution. Emissions of the three key GHGs are shown in Figure 2. The data shown in Figure 2 demonstrates ongoing increases in GHG emissions over the period 1970 to 2012.

1.5   Atmospheric CO₂ Concentrations

The trend in emissions of CO_2 shown in Figure 2 is reflected in the trend in levels of CO_2 recorded in the atmosphere. The longest continuous record of direct atmospheric CO_2 levels are measurements made at the Mauna Loa

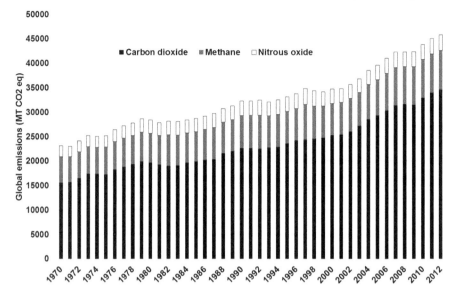

Figure 2 Global GHG emissions (adapted from ref. 8). (Note: calculated using 100 year time horizon GWP for methane of 21 and nitrous oxide of 310.[9] Updated GWP values for methane of 23 and nitrous oxide of 298 were subsequently published by the IPCC.[10])

observatory, Hawaii by the US National Oceanic and Atmospheric Administration (NOAA). Levels of CO_2 recorded at this location between 1960 and 2014 are shown in Figure 3.

The annual increase in atmospheric CO_2 concentration measured at the Mauna Loa observatory is calculated to be between 0.28 and 2.93 ppm per year.[11]

1.6   The Consequences of Climate Forcing

These GHG emissions, together with other natural and anthropogenic (man-made) influences on the global atmosphere, result in a range of consequences for the global environment. Some of the key observations from the IPCC[1] are:

(a) It is extremely likely that more than half of the observed increase in global average surface temperature from 1951 to 2010 was due to the combination of man-made increases in GHG concentrations and other anthropogenic forcings (see Figure 4 below).
(b) Anthropogenic forcings, dominated by well-mixed GHGs, are likely to have contributed to the warming of the lower atmosphere since 1961.
(c) Anthropogenic forcings are very likely to have contributed to a rise in ocean temperature since 1970, as well as an increase in acidity (due to uptake of CO_2) and changes in salinity.

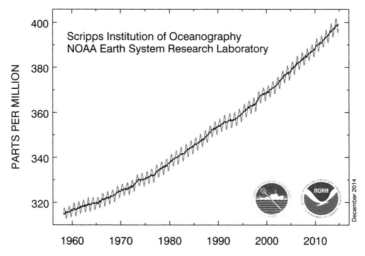

Figure 3 Atmospheric CO_2 levels recorded at the Mauna Loa observatory.[11]

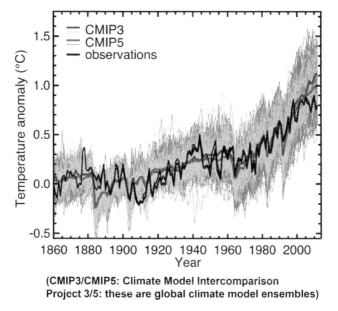

(CMIP3/CMIP5: Climate Model Intercomparison
Project 3/5: these are global climate model ensembles)

Figure 4 Global mean surface temperature and climate model simulations with anthropogenic and natural forcings.[4]

(d) Anthropogenic forcings are very likely to have contributed to Arctic sea ice loss since 1979.

(e) Climate changes are likely to have contributed to changes in the fresh water cycle since 1960.

(f) Anthropogenic forcings are very likely to have contributed to observed changes in the frequency and intensity of extreme temperature events since the mid-20th Century.

The uncertainty associated with forecasting future impacts of ongoing GHG emissions to the global environment is high. However, the IPCC[12] highlights a wide range of risks which may be triggered and/or exacerbated by human interference with the climate system:

(a) Risk of death, injury, ill-health, or disrupted livelihoods in low-lying coastal zones and small island developing states and other small islands, due to storm surges, coastal flooding, and sea level rise.

(b) Risk of severe ill-health and disrupted livelihoods for large urban populations due to inland flooding in some regions.

(c) Systemic risks due to extreme weather events leading to breakdown of infrastructure networks and critical services such as electricity, water supply, and health and emergency services.

(d) Risk of mortality and morbidity during periods of extreme heat, particularly for vulnerable urban populations and those working outdoors in urban or rural areas.

(e) Risk of food insecurity and the breakdown of food systems linked to warming, drought, flooding, and precipitation variability and extremes, particularly for poorer populations in urban and rural settings.

(f) Risk of loss of rural livelihoods and income due to insufficient access to drinking and irrigation water and reduced agricultural productivity, particularly for farmers and pastoralists with minimal capital in semi-arid regions.

(g) Risk of loss of marine and coastal ecosystems, biodiversity, and the ecosystem goods, functions, and services they provide for coastal livelihoods, especially for fishing communities in the tropics and the Arctic.

(h) Risk of loss of terrestrial and inland water ecosystems, biodiversity, and the ecosystem goods, functions, and services they provide for livelihoods.

As one specific example, the IPCC[1] reports that "The available evidence indicates that global warming beyond a threshold would lead to the near-complete loss of the Greenland ice sheet over a millennium or longer, causing a global mean sea level rise of approximately 7 m. Studies with fixed present-day ice sheet topography indicate that the threshold is greater than 2 °C but less than 4 °C (medium confidence) of global mean surface temperature rise above pre-industrial."

2   The International Response to Climate Change

2.1   *The United Nations Framework Convention on Climate Change*

The United Nations Conference on Environment and Development (UNCED)—also known as the Rio Summit, the Rio Conference, and the Earth Summit—was a major United Nations conference held in Rio de Janeiro in

1992. During this conference a treaty known as the United Nations Framework Convention on Climate Change (UNFCCC) was opened for signature, the ultimate objective of which is the "stabilization of greenhouse gas concentrations in the atmosphere at a level that would prevent dangerous anthropogenic interference with the climate system..."[13]

The Convention served as a call to action for the world to address the threat of climate change. Through this Convention, industrialised countries agreed to support developing countries with both financial and technical support for their burgeoning climate change activities. This support also recognized the important role of economic development to developing countries by acknowledging that greenhouse gas emissions produced by those nations would continue to grow and that the burden of effort to reduce emissions would fall on the shoulders of wealthier nations.

The UNFCCC secretariat supports the institutions involved in the international climate change negotiations, particularly the Conference of the Parties (COP) (which serves as the meeting of the Parties (CMP)), the subsidiary bodies (which advise the COP/CMP), and the COP/CMP Bureau (which deals mainly with procedural and organizational issues arising from the COP/CMP and also has technical functions).[13]

2.2   Industrialised and Developing Countries

One of the Convention's core principles is that of "common but differentiated responsibility" which addresses the fact that developed countries are largely responsible for the current high levels of GHG emissions in the atmosphere and are therefore most responsible for mitigating the problem. In keeping with this principle, Parties to the UNFCCC are divided into Annex I Parties ("industrialised" countries) and non-Annex I Parties ("developing" countries).

Within the UNFCCC, both Annex 1 and Non-Annex 1 Parties share reporting obligations to describe the efforts they are taking to implement the Convention to the COP. These reports are commonly known as National Communications (NCs). Annex I countries must submit an annual inventory of their greenhouse gas emissions and report on their climate change policies and measures, while Non-Annex I countries provide a less frequent and less detailed update on their actions both to address climate change and to adapt to its impacts.[14]

2.3   The Kyoto Protocol

Although the UNFCCC treaty came into effect in 1994, the global community had already recognized its shortcomings and thus launched negotiations to develop a more aggressive response to climate change. This led to the adoption of the Kyoto Protocol in 1997, with a goal to reduce the net GHG emissions of Annex I countries to 5.2% below 1990 levels in the first commitment period of 2008 to 2012. The development of the Kyoto Protocol was

an important achievement at the time, as it provided a co-operative mechanism for countries to legally bind themselves to emission reduction targets.
However, of the 192 Parties to the Kyoto Protocol today, only 37 industrialised countries are committed to such targets. The vast majority of United
Nations member states have ratified the Kyoto protocol. However, the USA
has signed but not ratified the protocol, and Canada ceased to be a member
in 2012.

The Kyoto Protocol was followed by a series of Conference of the Parties
(COP) meetings which led to the production of agreements known as the
Marrakesh Accords. These were designed to deal with unresolved issues
regarding the reporting of land use, land use change and forestry (LULUCF).
This process required compromise between the parties, but was necessary in
order to secure implementation of the overall Kyoto protocol. Although the
protocol contained compromises, and was inherently flawed,[15,16] it nevertheless came into force with its ratification by Russia in 2006.

2.4   Post-Kyoto Protocol

Following the adoption of the Kyoto protocol, UNFCCC Parties reconvened
in 2007 for further negotiations which led to a set of agreements that included new GHG reduction goals for Annex I countries, "nationally appropriate mitigation actions" (NAMAs) for non-Annex I countries, and other
targets for the second commitment period (2013–2020). Referred to as the
Bali Action Plan, these mandates failed to materialize in time for the 2009
COP in Copenhagen, Denmark and did not produce a legally binding treaty,
but rather a short, hastily developed political document called the Copenhagen Accord. Many of the elements of the Copenhagen Accord, the Bali
Action Plan, and the UNFCCC were adopted officially at the 2010 COP in
Cancún, however, which yielded several decisions collectively called the
Cancún Agreements,[17] described by the UNFCCC as follows:

> *"The Cancún Agreements are a set of significant decisions by the international
> community to address the long-term challenge of climate change collectively
> and comprehensively over time and to take concrete action now to speed up
> the global response."*

The Cancún Agreements state that future global warming should be limited
to below 2 °C compared to pre-industrial levels. Negotiations following the
Kyoto protocol have focused on agreeing emissions reductions pathways, targets and evaluation methods with the aim of limiting global temperature increase to 2 °C. Altogether, however, these negotiations have not been successful.

2.5   Unilateral and Bilateral Initiatives

The seriousness of the climate change challenge coupled with the lack of
production at the international level has led many states, authorities and

businesses to undertake unilateral initiatives to reduce GHG emissions. In the UK, for example, the Climate Change Act was passed in 2008 and established a framework to develop an economically credible emissions reduction path. It also strengthened the UK's leadership internationally by highlighting the role it would take in contributing to urgent collective action to tackle climate change under the Kyoto Protocol. The Climate Change Act includes the following elements:

(a) The Act commits the UK to reducing emissions by at least 80% in 2050 from 1990 levels. This target was based on advice from the Committee on Climate Change (CCC) report "Building a Low-carbon Economy".[18]

(b) The Committee on Climate Change was set up to advise the Government on emissions targets, and report to Parliament on progress made in reducing GHG emissions. It includes the Adaptation Sub-Committee (ASC) which scrutinises and advises on the Government's programme for adapting to climate change.

(c) The Act requires the Government to set legally binding "carbon budgets". A carbon budget is a cap on the amount of greenhouse gases emitted in the UK over a five-year period. The Committee provides advice on the appropriate level of each carbon budget. These budgets are designed to reflect cost effective paths to achieving the long term objectives. The first four carbon budgets have been put into legislation and run up to 2027.

(d) Alongside these emissions reductions programmes, the Government is required to produce a National Adaptation Plan which assesses the UK's risks from climate change, prepare a strategy to address them, and encourage critical organisations to do the same.

The European Union (EU) has made a commitment to reduce GHG emissions by 20% compared to 1990; to increase renewable sources to 20% of energy consumption; and to increase energy efficiency by 20%—all by 2020. The EU Effort Sharing Decision[19] establishes binding annual greenhouse gas emission targets for Member States for the period 2013–2020. These targets concern emissions from most sectors not included in the EU Emissions Trading System (EU ETS), such as transport (except aviation and international maritime shipping), buildings, agriculture and waste. The Effort Sharing Decision is designed to move Europe towards a low-carbon economy and increase its energy security.

In the US, over 20 States have established GHG reduction targets. For example, the California Global Warming Solutions Act of 2006 (Assembly Bill 32) requires a significant reduction in GHG emissions from the state of California. AB 32 is similarly designed to move California towards a lower carbon economy, while conserving resources and facilitating a strong economy. AB 32 requires California to reduce its GHG emissions to 1990 levels by 2020 (approximately equivalent to a 15% reduction compared to "business as usual").

Additionally, in November 2014, the US and China announced together a bilateral establishment of goals to reduce those nations' GHG emissions. The US announced it would reduce net GHG emissions by 26%–28% below 2005 levels by 2025 while China announced its goal was to "peak" CO_2 emissions around 2030 while increasing the non-fossil fuel portion of its energy mix to around 20% by 2030. The bilateral announcement came as a welcome surprise to the international community as the US and China together emit over one-third of global GHG emissions.[20]

2.6   *Mobilizing Climate Finance*

The UNFCCC has projected that an additional US\$ 200–210 billion of investments and financial flows in climate change mitigation and adaptation is needed annually until 2030 in both developed and developing countries in order to reduce global GHG emissions to sustainable levels.[21] However, because the bulk of GHG emissions growth is expected to come from resource-constrained middle and lower-income countries, international financial support is crucial if they are to be successful at combating climate change. Furthermore, these countries are not historically responsible for today's climate crisis and their priority is to change the trajectory of their economies and improve the livelihoods of their citizens.[22]

In response to this reality, investments in climate mitigation activities have been supplied largely by multilateral, bilateral and private financing sources. The World Bank and other Multilateral Development Banks (MDBs), such as the United Nations Development Programme (UNDP) and the United Nations Environment Programme (UNEP), are prime examples of multilateral financing sources. Other special financial institutions such as the Global Environment Facility (GEF) have been established specifically to provide monies for climate change mitigation activities. In addition, several carbon finance funds have been established to help facilitate the market for Certified Emission Reduction (CER) credits generated by GHG reduction projects for use in compliance and voluntary carbon markets.[21]

To give a sense of scale, the World Bank Group invested \$11.3 billion in 221 climate mitigation and adaption projects in over 60 countries during FY 2014. The Bank's private sector arm—the International Finance Corporation (IFC)—also invested \$1 billion in renewable energy generation during the same time period.[23] At the Copenhagen COP in 2009, industrialized countries set a joint goal to mobilize an additional \$100 billion a year by 2020 to address the needs of developing countries.[21]

Complementing these multilateral sources of climate finance is the newly operational, official financial mechanism of the UNFCCC, The Green Climate Fund (GCF). The GCF was proposed in the 2009 Copenhagen Accord and was formally adopted by countries during the COP in Durban, South Africa in 2011. The GCF's goal is to help developing countries address climate change "through the provision of grants and other concessional financing for mitigation and adaptation projects, programs, policies, and activities."[22] By the end of 2014, the GCF had received $7.5 billion in pledges, including Germany and France ($1 billion each), Sweden ($550 million), and Japan and the United States ($4.5 billion collective pledge ahead of a pledging meeting in November 2014).

3 GHG Emissions Data: Measurement, Reporting and Verification

3.1 Role of Emissions Inventories

Greenhouse gas inventories are the fundamental evidence base for linking public policy and actions to the scientific evidence for climate change. Consequently, regular GHG inventories are required as part of national reporting commitments under the Kyoto Protocol. The Inter-Governmental Panel on Climate Change (IPCC) has published methodological guidance to enable GHG inventories to be compiled on a consistent basis, and the UNFCCC conducts audits and reviews of signatory state inventories and National Communications. The longest-standing inventories (*e.g.* the UK GHG inventory) go back to the early 1970s and provide a valuable record of changes in sources and priorities for GHG emissions over this time period. This provides a useful resource for scientific analysis of climate change.

Inventories can be developed in a number of different ways. For example, inventories may be based on production or consumption metrics, whereby emissions are allocated to the activity giving rise to the emission (as for national inventories submitted to the UNFCCC), or to the final beneficiary of the carbon emitted (as for end user emission inventories and for carbon footprints). All approaches have advantages and disadvantages, and are designed to meet different requirements for different types of organisation. Emission inventories are primarily used for:

(a) Annual reporting of GHG emissions by source, to meet (*e.g.* UNFCCC) reporting requirements on a consistent, comparable basis between parties.
(b) Setting emission reduction targets, at the national, local and/or sector level, and then measuring progress towards those targets through time.
(c) Providing the historical evidence base for analysis of emissions through time by source, by region/organisation, enabling some degree of evaluating the impacts of policies and measures, economic drivers, climatic conditions.

Emission inventories serve as an invaluable resource for providing the historic evidence base for any *ex ante* GHG emissions policy appraisal. The historic data from emission inventories provide an insight in to the effectiveness of current and past policies. Analysis of policy objectives in relation to historic emission inventory data provides a means of measuring the successes of a policy. However, in many cases more detailed data is needed to assess the specific policy impacts, to understand the precise contribution of policies and measures to the observed inventory trend, amidst all of the other social, economic, financial and climatic factors influencing activities and emissions.

This requires the development of emissions mitigation actions and projects which are measurable, reportable and verifiable (referred to as "MRV"). Publishing of emissions inventories to support MRV mitigation actions is important, but there are many approaches to developing such inventories. These approaches are typically not mutually consistent, and make it hard to verify actual achievements in emissions reductions.

3.2   Policy and Inventories

Highlighting the specific areas in which a policy was effective provides the framework against which future policies can be can be designed. The European Environment Agency (EEA)'s publication on the *Trends and projections in Europe 2013*[24] shows the progress the EU 28 countries have made towards their 2020 Kyoto target, as well as providing sectoral breakdowns of emission reductions achieved, and the mechanism by which reductions were achieved. Looking at the technical improvement as well as historic trends of GHG emissions, inventories can provide the evidence for interpreting/ determining/ evaluating the effectiveness of national and international initiatives, and the basis for identifying and prioritising mitigation actions across an economy.

3.3   Sub-national and City Inventories

Sub-national emission inventories have been developed in many regions globally, often driven by specific local circumstances and priorities.[25] In the UK since the late 1990s, GHG and air emissions inventories have been developed for Scotland, Wales, Northern Ireland and England, and these inventories have underpinned the development of country-specific climate change strategies, policies and emission reduction targets.

These sub-national inventories reflect the need to tailor local and regional strategies and policy actions to the specific local opportunities, as mitigating GHGs to meet national-level targets is benefited by a strong, informed and motivated "bottom-up" climate change policy agenda driven by engaged local organisations. Experience from the UK system demonstrates that research to improve the evidence-base for sub-national inventories plays an increasing role in improving the available dataset for the national inventory, reducing uncertainties, improving policy sensitivity of the inventories and influencing the national-level decision-making. Access to more detailed local-level data to challenge and improve the national statistics ensures that the local to national-level climate change strategy can become increasingly joined-up and complementary in achieving prioritised, effective and cost-effective GHG mitigation nationally.

3.4 Inventory Verification

Independent verification of GHG inventories is crucial to provide confidence in the findings, and to enable inventories to be used as the basis for GHG policies and measures. Verification is normally carried out *via* an independent audit process and provides information for countries, companies and other sub-national entities to improve their inventories. There are many different considerations involved before embarking on the verification of a GHG inventory such as the expertise needed and expense of carrying out a full verification, the availability of reliable data, and the required level of accuracy and precision.[5]

4 Science and Policy Challenges

4.1 Benchmarks

Despite the considerable international response to climate change from both the public and private sectors, global mean surface temperatures continue to rise as a result of the persistent contribution of anthropogenic GHG emissions to the atmosphere (see Section 1). It is also expected that this persistent growth in emissions will continue as the world's population and economy continues to expand, unless considerable effort is made to scale-up GHG mitigation efforts to date. In fact, according to the IPCC, "Baseline scenarios, those without additional mitigation, result in global mean surface temperature increases in 2100 from 3.7 °C to 4.8 °C compared to pre-industrial levels".[7]

A temperature change caused by anthropogenic GHG emissions that exceeds 2 °C compared to pre-industrial levels is considered a critical threshold by the scientific community (see Section 2). More specifically, the international scientific community suggests that 2 °C is a temperature level which society should endeavour not to exceed in order to limit the dangerous effects of climate change. This threshold has become an integral statistic within the international response to climate change as evidenced by its use as a target in the 2009 Copenhagen Accord and the 2010 Cancún declaration.

Alongside this statistic is another important figure to the international community: 450 ppm. 450 ppm (parts per million) refers to the atmospheric concentration levels of GHGs at which temperature change can still be limited to 2 °C, and thus has become another threshold which both the scientific community as well as policy makers utilise to develop emissions scenarios and mitigation pathways.

Together, 2 °C and 450 ppm serve as critical parameters for decision-making within international efforts to combat climate change. To help illustrate their importance, the IPCC states that "Scenarios reaching atmospheric concentration levels of about 450 ppm CO_2eq by 2100 (consistent with a likely chance to keep temperature change below 2 °C relative to pre-industrial levels) include substantial cuts in anthropogenic GHG emissions by mid-century through large-scale changes in energy systems and potentially land use".[7]

A multitude of mitigation scenarios and actions have been considered and are being carried out globally. However, there is no silver bullet to solving the climate crisis and global consensus on the most cost-effective and equitable path towards reaching the aforementioned science-based targets remains elusive. Why, then, if climate change does indeed constitute a global crisis are consensus-based solutions so difficult to achieve?

4.2   Growth

According to the World Bank, over 1.2 billion people still do not have access to electricity and roughly 2.8 billion more depend on traditional solid fuels for their everyday cooking and heating needs.[26] Sustainable development requires that these people are afforded access to modern energy services to not only improve the health of their families and communities, but also to help lift them out of poverty and enhance their livelihoods. As energy is a major source of greenhouse gas emissions, the principle of sustainable development challenges us to find a way to meet the energy needs of a growing global population and economy with climate-friendly sources of energy.

4.3   Short-lived Climate Forcing Agents

Some climate forcing agents have relatively short atmospheric lifetimes compared to that of CO_2, which is up to around 20 years.[4] This opens up the possibility of targeted action to address short-lived GHGs, with the aim of limiting impacts of climate change over a time horizon of up to two decades. Ongoing action would also be needed to reduce emissions of long-lived climate forcing agents, principally carbon dioxide. The key short-lived climate forcing agents are:[4,27]

(a) Methane (atmospheric lifetime of about 12 years)
(b) Tropospheric ozone (atmospheric lifetime of 4 to 18 days)
(c) Stratospheric ozone (atmospheric lifetime from days to several years; varies through stratosphere)
(d) Black carbon (BC) (atmospheric lifetime of 3 to 8 days)
(e) Hydrofluorocarbons (HFCs) (atmospheric lifetime of HFC-134a of about 13 years)
(f) Stratospheric water vapour.

Initiatives such as the Climate and Clean Air Coalition to Reduce Short-Lived Climate Pollutants and the Global Methane Initiative are focused on securing reductions in emissions of short-lived climate forcing agents. Addressing emissions of these pollutants may be more tractable than securing reductions in emissions of carbon dioxide. One reason for this is that it is often possible to address local air quality and amenity impacts at the same time as reducing emissions of climate forcing agents. It may also be possible to use the mechanism of the Montreal Protocol to develop further controls of HFCs, focused on limiting climate change impacts.

4.4 Credibility

For those people who do not work in the field of climate change on a day-to-day basis the topic often remains a peripheral subject of speculation or debate in their lives. This applies to many of those responsible for making important legal, financial and political decisions—decisions that can impact each opportunity to build economic and environmental resilience into our cities, states and regions. There also remains a section of the public—a small yet vocal one—that continues to question the validity and certainty of the science of climate change despite the overwhelming consensus among the scientific community. These examples represent a small sample of the types of non-scientific uncertainty that serve as one hurdle to addressing climate change.

4.5 Political Economy

International cooperation on climate change can become a lower priority than, or be undermined by, the economic realities and social plans of participating countries. The complexity of the science is easily matched by the myriad of social, political and financial challenges that a massive environmental issue like climate change presents—not to mention the sheer number of stakeholders involved. The outcomes of a country's elections, for example, can fundamentally alter the direction of its stance on climate change and thus the entire shape of its energy and environmental policy. A sudden downturn in the global economy can quickly change a country's priorities and distribution of scarce resources.

4.6 Outsourcing Emissions

Even as developed countries make progress towards reducing their emissions, their climate gains are being cancelled out by the considerable increase of emissions from countries which are heavy manufacturers and exporters of carbon-intensive goods, most notably China. Depending on the producer and consumer of the goods and the corresponding GHG emissions this can be referred to as "leakage." Because a country's emissions are typically associated with what it produces, rather than what it imports and consumes, countries can claim to be making progress on their climate targets when in fact they have simply transferred a generation of emissions to another country. There is a need to apply carbon footprinting techniques to emissions inventories more widely, to enable GHG mitigation policies to focus on the final beneficiary of the carbon emitted.

5 The Energy Sector and Technology

The energy sector is responsible for approximately two-thirds of global GHG emissions and therefore plays an especially important role in the emissions mitigation challenge. According to the International Energy Agency (IEA) and the US Energy Information Administration (EIA), worldwide demand for energy is expected to grow by 37% and 56% by 2040, respectively.[28,29] This demand will be driven largely by economic growth in India, Southeast Asia, the Middle East and sub-Saharan Africa. Hence, supplying this growth with cleaner, more efficient energy sources and technologies—while decarbonizing existing energy systems in developed countries—is critically important if dangerous climate change is to be avoided.

5.1   Technology Types

The energy sector has been dominated historically by technologies that utilize fossil fuels to generate power—for example, oil used to fuel various modes of transportation or coal combusted to produce electricity. In fact fossil fuels account for almost 80% of total energy consumption, a figure that continues to increase gradually (see ref. 30, page 15). However, there is a wide range of climate-friendly sources of power, such as renewable energy from wind, solar, hydro, geothermal, and certain biomass fuels as well as non-renewable sources such as nuclear energy. Policy makers are using these technologies—along with policies and measures to use energy more efficiently—to provide large and small-scale solutions to decarbonize the energy sector and change the trajectory of the world's GHG emissions profile. Although not a source of energy, the capture and permanent storage of carbon dioxide from traditional coal-fired power plants is another energy-related technology that has the potential to provide considerable benefit to the climate, but is still largely in the research, development and pilot stage.

5.2   Market Potential and Challenges

Renewable energy is the fastest growing source of power according to the IEA and in 2013 expanded at its most rapid pace to date reaching 22% of world energy supply (compared with 21% in 2012) and is now on par with electricity generated by natural gas. The IEA has also reported that more than $250 billion was invested in new renewable power capacity worldwide and projects that levels of investment are likely to remain high through 2020.[31]

Despite this growth, renewable energy historically has not been cost-competitive compared to fossil fuel energy sources, with the exception of a few niche regions and markets. Like most infrastructure investments renewable energy projects are capital-intensive, and, while economic subsidies have been an available and important resource to the renewables sector as a whole, subsidies for fossil fuels are six times the level of support to renewable energy, amounting to roughly $523 billion in 2011 (see ref. 30, page 11). Even in situations with a more level playing field, policy, market and technology risks abound and often undermine a renewable energy project's feasibility.

Factors such as grid integration challenges, and macroeconomic and currency risks can alter financing conditions negatively. For example, utilities and grid operators have often been reluctant to support significant scale-up of technologies like wind and solar simply because of their intermittent nature. This kind of variability is not unusual for grid operators, but an abrupt and significant switch of the grid's energy supply from traditional fuels to renewables would likely stress existing power systems. Options to help balance the supply and demand for energy do exist and a include storage and load-management enabled by smart grids to name a few.[31]

Nuclear power is not considered a renewable form of energy but it does constitute almost 11% of world electricity production while producing few GHG emissions.[32] Because of its ability to produce large amounts of low carbon power, nuclear energy plays an important role in most climate mitigation scenarios where society meets the 2 °C and 450 ppm benchmarks. The nuclear industry, however, has many detractors due to safety concerns associated with several very serious incidents (*e.g.* Chernobyl, 1986; Three Mile Island, 1979; Fukushima Daiichi, 2011) as well as the production, use and disposal of its fuel and facilities. In addition to high capital and maintenance costs, decommissioning old plants and building new ones face considerable regulatory and political hurdles in virtually every country where it is employed.

6   What does the Future Hold?

6.1   GHG Emissions

In May 2013, CO_2 levels in the atmosphere exceeded 400 ppm for the first time in several hundred millennia (see ref. 30; Figure 3). This is a consequence of the ongoing and increasing emissions of CO_2, as illustrated in Figure 2. However, recent trends are for a slow-down in global CO_2 emissions (see ref. 33; Figure 5), mainly due to a slowdown in the Chinese economy—the world's largest emitter of greenhouse gases. Global CO_2 emissions increased by only 1.1% in 2012, less than half the average annual increase of 2.9% seen over the last decade. Downward trends in emissions were observed in the USA and Europe, with increased emissions in economies including China, Japan and India.

6.2   The Global Environment

The IPCC proposed a number of Representative Concentration Pathways (RCPs). These are trajectories of GHG concentrations leading to a given Radiative Forcing Value in the year 2100 relative to the pre-industrial (1750) climate. The four scenarios cover the range 2.5 to 8.5 Wm^{-2}. For context, anthropogenic radiative forcing was estimated to be 2.3 Wm^{-2} in 2011. This range of scenarios is forecast to lead to a rise in global surface temperature of 1.5 to 4.6 °C.[12]

Increasing radiative forcing results in increasing severity of impacts and risks, including those highlighted in Section 1. Conversely, reducing GHG emissions would be effective in reducing (although not eliminating) impacts and risks due to climate change, particularly in vulnerable communities in the second half of the 20th Century.

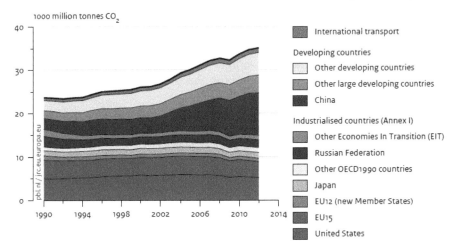

Figure 5   Global CO_2 emissions per region from fossil-fuel use and cement production.[33]

For example, if drastic emissions cuts are agreed and implemented from 2020 (represented by RCP2.6), sea levels are projected to rise by between 26 and 54 cm from 1986–2005 levels by the end of the century. If emissions continue to rise rapidly (represented by RCP8.5), sea levels are projected to rise by between 45 and 82 cm. The IPCC estimates that the higher RCP scenarios would be associated with impacts including harm or elimination of unique ecosystems, extreme weather events, global aggregate impacts and large-scale singular events.

The IPCC forecasts indicate that, whichever scenario is followed, we are likely to see rising temperatures and sea levels in 40 years' time. A rise in sea levels at least to the extent predicted under RCP2.6 appears almost inevitable.

6.3   Overcoming the Barriers to Effective Action

There are a number of global environmental issues for which effective international action has been carried out. For example, the Antarctic ozone hole first became apparent in the mid-1980s.[34] An extensive programme of co-operative scientific research was carried out, which confirmed that the active agent was emissions of chlorofluorocarbons. Production and use of the key chemicals has been phased out under the Montreal Protocol of 1987, with atmospheric concentrations now reducing. Consequently ozone depletion is forecast to reduce significantly over the coming decades.

Experience since the agreement of the Kyoto Protocol confirms that securing effective action to deal with climate change is extremely difficult. A number of factors can be identified which contribute to this difficulty:

(a) There are substantial vested interests in continuing to burn fossil fuels to maintain quality of life and economic activity. These vested interests operate at all levels: individual, corporate and national. All economic activity depends to a greater or lesser extent on energy use, which often means combustion of fossil fuels. In contrast, in the case of ozone depletion, while there were vested interests in producing the precursor chemicals, the economic benefit of these chemicals was relatively small-scale, and viable less harmful alternatives were found.

(b) The importance of energy to people and their economies cannot be understated. Yet the perils of climate change leaves the global community with little choice but to reshape a remarkably complex and dynamic sector that mixes technological, financial, development, geopolitical and security challenges. The introduction of a cost of carbon through market mechanisms such as the EU's Emissions Trading scheme or renewable portfolio standards make renewable energy sources more affordable, but technology and financial hurdles remain entrenched in a sector that historically has been slow to change. In many countries, energy and economic development policy is compartmentalised from climate policy. The UK's Department of Energy and Climate Change is a worthwhile first attempt to integrate energy and climate policy. In developing countries in particular, there are opportunities to introduce low-carbon energy solutions into development and energy provision policies and projects, but this can only happen if there is a joined-up approach to ensuring that new development follows low carbon pathways.

(c) There is often a mismatch between those who benefit from ongoing emissions of GHG (such as companies and shareholders in energy-intensive industries) and those most affected by climate change (such as those living in vulnerable developing countries). In many cases, the states most likely to be adversely affected by climate change possess neither the economic strength nor the political leverage to steer others towards fundamental change.

(d) Dealing with GHG emissions requires action at an individual level, as well as at governmental, corporate and international level. It is the aggregation of many such individual actions which could potentially result in a significant reduction in GHG emissions. Yet, an individual's minuscule contribution to reducing GHG emissions may be demotivating for that individual and potentially discourage individual actions.

There are many priorities which compete for attention from policy-makers and decision-makers. While climate change is arguably the biggest single issue facing mankind, other issues may be seen as more urgent, more evident, or more relevant to a particular local community. Local action is important in both securing reductions in GHG emissions, and also in ensuring that the global climate remains a key priority for action for politicians. The time may be right for sub-national groups such as cities and regions to take a lead in mitigating climate change.

Improved, robust engagement between public authorities and private sector is also essential in order to secure a sustainable future. In view of the evidence for climate change, businesses do not need to look far ahead to realise that they have a strong interest in preventing or mitigating climate change.

The scientific community has an ongoing role to produce data and analysis to link individual, corporate and state actions and policies to evaluations of GHG emissions, and analysis of evidence for changes in the global climate. Ongoing positive engagement between the scientific community, policy- and decision-makers is vital to ensure that decisions have access to the best scientific information. At a time when there are many competing issues, it is essential that the scientific and policy communities continue to engage with the media, to ensure that the public is properly informed about the vitally important issues associated with global climate change. Public support for positive action and investment to mitigate climate change is essential to give the politicians the mandate they need to take the difficult decisions which they will continue to face in the coming decades.

References

1. Intergovernmental Panel on Climate Change, Glossary in Climate Change 2013: The Physical Science Basis. Contribution of Working Group I to the Fifth Assessment Report of the Intergovernmental Panel on Climate Change, ed. T. F. Stocker, D. Qin, G.-K. Plattner, M. Tignor, S. K. Allen, J. Boschung, A. Nauels, Y. Xia, V. Bex and P. M. Midgley, Cambridge University Press, Cambridge, United Kingdom and New York, NY, USA.
2. Intergovernmental Panel on Climate Change, Summary for Policymakers in Climate Change 2013: The Physical Science Basis. Contribution of Working Group I to the Fifth Assessment Report of the Intergovernmental Panel on Climate Change, ed. T. F. Stocker, D. Qin, G.-K. Plattner, M. Tignor, S. K. Allen, J. Boschung, A. Nauels, Y. Xia, V. Bex and P. M. Midgley, Cambridge University Press, Cambridge, United Kingdom and New York, NY, USA.
3. National Research Council, *America's Climate Choices: Advancing the Science of Climate Change*, NRC Panel on Advancing the Science of Climate Change, Board on Atmospheric Sciences and Climate, Division on Earth and Life Studies, The National Academies Press, 2010, http://www.nap.edu/catalog.php?record_id=12782.
4. IPCC, Technical Summary in Climate Change 2013: The Physical Science Basis. Contribution of Working Group I to the Fifth Assessment Report of the Intergovernmental Panel on Climate Change, ed. T. F. Stocker, D. Qin, G.-K. Plattner, M. Tignor, S. K. Allen, J. Boschung, A. Nauels, Y. Xia, V. Bex and P. M. Midgley, Cambridge University Press, Cambridge, United Kingdom and New York, NY, USA.

5. Intergovernmental Panel on Climate Change, 2006 IPCC Guidelines for National Greenhouse Gas Inventories, Volume 1: General Guidance and Reporting, Chapter 6, Quality Assurance/Quality Control and Verification, Intergovernmental Panel on Climate Change, 2006.

6. United Nations Environment Programme, *The Emissions Gap Report 2013*, UNEP, Nairobi, 2013.

7. Intergovernmental Panel on Climate Change, Summary for Policymakers: Climate Change 2014, Mitigation of Climate Change. Contribution of Working Group III to the Fifth Assessment Report of the Intergovernmental Panel on Climate Change, ed. O. Edenhofer, R. Pichs-Madruga, Y. Sokona, E. Farahani, S. Kadner, K. Seyboth, A. Adler, I. Baum, S. Brunner, P. Eickemeier, B. Kriemann, J. Savolainen, S. Schlömer, C. von Stechow, T. Zwickel and J. C. Minx, Cambridge University Press, Cambridge, United Kingdom and New York, NY, USA.

8. European Commission, Emissions Database for Global Atmospheric Research (EDGAR), data published 2014 on http://edgar.jrc.ec.europa.eu, accessed 2015.

9. Intergovernmental Panel on Climate Change 1996, Climate Change 1995: A report of the Intergovernmental Panel on Climate Change, Second Assessment Report of the Intergovernmental Panel on Climate Change.

10. Intergovernmental Panel on Climate Change, *Climate Change 2007: Synthesis Report, Contribution of Working Groups I, II and III to the Fourth Assessment Report of the Intergovernmental Panel on Climate Change*, 2007, ISBN 92-9169-122-4.

11. P. Tans and R. Keeling, *Trends in Atmospheric Carbon Dioxide*, National Oceanic and Atmospheric Administration Earth System Research Laboratory and Scripps Institution of Oceanography, 2014, available from http://www.esrl.noaa.gov/gmd/ccgg/trends/, accessed 2015.

12. Intergovernmental Panel on Climate Change, Summary for policymakers in Climate Change 2014: Impacts, Adaptation, and Vulnerability. Part A: Global and Sectoral Aspects. Contribution of Working Group II to the Fifth Assessment Report of the Intergovernmental Panel on Climate Change, ed. C. B. Field, V. R. Barros, D. J. Dokken, K. J. Mach, M. D. Mastrandrea, T. E. Bilir, M. Chatterjee, K. L. Ebi, Y. O. Estrada, R. C. Genova, B. Girma, E. S. Kissel, A. N. Levy, S. MacCracken, P. R. Mastrandrea, and L. L. White, Cambridge University Press, Cambridge, United Kingdom and New York, NY, USA, pp. 1–32.

13. United Nations Framework Convention on Climate Change, 2014, First steps to a safer future: Introducing The United Nations Framework Convention on Climate Change, http://unfccc.int/essential_background/convention/items/6036.php.

14. United Nations Framework Convention on Climate Change, 2014, National Reports, http://unfccc.int/national_reports/items/1408.php.

15. I. Fry, Twists and Turns in the Jungle: Exploring the Evolution of Land Use, Land-Use Change and Forestry Decisions within the Kyoto Protocol, *RECIEL*, 2002, **11**(2), 159.

16. I. Fry, More Twists, Turns and Stumbles in the Jungle: A Further Exploration of Land Use, Land-Use Change and Forestry Decisions within the Kyoto Protocol, *RECIEL*, 2007, **16**(3), 341.

17. Congressional Research Service 2013, International Climate Change Financing: The Green Climate Fund (GCF), Congressional Research Service, Lattanzio, R., April, 2013. http://fas.org/sgp/crs/misc/R41889. pdf.

18. Committee on Climate Change, Building a low-carbon economy – the UK's contribution to tackling climate change, 2008, available from http://www.theccc.org.uk/publication/building-a-low-carbon-economy-the-uks-contribution-to-tackling-climate-change-2/.

19. European Commission, 23/04/2009 – Decision No 406/2009/EC of the European Parliament and of the Council of 23 April 2009 on the effort of Member States to reduce their greenhouse gas emissions to meet the Community's greenhouse gas emission reduction commitments up to 2020.

20. White House 2014, Fact Sheet: U.S.-China Joint Announcement on Climate Change and Clean Energy Cooperation, The White House, November 2014. http://www.whitehouse.gov/the-press-office/2014/11/11/fact-sheet-us-china-joint-announcement-climate-change-and-clean-energy-c.

21. United Nations Environment Programme, 2012, Accessing International Financing for Climate Change Mitigation: A Guidebook for Developing Countries, TNA Guidebook, Series, page xvi. Limaye, D., Zhu, X. UNEP Risoe Centre on Energy, Climate and Sustainable Development. http://tech-action.org/media/k2/attachments/TNA_Guidebook_MitigationFinancing_13.pdf.

22. Congressional Research Service, International Climate Change Financing: The Green Climate Fund (GCF), Congressional Research Service, Lattanzio, R., April, 2013. http://fas.org/sgp/crs/misc/R41889.pdf.

23. World Bank, http://www.worldbank.org/en/topic/climatechange/overview#2.

24. European Environment Agency, Trends and projections in Europe 2013, Report No 10/2013, available from http://www.eea.europa.eu/publications/trends-and-projections-2013.

25. World Resources Institute, *Global Protocol for Community-Scale Greenhouse Gas Emission Inventories*, 2014, available from http://www.ghgprotocol.org/city-accounting.

26. World Bank, 2013a, Toward a Sustainable Energy Future for All: Directions for the World Bank Group's Energy Sector, Board Report, The World Bank, July 2013.

27. United Nations Environment Programme, 2011, Near-term Climate Protection and Clean Air Benefits: Actions for Controlling Short-Lived Climate Forcers – A UNEP Synthesis Report, available from http://hqweb.unep.org/publications/ebooks/slcf/.

28. International Energy Agency World Energy Outlook 2014, Executive Summary, page 1, International Energy Agency, November 2014.

29. Energy Information Administration, International Energy Outlook 2013, U.S. Energy Information Administration, July 2013.
30. International Energy Agency, Redrawing the Energy-Climate Map. World Energy Outlook Special Report, International Energy Agency, June 2013.
31. International Energy Agency 2014, Renewable Energy Medium-Term Market Report 2014, Executive Summary, International Energy Agency, August 2014.
32. International Energy Agency 2014, Key World Energy Statistics 2014, International Energy Agency, September 2014.
33. J. G. J. Oliver, G. Janssens-Maenhout, M. Muntean and J. A. H. W. Peters, *Trends in global CO_2 emissions*, 2013 Report, PBL Netherlands Environmental Assessment Agency, The Hague, Joint Research Centre, Ispra, 2013.
34. J. C. Farman, B. G. Gardiner and J. D. Shanklin, Large losses of total ozone in Antarctica reveal seasonal ClOx/NOx interaction, *Nature*, 1985, **315**, 207–210.

Trends in Local Air Quality 1970–2014

ROY M. HARRISON,* FRANCIS D. POPE AND ZONGBO SHI

ABSTRACT

Trends in air pollutant emissions from 1970 to the present day are described and discussed for the United Kingdom and the United States. These are compared with trends in ambient concentrations and the similarities and divergences are discussed. There have been notable success stories in terms of reductions in smoke and sulfur dioxide concentrations, although the current concentrations are still a matter of concern. Concentrations of carbon monoxide and many volatile organic compounds have decreased very substantially in recent years. Nitrogen dioxide, particulate matter and ozone still give cause for concern. The trends in the United Kingdom and United States are contrasted with those in less developed countries for which China is taken as a case study. The trends in emissions and airborne concentrations are reviewed and contrasted with those in the more developed world.

1 Introduction

Air pollution has been a highly topical subject in the public mind over the 40 years considered by this volume. The classical London smogs caused by black smoke and sulfur dioxide emissions arising largely from domestic coal combustion persisted into the 1960s but were disappearing with the arrival of the 1970s. This period, however, was characterised by a huge growth in

---

*Corresponding author.

Issues in Environmental Science and Technology No. 40
Still Only One Earth: Progress in the 40 Years Since the First UN Conference on the Environment
Edited by R.E. Hester and R.M. Harrison
© The Royal Society of Chemistry 2015
Published by the Royal Society of Chemistry, www.rsc.org

the volumes of motor traffic and therefore a transformation of the pollution climate into one dominated by road vehicle emissions. By 1991, a pollution episode in London in early December[1] was notable mainly for the highly elevated concentrations of nitrogen dioxide rather than sulfur dioxide, and particle loadings although high, were very much lower than those which characterised the smogs of the 1950s.

The United Kingdom can be taken as fairly typical of western Europe, but other countries were progressing at a different pace. In the United States, the photochemical smog problems most notably afflicting California were leading to action on traffic-generated pollutants somewhat earlier than in Europe. In North America, oxidation catalysts were introduced on gasoline vehicles in 1975 with three-way catalysts becoming mandatory on new cars in 1981. In Europe, however, the first move was to three-way catalysts which became mandatory in 1993. On the other hand, most of the less developed world has lagged behind developments in North America and Europe, and as a consequence many major cities in the less developed world suffer problems currently both with classic smoke and sulfur dioxide smogs but accompanied by high concentrations of traffic-generated pollutants.

A pollutant which gave rise to concern worldwide in the early part of the period was lead which was added to motor fuel in the form of alkyllead additives, causing emission of inorganic lead salts into the atmosphere.[2] These lead salts led to human exposure not only through direct inhalation but by incorporation in surface dusts which could be ingested by children and by deposition on food crops. Allegations of retarded IQ development in children were highly controversial, but most developed countries took pre-cautionary action first to limit and then remove lead from motor fuels. This has been highly effective in reducing airborne concentrations of lead, and has also proved to be soundly based in the light of more recent epidemiology.[3] The use of lead in motor fuels has been banned in the European Union since the year 2000.

2 The United Kingdom

The great smog of December 1952 was associated with at least 4000 add-itional deaths. This led to detailed investigation of the causes which was followed by the passage of Clean Air Acts in 1956 and 1968. These led to the establishment of smokeless zones which were areas in which it was required to burn only smokeless fuels or to use appliances capable of burning fuel smokelessly. These legislative developments took place at around the same time that natural gas was becoming widely available as a fuel and many households changed over to gas-fired central heating, dispensing with do-mestic solid fuel combustion altogether. The outcome was a huge im-provement in urban air quality with respect to black smoke and sulfur dioxide. At the same time, urban electricity generating power stations were closing with the construction of large coal-fired stations in rural areas emitting through high stacks. There was a consequent convergence of urban and rural sulfur dioxide concentrations at levels far below those which had prevailed previously in urban areas.[4] Similarly, urban concentrations of black smoke declined rapidly and far faster than national emissions.

2.1 Trends in Emissions in 1970

2.1.1 Particulate Matter (Including Black Smoke). Routine monitoring of particulate matter by mass (as PM_{10}; *i.e.* the mass concentration of particles <10 μm aerodynamic diameter) did not start in the UK until the early 1990s. Prior to that, airborne concentrations had been measured for many years as black smoke which was measured by the blackness of filter papers using a calibration based upon coal smoke.[5] During the time that the measurement networks were analysing black smoke, emissions inventories were available also for black smoke. That is now no longer the case and the most appropriate recent inventory is that for $PM_{2.5}$, referring to particles <2.5 μm aerodynamic diameter which appears in Figure 1.[6] It is important to emphasise that emissions inventories can

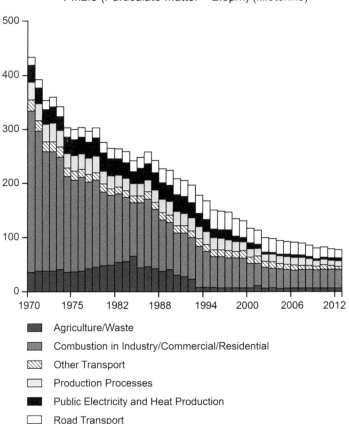

PM2.5 (Particulate Matter < 2.5μm) (kilotonne)

Figure 1   Trends in UK emission of primary $PM_{2.5}$ (kilotonnes) according to major source category from 1970–2012.[6]

include only primary pollutants (*i.e.* those directly emitted) and do not include secondary pollutants which are formed in the atmosphere through atmospheric chemistry. This is one of the many reasons why trends in ground-level concentrations may not well reflect the trends in emissions. In the case of $PM_{2.5}$, there has been a huge reduction in the category entitled "combustion in industry/commercial/residential" which includes coal smoke. The banning of agricultural waste burning in the early 1990s had a major impact, and currently the main sources are road transport (mainly diesel vehicles) and combustion of solid fuels, including wood as well as coal.

A significant proportion of primary particulate matter is in the form of coarse particles defined as those between 2.5 and 10 μm aerodynamic diameter. PM_{10} includes both the coarse and fine ($PM_{2.5}$) particles and the National Inventory for the UK for PM_{10} is shown in Figure 2. Much of this is reflective of the $PM_{2.5}$ inventory which makes up a substantial proportion,

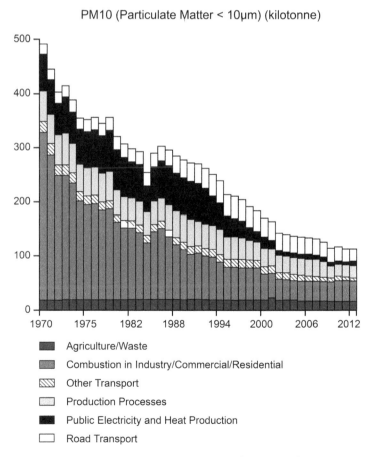

Figure 2   Trends in UK emission of primary PM_{10} (kilotonnes) according to major source category from 1970–2012.[6]

but some differences in contributions are notable. In particular, production processes which include some dusty processes such as quarrying and open cast coal mining represent a bigger proportion of PM_{10} emissions than for $PM_{2.5}$. Road transport emissions of PM_{10} also exceed those for $PM_{2.5}$ because of the relatively coarse particles emitted from abrasion of brakes and tyres.

2.1.2 Sulfur Dioxide.   Emissions of sulfur dioxide arise predominantly from the combustion of fuels containing sulfur which is in large proportion converted to sulfur dioxide in the combustion process. Some industrial processes such as metal smelting and sulfuric acid manufacture are also sources of sulfur dioxide emissions. Trends in sulfur dioxide emissions in Figure 3 show a huge reduction in the categories referred to as "combustion in energy and transformation industry" and "combustion in manufacturing industry". The larger of the two is the former which refers primarily to combustion of coal in power stations, from which emissions have reduced very substantially since the early 1990s due to a steady

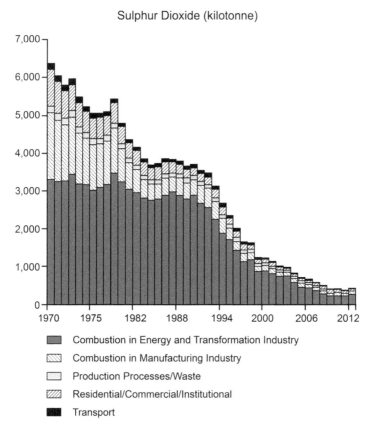

Figure 3   Trends in UK emission of primary sulfur dioxide (kilotonnes) according to major source category from 1970–2012.[6]

reduction in the use of coal for generating electricity (in favour of natural gas) and the installation of flue gas desulfurisation abatement equipment on major power plants. Emissions from manufacturing industry have also reduced substantially due to the use of cleaner fuels and more effective abatement plant. Road traffic has been a relatively minor contributor through this period and progressive reductions in the sulfur content of motor fuels have caused reductions in this source.

2.1.3 Oxides of Nitrogen. Oxides of nitrogen are emitted predominantly from combustion processes, and arise both from the combustion of nitrogen contained in some fuels (*e.g.* coal) and the combination at high temperatures of atmospheric oxygen and nitrogen. Consequently, a wide range of combustion source shown in Figure 4 make a contribution to emissions of oxides of nitrogen. All have shown some reduction over the period 1990–2012, with the most notable reductions coming from public

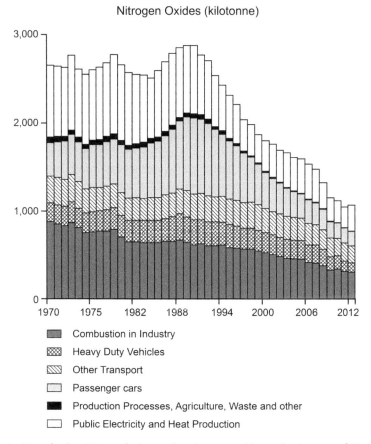

Figure 4 Trends in UK emission of primary oxides of nitrogen (kilotonnes) according to major source category from 1970–2012.[6]

electricity and heat production, passenger cars and combustion in industry. The emissions reductions from heavy duty vehicles and other transport (*e.g.* railways, canals *etc.*) have been relatively modest. A major limitation on the reduction in emissions from road traffic has been the small reduction in emissions from diesel engines[7] whereas petrol engines, which have been fitted with three-way catalysts since the early 1990s, have seen a very substantial reduction in emissions of oxides of nitrogen which accounts for a large part of the emissions reduction from passenger cars seen in Figure 4. In more recent years, increased sales of diesel passenger cars have led to a slowing of that reduction. A decline in energy intensive industries such as iron and steel contributed to the fall in industrial combustion emissions between 1970 and 1985, and the fitting of low-NO_x burners in large power plants from 1988 onwards led to a reduction in emissions.

2.1.4 Volatile Organic Compounds. The UK National Atmospheric Emissions Inventory accounts separately for methane and for non-methane volatile organic compounds (VOCs). Methane is of little significance for local air quality but is important as a greenhouse gas with a global warming potential (per unit mass) very much greater than that of carbon dioxide. The inventory for non-methane volatile organic compounds shown in Figure 5 shows a multiplicity of sources with "solvent and other product use" dominating emissions in 2012 and having decreased to only a relative small degree since 1970. Some benefits have accrued from the reduction in organic solvent content of a number of products such as paints in recent years. The most notable decline has been in emissions from transport and particularly road transport occasioned by the fitting of three-way catalytic converters to petrol cars and of oxidation catalysts on many diesels. Reductions have also occurred in the category of "extraction and distribution of fossil fuels" due to implementation of a variety of measures including the fitting of vapour recovery on vehicle fuel tanks. Overall, the emissions reduction for non-methane volatile organic compounds has been significantly less successful than for some other pollutants such as sulfur dioxide and primary particulate matter.

2.1.5 Carbon Monoxide. Emissions of carbon monoxide arise from inefficient combustion of fossil fuels, and for the majority of the period since 1970, the predominant source has been from road transport and especially from petrol engines which operate at a higher fuel-to-air ratio than diesel engines.[8] However, the introduction of catalytic converters in the early 1990s has led to a dramatic reduction in emissions from the road transport source. The other source to have decreased substantially over this period is residential combustion primarily due to much reduced use of solid fuels such as coal in favour of gas and electricity for heating (see Figure 6).

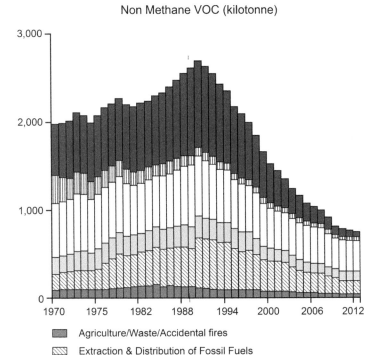

Non Methane VOC (kilotonne)

Agriculture/Waste/Accidental fires

Extraction & Distribution of Fossil Fuels

Production Processes

Solvent & Other Product Use

Stationary Combustion

Transport

Figure 5 Trends in UK emission of primary non-methane volatile organic compounds (kilotonnes) according to major source category from 1970–2012.[6]

2.1.6 Toxic Organic Micropollutants. This term describes a range of pollutant classes, all of significance for human health and referred to as micropollutants because of their generally low concentrations.

The polycyclic aromatic hydrocarbons (PAHs) are an important class of organic micropollutants emitted from all inefficient combustion of carbonaceous fuels and also from the anode-baking process used in aluminium smelting. The PAHs include a very wide range of compounds, several of which are carcinogenic, but much of the carcinogenic risk is associated with exposure to one compound, benzo[*a*]pyrene.[9] Emissions of this compound since 1990 are shown in Figure 7 and show a dramatic reduction. The cessation of burning agricultural stubble waste in the early 1990s caused a dramatic reduction in emissions from this source, and the other category to have shown a major reduction is that of production processes, primarily referring to the anode-baking process in aluminium smelting. The progressive closure of aluminium smelters in the UK has been responsible for

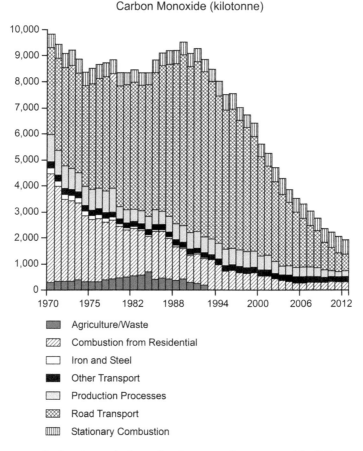

Figure 6   Trends in UK emission of primary carbon monoxide (kilotonnes) according to major source category from 1970–2012.[6]

the major part of the reduction from this source category. Residential/ commercial/ institutional combustion has remained a small but significant source which now dominates emissions.

A second important class of organic micropollutants are the dioxins and furans. These have given rise to substantial public concern, although exposures are actually very small. Emissions since 1990 expressed as the toxic equivalent of the sum of many compounds shown in Figure 8 show a substantial reduction in the earlier periods since 1990 but a levelling off over the past decade. The residential/commercial/institutional combustion source reduced substantially in the early years of this period due to a change-over from coal to cleaner fuels such as gas and electricity. However, a reversal of trends is seen in recent years due primarily to increased combustion of wood. Emissions from waste incineration were cut substantially mainly due to implementation of the EU Waste Incineration Directive which placed strict limits

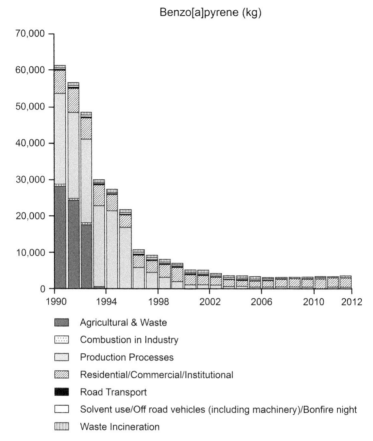

Figure 7   Trends in UK emission of primary benzo[*a*]pyrene (kilograms) according to major source category from 1990–2012.[6]

on dioxin emissions from incinerators from 1995 onwards leading to closure of plant, or fitting of more efficient abatement plant to existing incinerators.

2.2   Ambient Air Monitoring

Major changes have taken place both in the monitoring networks and the availability of public information in the UK in the period since 1970. The National Survey of Smoke and Sulfur Dioxide was established in 1961, and at its peak had around 2000 monitoring stations. These were operated in the main by local government authorities and used very basic technology to measure black smoke from the change of reflectance of a filter paper and sulfur dioxide by the increase in acidity of a hydrogen peroxide solution used to trap and oxidise sulfur dioxide. These techniques were designed essentially for the monitoring of coal smoke and of high sulfur dioxide levels and were widely used until the 1990s, by which time it had become very clear that the main source of black smoke at most locations was diesel vehicles rather

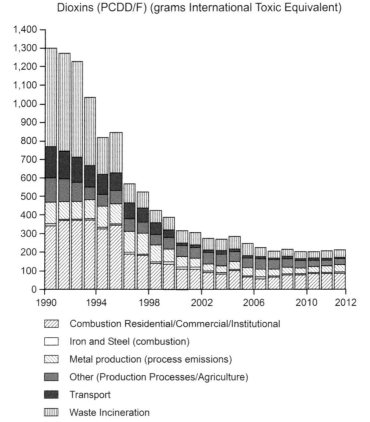

Figure 8   Trends in UK emission of primary dioxins (polychlorinated dioxins/furans; PCDD/F) (grams international Toxic Equivalent) according to major source category from 1990–2012.[6]

than coal smoke, and sulfur dioxide concentrations had declined to the point where the method produced unreliable data. This led to a major change which started in 1992 with the opening of the first comprehensive automatic monitoring stations which now form the backbone of the Automatic Urban and Rural Network. These stations used automatic instruments to measure concentrations of sulfur dioxide, particulate matter (as PM_{10}), oxides of nitrogen (as NO, NO_2 and NO_x), carbon monoxide and ozone. Data were collected continuously with fast-response monitors and data were made available to the public in near real-time linked subsequently to health advice for those susceptible to air pollutant exposures. The number of automatic monitoring stations increased steadily with time, although not all stations were capable of measuring the full suite of pollutants, and many now also measure $PM_{2.5}$. A smaller number of automated instruments were installed for measuring volatile organic compounds (predominantly hydrocarbons) on a semi-continuous basis. These instruments sample over almost one hour, collecting an integrated sample which is then desorbed into a gas chromatograph for separation and quantification of individual volatile organic compounds. The number of sites at which such instruments are operated has declined significantly in recent years but this has been accompanied by the introduction of less sophisticated instrumentation to measure key compounds such as benzene on a time-integrated basis.

One of the pollutants which has consistently exceeded health-based standards at many locations is nitrogen dioxide, and in order to obtain better spatial coverage of data for nitrogen dioxide, many local government authorities deployed diffusion tubes which are a low cost method of obtaining long term average concentrations of nitrogen dioxide.[10] Due to the relatively low cost of the tubes and their analysis, these could be deployed at relatively high spatial density, but the data generated are of low temporal resolution and have a tendency to be somewhat imprecise.[11]

The spatial distribution of automatic monitors has been determined largely by rules established by the European Commission. These are focussed very largely upon centres of population and there have been only very small numbers of stations monitoring concentrations in rural areas. Although such datasets are of limited value for the protection of human health, they can be extremely valuable in elucidating the sources and atmospheric processes which determine measured concentrations.

In common with all European Union countries, the UK has to comply with European Union air quality Limit Values, which are listed in Table 1. There are also requirements relating to the Average Exposure Indicator, which is the population-weighted mean concentration across the entire country (see Table 2).

2.3 Trends in Airborne Concentrations from 1970

2.3.1 Smoke and Sulfur Dioxide. Data are available from stations in the West Midlands conurbation of the UK going back to 1955 with comprehensive data from around 1965. Figures 9 and 10 show the trends in black smoke and sulfur dioxide at monitoring stations within the West Midlands conurbation between 1955 and 2005, respectively. These indicate a huge reduction in concentrations over this period largely reflective of the trend towards cleaner fuels and away from coal for domestic heating.

2.3.2 Particulate Matter as PM_{10}. The advent of the automatic network from 1992 onwards led to a rapid reduction in the monitoring of black smoke and the availability of data from continuous instruments based upon the Tapered Element Oscillating Microbalance (TEOM) principle. Data from 1992 to 2006 appear in Figure 11.[12] These show a modest but steady decrease in concentrations between 1992 and 2000, with a levelling of concentrations after that except for the relatively high pollution year of 2003. The TEOM instruments used up to that time had an inlet heated to 50 °C which led to the loss of semi-volatile constituents such as

Table 1   European Union air quality limit values and target values.

Pollutant	Concentration	Averaging period	Date entered	Permitted exceedances each year
Fine particles (PM$_{2.5}$)	25 µg m^{-3}	1 year	2015	n/a
Sulfur dioxide (SO$_2$)	350 µg m^{-3}	1 hour	2005	24
	125 µg m^{-3}	24 hours	2005	3
Nitrogen dioxide (NO$_2$)	200 µg m^{-3}	1 hour	2010	18
	40 µg m^{-3}	1 year	2010	n/a
PM$_{10}$	50 µg m^{-3}	24 hours	2005	35
	40 µg m^{-3}	1 year		n/a
Lead (Pb)	0.5 µg m^{-3}	1 year	2005	n/a
Carbon monoxide (CO)	10 mg m^{-3}	Maximum daily 8 hour mean	2005	n/a
Benzene	5 µg m^{-3}	1 year	2010	n/a
Ozone	120 µg m^{-3}	Maximum daily 8 hour mean	2010	25 days averaged over 3 years
Arsenic (As)	6 ng m^{-3}	1 year	2012	n/a
Cadmium (Cd)	5 ng m^{-3}	1 year	2012	n/a
Nickel (Ni)	20 ng m^{-3}	1 year	2012	n/a
Polycyclic aromatic hydrocarbons	1 ng m^{-3} (expressed as concentration of benzo[*a*]pyrene)	1 year	2012	n/a

Table 2   European Union exposure reduction requirements.

Title	Metric	Averaging period	Legal nature	Permitted exceedances each year
PM$_{2.5}$ exposure concentration obligation	20 µg m^{-3}	Based on 3 year average	Legally binding in 2015 (years 2013, 2014, 2015)	n/a
PM$_{2.5}$ exposure reduction target	Percentage reduction^a + all measures to reach 18 µg m^{-3} (AEI)	Based on 3 year average	Reduction to be attained where possible in 2020, determined on the basis of the value of exposure indicator in 2010	n/a

^{a}Depending on the value of AEI in 2010, a percentage reduction requirement (0%, 10%, 15% or 20%) is set in the Directive. If AEI in 2010 is assessed to be over 22 µg m^{-3}, all appropriate measures need to be taken to achieve 18 µg m^{-3} by 2020.

Note: Average exposure indicator (AEI) is determined as a 3 year running annual mean PM$_{2.5}$ concentration averaged over the selected monitoring stations in agglomerations and larger urban areas, set in urban background locations to best assess the PM$_{2.5}$ exposure to the general population.

ammonium nitrate, and in order to "correct" the data, an adjustment was made by multiplying by 1.3 before comparison of data with Air Quality Objectives and Limit Values.[13] After 2006, instruments became available

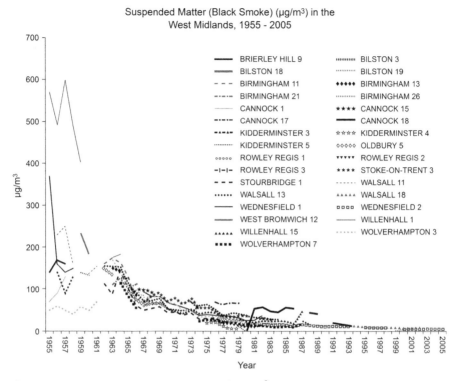

Figure 9 Measurements of black smoke (μg m^{-3}) at sites in the UK West Midlands between 1955 and 2005.

which were designed to compensate automatically for the loss of semi-volatile constituents and this fact, together with the movement of monitoring station locations, made it impossible to extend the datasets shown in Figure 11 with further comparable data. It appears likely, however, that there has been some modest reduction in PM$_{10}$ concentrations since 2006.

Widespread monitoring of PM$_{2.5}$ was instigated only following the introduction of the updated Directive on Ambient Air Quality and Cleaner Air for Europe (2008/50/EC) and high quality data on temporal trends are not yet available.

2.3.2.1 Secondary Particulate Matter

The trends in emissions of particulate matter discussed in Section 2.1.1 related only to primary particulate matter. A large proportion of airborne particulate matter in most of the developed world arises from secondary production, and the main components are sulfates, nitrates and secondary organic compounds. Both nitrates and secondary organic compounds have semi-volatile components and are therefore subject to measurement artefacts and temporal trend data are in the main rather poor. Additionally, estimation of secondary organic aerosol mass is difficult as discriminating

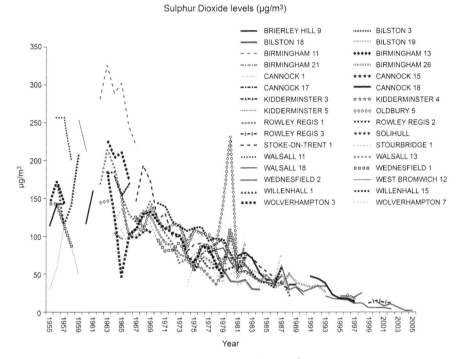

Figure 10   Measurements of sulfur dioxide ($\mu g\ m^{-3}$) at sites in the UK West Midlands between 1955 and 2005.

between primary and secondary material involves assumptions and/or measurement artefacts. Consequently, the best data are for sulfates and an analysis of trends in sulfates at UK sites has shown a small downward trend at all sites generally within the range of -0.1 to $-0.2\ \mu g\ m^{-3}$ per year.[14] This is a rate of decline very much slower than that in sulfur dioxide emissions or sulfur dioxide airborne concentrations. This appears to be a result of the atmospheric chemistry of conversion of sulfur dioxide to sulfate and appears also to apply to nitrogen oxides. Consequently, abatement of secondary particulate matter is likely to progress far more slowly than that of the precursor gases.

2.3.3 Oxides of Nitrogen.   Comprehensive continuously collected high quality monitoring data for the UK have been available only since 1992. An example of data for nitrogen dioxide appears in Figure 12 which also includes calculated NO_x emissions from road traffic in the UK. Ground-level concentrations of NO_x (which is the sum of $NO + NO_2$) have declined in a similar manner to the NO_x emissions from road traffic shown in Figure 12. However, the pollutant to which ambient air quality standards apply is nitrogen dioxide, and it may be seen from Figure 12 that at the London Bloomsbury site, although there has been some inter-annual variability, overall there has been very little change in concentrations between

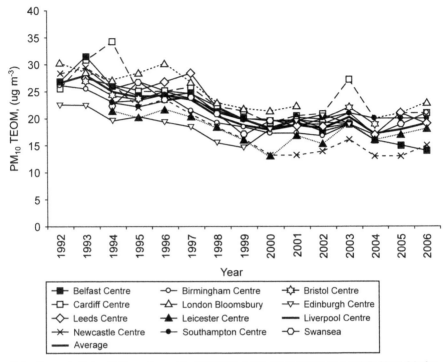

Note: Concentrations of PM_{10} measured by TEOM (as in this case) should be multiplied by 1.3 before comparison with EU Limit Values

Figure 11 Time series of annual mean PM_{10} concentrations at the long-running urban background sites in the United Kingdom from 1992 to 2006.[12]

1992 and 2010. This pattern of behaviour is reflected at most heavily trafficked sites across the UK,[7] and western Europe more generally. The reasons are two-fold. Firstly, the major component of NO_x as emitted from combustion sources is nitric oxide which is rapidly converted to nitrogen dioxide by reaction with ozone. However, the molar concentrations of nitric oxide at urban sites exceed those of available ozone and consequently the conversion to nitrogen dioxide is limited by the oxidant availability. Consequently, a reduction in nitric oxide emissions has only a very minor effect on nitrogen dioxide under such circumstances. However, some reduction in concentrations would be expected and a second factor has proved to be highly influential. As emissions standards for particulate matter from diesel vehicles have become stricter, vehicle technologies have been developed which reduce particle emissions but which increase the fraction of nitrogen dioxide in the exhaust, such that this has increased from around 5% in the early 1990s to 20–30% in the current decade. The emissions of primary nitrogen dioxide have strongly influenced ambient concentrations and provide the major explanation as to the lack of significant decline in ambient concentrations of nitrogen dioxide. Emissions from

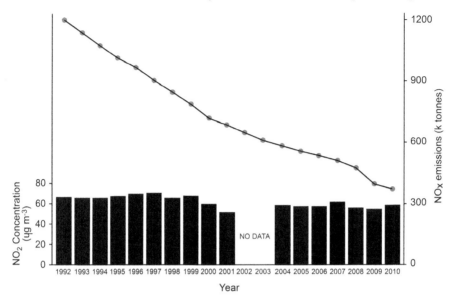

Figure 12   UK emissions of NO$_x$ from road traffic and concentrations of nitrogen
dioxide in London, Bloomsbury, 1992–2010.

other sources such as power stations are generally at a substantial altitude
and are highly dispersed by the time they reach ground level, having only a
relatively minor influence upon urban concentrations.

2.3.4   Volatile Organic Compounds.   There is considerable evidence of a
strong decline in the ground level concentrations of volatile organic compounds resulting particularly from the use of catalytic converters on petrol
cars. A time series for benzene appears in Figure 13 which shows a fairly
close parallel between the estimates of the emissions inventory for total
emissions and the decline in the average annual mean concentration
across 20 long running sites.[15] These data go back only to 2002 and are
part of a long running downward trend.

2.3.5   Carbon Monoxide.   There have been vast reductions in airborne concentrations of carbon monoxide resulting primarily from the use of catalytic
converters on petrol vehicles, which were previously far the major source of
carbon monoxide emissions. While UK emissions have fallen from around
9000 kilotonnes per year in 1990 to around 2000 kilotonnes per year in
2012, ambient concentrations expressed as the average maximum 8 hour
running mean carbon monoxide concentration have declined from 12 to
below 2 mg m^{-3}.[15] Thus, the decline in ground level concentrations was a
little larger than that in total emissions presumably because emissions
from road traffic fell faster over this period than total national emissions.

2.3.6   Toxic Organic Micropollutants.   Due to changes in network configuration and monitoring methods, there are no good homogeneous

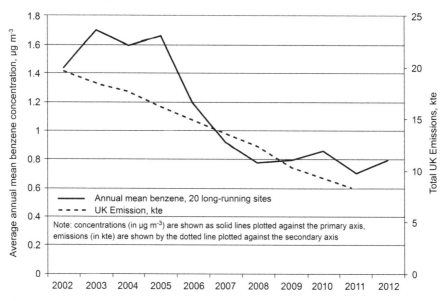

Figure 13 Annual mean benzene concentration, mean of 20 long-running non-automatic sites.

datasets from which to evaluate temporal trends comprehensively. However, there are sampling campaign-based measurements which provide superficial evidence for a very substantial decline in airborne concentrations of polycyclic aromatic hydrocarbons. For example, Lawther and Waller[16] found a reduction of 90% in London's concentration of benzo[a]-pyrene between 1949 and 1973, and Smith and Harrison[17] report a decline in PAH concentrations of 2–10 fold in Birmingham, UK between 1977/8 and 1992. Since 1992, mean concentrations of PAH in Birmingham have declined further by a factor of more than two.[18]

2.3.7 Ground-level Ozone. There has been a general trend across Europe in which peak episodic ozone concentrations have been declining, but the annual average at most sites has been slowly increasing. The reduction in episodic peak concentrations is attributed to reductions in the precursors of ozone formation, *i.e.* oxides of nitrogen and volatile organic compounds. The reasons for the increase in baseline concentrations and hence the annual average are more complex. Atmospheric chemical processes are responsible not only for the formation of ozone, but also for its destruction. In particular, the reaction of ozone with nitric oxide leads to ozone destruction and this process predominates over the UK. North Atlantic background air contains higher concentrations of ozone than are typically measured over the UK due to net ozone destruction. As emissions of NO_x have declined, so the primary nitric oxide responsible for ozone destruction has fallen in concentration, and therefore the extent of ozone destruction has also declined. This has led to an increase in background ozone which has been greatest at urban sites where emissions of NO_x are most influential. There has consequently been a tendency for urban and rural concentrations to converge as urban concentrations have risen more rapidly than rural concentrations. A recent study of the period 1990–2010 has shown annual mean concentrations across EU rural sites, increasing from 15% below the 24 year mean to 10% above the 20 year mean.[19] For rural sites, the increase has been about half as rapid. On the other hand, trends in the average maximum hourly concentration within a year for European sites showed a decline from around 12% above the 20 year mean in 1990 to 8% below the 20 year mean in 2010. For rural sites, the decline in average maximum concentrations was slightly higher for urban than rural sites. Overall, a broadly similar pattern of behaviour was observed also at sites in the USA.[19]

3   The United States of America

3.1   Introduction

The United States of America (USA) is a large country in terms of size and population. The total land area of the country is 9.9 million km^2 and as of 2014 the population is 319 million. The population density is relatively small (32 people per km^2) in comparison to other countries of similar development, but the population density is highly heterogeneous within the country and great variation occurs. Very broadly, the eastern half of the country has much higher population density than the western half with the exceptions being the west coast states of California and Washington. Both the east and west coasts contain one megacity (population greater than 10 million) each, namely New York and Los Angeles. The large size and varying population densities within the USA, in combination with differing regional climates, result in a non-uniform pollution environment throughout the country. To aid the understanding of regional pollution trends in the USA, the country is often divided into nine regions: Northwest, West, Northern Rockies and Plains (West North Central), Upper Midwest (East North Central), Northeast, Ohio Valley (Central), South, Southeast and Southwest.[20]

Like the 1950s London smog events in the UK, the US suffered from several major air pollution events which became emblematic and focused the attention of US legislators on the dangers of air pollution on human health. The first deadly air pollution event in the USA which generated substantial attention was the Donora smog disaster.[21] Donora is situated in an industrial region of Pennsylvania, and in October 1948 was engulfed by an air pollution episode that emanated from local industrial activity, including zinc and steel factories. The event lasted for several days and resulted in 20 deaths and illness in at least 6000 other residents. The cause of this severe pollution episode was a temperature inversion which trapped toxic chemicals, such as carbon monoxide, sulfur dioxide and metals, in the breathable air.

Another example of air pollution, that covered a much greater geographical area, compared to Donora, was recognized in Southern California in 1950s. In particular the metropolitan Los Angeles region and the surrounding urbanized areas were and are still affected to present. The geography, meteorology and climatology of the region, in part, help the formation of smog by trapping the pollution between the ocean and mountains. Smog episodes, distinct in nature to the London smogs, were found to occur under conditions of high nitrogen oxides (NO_x) and volatile organic compounds (VOCs) in the presence of sunshine. To differentiate between the London-type smogs which were dominated by black smoke and sulfur dioxide, the Southern Californian smogs are referred to as "photochemical smogs". These photochemical smogs produce a distinctly brown haze which can much diminish visibility. Photochemical smogs also contain chemical species which are harmful to human health.[22]

Federal legislation regarding air pollution was initiated in 1955 with the Air Pollution Control Act which enabled research into the sources of air pollution. This was followed in 1963 with the original Clean Air Act (CAA) which started a national program to address air pollution. In 1970 the CAA was significantly expanded and strengthened. In particular, the National Ambient Air Quality Standards (NAAQS) were first implemented. The Environmental Protection Agency (EPA) was setup soon after the 1970 CAA with a remit to ensure environmental protection at a federal level. Subsequent to 1970 the CAA has been revised several more times, notably in 1990 when the EPA was given considerably greater powers.[23]

Since 1970, the EPA have designated six pollutants as criteria pollutants: carbon monoxide (CO), ground-level ozone (O_3), lead (Pb), particulate matter (PM), nitrogen oxides ($NO_x = NO + NO_2$) and sulfur dioxide (SO_2).[24] Each of these criteria pollutants have a permissible level designated to them. The current NAAQS for these criteria pollutants are given in Table 3. NAAQS are designated either primary or secondary. The primary standards are for human health protection, and the secondary standards are for non-health related protection including the reduction in damage to vegetation, crops and buildings, and reduced visibility. The EPA designates areas as meeting ("attainment") or not meeting ("non-attainment") the standard.

Another major driver for the reduction of atmospheric pollutants is their effect on visibility and in particular haze. Towards these aims the EPA established the Regional Haze Rule (RHR) which aims to remove anthropogenic visual impairment in national parks and other locations of special value to the nation.[25]

Emissions reductions in the USA have had to occur under the difficult conditions of increasing population, increasing vehicle miles travelled, increasing gross domestic product (GDP) and increased energy consumption.[26] It is noted that the global economic downturn which started in 2007 has led to decreases in vehicle miles travelled, GDP and energy consumption but these are now starting to rebound.

Table 3 Current EPA National Ambient Air Quality Standards (NAAQS). http://www. epa./gov/ttn/naaqs/.

Pollutant	Standard: primary or secondary	Averaging time	Concentration	Criteria for exceedance
Carbon monoxide	Primary	8 hours	9 ppm	Not to be exceeded more than once per year
		1 hour	35 ppm	Not to be exceeded more than once per year
Lead	Primary and secondary	Rolling 3 month average	$0.15 \ \mu g \ m^{-3}$	Not to be exceeded
Nitrogen dioxide	Primary	1 hour	100 ppb	98th percentile, averaged over 3 years
	Primary and secondary	Annual	53 ppb	Annual mean
Ozone	Primary and secondary	8 hours	75 ppb	Annual fourth highest daily maximum 8 hour concentration, averaged over 3 years
$PM_{2.5}$	Primary	Annual	$12 \ \mu g \ m^{-3}$	Annual mean, averaged over 3 years
	Secondary	Annual	$15 \ \mu g \ m^{-3}$	Annual mean, averaged over 3 years
	Primary and secondary	24 hours	$35 \ \mu g \ m^{-3}$	98th percentile, averaged over 3 years
PM_{10}	Primary and secondary	24 hours	$150 \ \mu g \ m^{-3}$	Not to be exceeded more than once per year on average over 3 years
Sulfur dioxide	Primary	1 hour	75 ppb	99th percentile of 1 hour daily maximum concentrations, averaged over 3 years
	Secondary	3 hours	500 ppb	Not to be exceeded more than once per year

Various policy options are used by the EPA to reduce pollutant emissions. These can be broadly broken down into three categories: command and control measures, market based approaches, and non-regulatory approaches. Command and control works through setting technology and emissions standards, for example, specifying the amount of lead content allowed within gasoline. Market based approaches such as cap and trade allow for the market to decide how best to achieve the restriction of pollutants.[27] Non-regulatory approaches include methods such as badging the environmental credentials of products and improving education.

3.2 Trends in Emission

The data for this section is taken from the EPA's 2011 National Emissions Inventory (NEI)[28] for the time period from 1970 to 2013. It should be noted that the emissions data are not measured but estimated and estimation methods contain assumptions. Variations in the emissions estimates from year to year can be influenced by changes in the assumptions and estimation techniques employed in addition to the actual changes in emissions.

Emissions are typically broken down into four major categories: stationary fuel combustion, industrial and other processes, transportation, and miscellaneous. Furthermore, each category can be further subdivided to provide more detail when required. Stationary fuel combustion emission sources include power plants used to generate electricity and other industrial activities that generate steam, heat and energy. The major categories for emissions from industrial and other processes include chemical and allied product manufacture, metal processing, petroleum and related industries, solvents, storage and transport, waste disposal and recycling and other processes. Transportation emissions can be broken down into two major categories: road and non-road emissions. Road emissions are defined as those coming from vehicles that are used on roads to carry passengers or freight. Non-road emissions include aircraft, boats, ships, trains, industrial and construction vehicles, lawn and garden equipment, and recreational vehicles. Transportation emissions are typically further subdivided by fuel use: diesel, gasoline or other fuel.

3.2.1 Carbon Monoxide. Emissions of carbon monoxide (CO) have significant anthropogenic and biogenic sources.[29] The direct anthropogenic emissions are given in Figure 14 by source category. Overall, the total CO emissions have dramatically reduced since the 1970s by almost a factor of four. CO emissions are dominated by the transport sector, which includes both road transport and non-road transport which includes rail, air, boat and other off road and recreational vehicles.[30] This large vehicular source results in high CO levels in urban areas where car use is typically greatest.

Since the introduction of catalytic converters in the 1970s, which convert CO to CO_2, the emissions of CO from road transport have rapidly declined in a near linear fashion. The decrease in road transport emissions has much increased the relative contribution of non-road vehicles to total emissions. In 1970 highway vehicles accounted for $\sim$90% of transport emissions, whereas in 2013 they account for $\sim$60% of emissions. Since virtually all new road vehicles are fitted with catalytic converters, further decreases in CO emissions will become more difficult. In addition to catalytic converters some cities with high winter CO concentrations are mandated to use gasoline with a greater oxygenated content. The oxygenated content increases combustion efficiency in cold weather, thus reducing CO emissions. Whilst the industrial sector has always been a small contributor to total CO emissions, this sector has managed to decrease emissions by about 80% from 1970 levels to present.

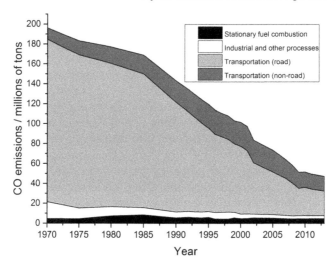

Figure 14　Annual anthropogenic emissions of carbon monoxide (CO) in the USA, by source category, from 1970–2013. Miscellaneous CO sources are not included in this analysis.
(Data source: http://www.epa.gov/ttn/chief/trends/index.html#tables).

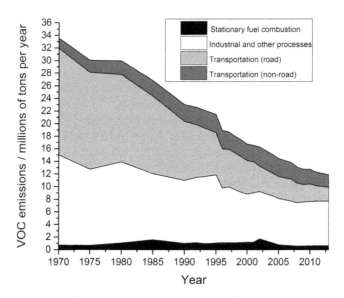

Figure 15　Annual anthropogenic emissions of volatile organic compounds (VOCs) in the USA, by source category, from 1970–2013. Miscellaneous VOC sources are not included in this analysis.
(Data source: http://www.epa.gov/ttn/chief/trends/index.html#tables).

3.2.2　Volatile Organic Compounds.　Volatile Organic Carbon (VOC) emissions in the USA are dominated by biogenic emissions with only around one quarter of emissions coming from anthropogenic sources. Figure 15

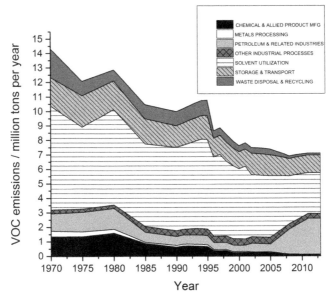

Figure 16 Annual industrial emissions (non-fuel combustion of volatile organic compounds (VOCs) in the USA, by source category, from 1970–2013. Miscellaneous VOC sources are not included in this analysis. (Data source: http://www.epa.gov/ttn/chief/trends/index.html#tables).

provides the direct anthropogenic emissions of VOCs. It should be noted that the definition of VOC used by the EPA is distinct from that used in the UK. Most notably, USA VOC emissions contain methane whereas the UK compiles separate emission inventories for non-methane VOC (NMVOC) and for methane.

USA anthropogenic emissions of VOCs have two major sources: transportation and industry. Major reductions in anthropogenic emissions have been achieved – since 1970 total emissions have been almost halved. This reduction is largely due to the remarkable reductions in VOC emissions in the transportation sector which have undergone a reduction of 80% in the same time frame. The decrease in the transportation has been near linear over the last 40 years and is in large part due to the introduction of catalytic converters.

The industrial sector has also seen major reductions since 1970. Figure 16 provides the non-fuel combustion industrial emissions of VOCs split into different industries. Overall the industrial sector emissions have reduced by about half since 1970 with most industry types showing reductions. The waste disposal and chemical industries have shown particularly impressive reductions in their emissions.

3.2.3 Lead. Huge reductions in the emissions of lead have been achieved over the last forty years. Figure 17 provides the annual anthropogenic emissions of lead in the USA since 1980. The reduction in emissions has been

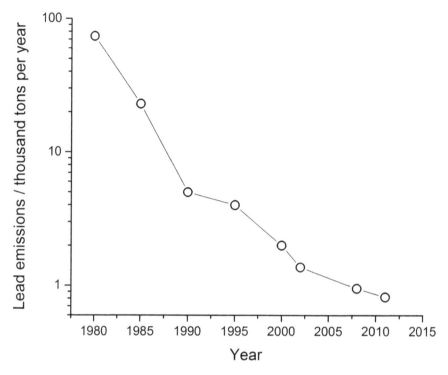

Figure 17   Annual anthropogenic lead emissions in the USA, from 1980–2012 (fire
and dust emissions excluded).
(Data sources: NEI 2011).[27]

mostly driven by the introduction of non-leaded gasoline. Lead was initially
added in the form of tetraethyl lead as an anti-knocking agent to improve
engine performance in the 1920s. Leaded fuel was found to poison catalytic
converters and hence the increase in catalytic converters since the 1970s
has been mirrored by a reduction in leaded gasoline and hence reductions
in lead emissions. A complete ban on lead in automobile gasoline was
achieved in 1996. Prior to the ban of leaded gasoline the predominant
source of airborne lead was through road vehicle use.

The most important sources of lead emissions are now from the following:
aircraft emissions (59%), industrial processes – predominantly metallurgical
in nature (28%), fuel combustion (10%) and waste disposal (1%).[28] It should
be noted that the NEI databases do not consider the resuspension of road
dust as an emission source of lead; it is likely to be an important source.[31]

3.2.4   Particulate Matter.   As in the UK, particulate matter is measured as
either PM_{10} or $PM_{2.5}$ in the USA. Figure 18 provides the annual USA emissions of $PM_{2.5}$ and PM_{10} from 1990 to 2013. Overall the emissions of PM_{10}
have decreased slightly since 1990 levels, whereas the $PM_{2.5}$ emissions have

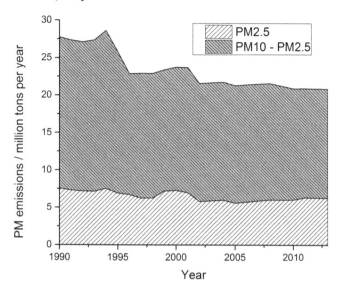

Figure 18 Annual emissions of particulate matter (PM) in the USA, from 1990–2013.
Note that $PM_{2.5} + (PM_{10}-PM_{2.5}) = PM_{10}$.
(Data source: http://www.epa.gov/ttn/chief/trends/index.html#tables).

remained rather static. The ratio of PM_{10} to $PM_{2.5}$ has also remained relatively constant with the proportion of PM_{10} which is in the fine $PM_{2.5}$ size range varying between 26–31% from 1990 to the present date.

Figure 19 provides the breakdown of $PM_{2.5}$ and PM_{10} emission categories for 2011.[27] The emission sources for $PM_{2.5}$ and PM_{10} are broadly similar but do show distinct differences. For example, the major sources of $PM_{2.5}$ and PM_{10} are fires and dust, respectively. Much of the PM_{10} emission sources can be classified as fugitive emissions which include material generated through the mechanical disturbance of particulate material settled on the ground. Major sources include unpaved roads, agricultural activities and construction. Both $PM_{2.5}$ and PM_{10} have fire sources which are both natural and anthropogenic in origin. Biogenic wildfires provide a slightly greater source, in both fine and coarse particle categories, compared to anthropogenic prescribed fires and agricultural field burning.

3.2.5 Nitrogen Oxides. Nitrogen oxide (NO_x) emissions are dominated by two sources: stationary and mobile source fuel combustion. Substantial reductions in the total emissions of NO_x have been achieved since 1970 with present levels roughly half of 1970. The annual USA emissions by source category are given in Figure 20. Most of the emission reductions were achieved from the mid-1990s onwards subsequent to the 1990 Clean Air Act Amendments (CAAA) which targeted in particular NO_x emissions from power stations. Notably, the stationary fuel combustion sources, which include power stations, have halved since 1990. Conversely the contribution from mobile sources, although reducing significantly in absolute magnitude, has

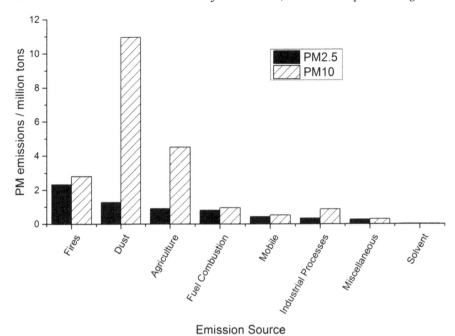

Figure 19   Source categories for particulate matter (PM) emissions of in the USA for 2011. (Data source: National Emissions Inventory (NEI) 2011 http://www.epa.gov/ttn/chief/eiinformation.html).

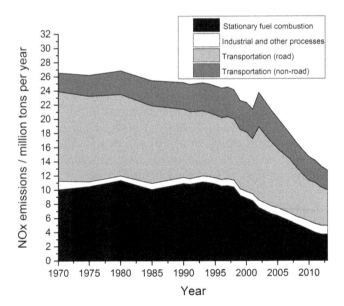

Figure 20   Annual anthropogenic emissions of nitrogen oxides (NO_x) in the USA, by source category, from 1990–2013. Miscellaneous NO_x sources are not included in this analysis.
(Data source: http://www.epa.gov/ttn/chief/trends/index.html#tables).

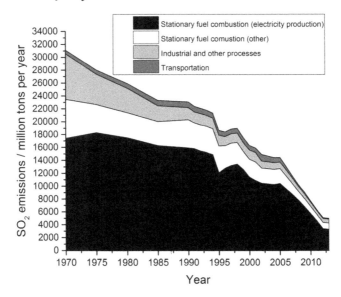

Figure 21 Annual anthropogenic emissions of sulfur dioxide (SO₂) in the USA, by source category, from 1990–2013. Miscellaneous SO₂ sources are not included in this analysis.
(Data source: http://www.epa.gov/ttn/chief/trends/index.html#tables).

increased their percentage of the total emissions. Significant absolute reductions in NO_x from road vehicles have been achieved since the introduction of the 3-way catalyst in the 1980s. The large decrease in the road transport sector has not been achieved by the off road sector. For example, in 1970 off-highway sources contributed approximately one fifth of the mobile emissions, whereas in 2010 it was approximately one third.

3.2.6 Sulfur Dioxide. Large total decreases in sulfur dioxide (SO_2) emissions have been achieved in the last 40 years. Reductions have occurred within all major sectors: fuel combustion, non-combustion industrial activities, and transportation. The total emissions are dominated by stationary fuel combustion sources predominantly from coal power stations but there are also significant amounts emitted from fuel combustion in industrial and other sources. Figure 21 provides the Annual Emissions Inventory of SO_2. Reductions in the stationary fuel combustion sources, including power stations, are large due to the introduction of flue gas desulfurization technologies and the use of cleaner fuels. Lower sulfur fuels have significantly reduced the SO_2 emissions from transportation.

3.3   Trends in Airborne Concentrations

3.3.1 Carbon Monoxide. Large decreases in the annual atmospheric concentrations of carbon monoxide (CO) have been measured since 1980, see

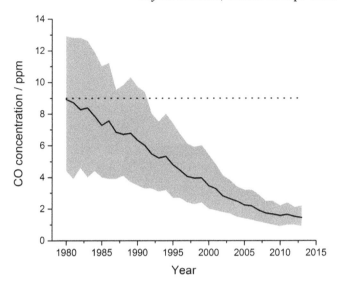

Figure 22   Annual atmospheric concentration of carbon monoxide (CO) in the USA, from 1980–2013. Data is generated from the measurements taken at 82 independent sites. The black line is the annual second highest non-overlapping 8 hour average. The dashed line is the National Ambient Air Quality Standard. The grey area is banded by the 10th and 90th percentile values.
(Data source: http://www.epa.gov/airtrends/carbon.html).

Figure 22. These decreases have easily satisfied the NAAQS of 9 ppm since the early 1990s. The decrease in atmospheric concentration has largely mirrored the reduction in direct anthropogenic emissions (see Figure 14). Since 1980 to present day, direct emissions have reduced by ~70% whilst atmospheric concentrations have reduced by ~80%.

3.3.2   Lead.   Atmospheric concentrations of lead have decreased significantly since measurements began in the 1980s, see Figure 23, with the average national average decreasing by ~90%. Since 2009 the lead NAAQS of 0.15 µg m^{-3} has been achieved by the nationwide average concentration but many measurement locations still regularly exceed the NAAQS limit. The reductions in atmospheric concentrations of lead lag significantly behind the emission rates. This is somewhat surprising since the major emissions category was leaded gasoline which is now completely banned apart from specialist uses. The lingering amounts of lead are now likely to be mainly due to resuspension of lead-containing particles that were emitted directly into the atmosphere from historic vehicle use and industrial sources.[29]

3.3.3   Particulate Matter.   Atmospheric concentration data for PM$_{10}$ and PM$_{2.5}$ have only been available, on a national scale, since 1990 and 2000, respectively. Figures 24 and 25 provide the annual atmospheric

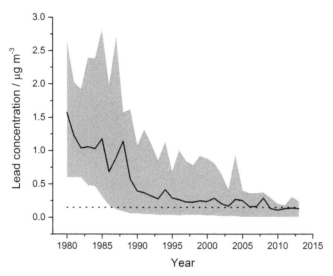

Figure 23 Annual atmospheric concentration of lead (Pb) in the USA, from 1980–2013. Data is generated from the measurements taken at 12 independent sites. The black line is the annual maximum 3 month average. The dashed line is the National Ambient Air Quality Standard. The grey area is banded by the 10th and 90th percentile values.
(Data source: http://www.epa.gov/airtrends/lead.html).

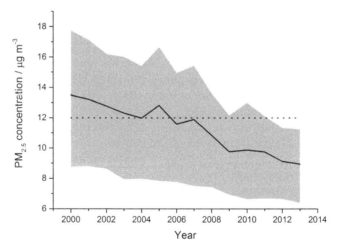

Figure 24 Annual atmospheric concentration of PM$_{2.5}$ in the USA, from 2000–2013. Data is generated from the measurements taken at 537 independent sites. The black line is seasonally weighted annual average. The dashed line is the National Ambient Air Quality Standard. The grey area is banded by the 10th and 90th percentile values.
(Data source: http://www.epa.gov/airtrends/pm.html).

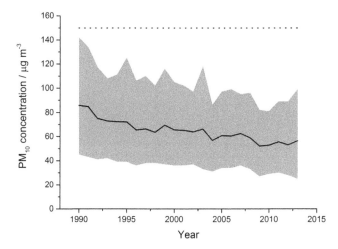

Figure 25 Annual atmospheric concentration of PM_{10} in the USA, from 1990–2013. Data is generated from the measurements taken at 207 independent sites. The black line is the annual 2nd maximum 24 hour average. The dashed line is the National Ambient Air Quality Standard. The grey area is banded by the 10th and 90th percentile values. (Data source: http://www.epa.gov/airtrends/pm.html).

concentrations of PM_{10} and $PM_{2.5}$, respectively. Significant reductions in PM_{10} of $\sim 35\%$ have been achieved since 1990 to present. A similar reduction of $\sim 35\%$ has been achieved in $PM_{2.5}$ albeit from 2000 to present. The changes in atmospheric concentration far exceed the modelled changes in emissions, which suggest that the emissions estimates are lacking or reflect siting of monitoring stations close to controllable ground-level sources. It may also reflect the emission estimates only including primary PM emissions whereas airborne concentrations also include a substantial contribution from secondary components formed by atmospheric chemical processes. The PM_{10} NAAQS has been easily achieved on the national level since the 1990s whereas for $PM_{2.5}$ it took until the mid-2000s until the NAAQS was routinely achieved. Until very recently many locations for $PM_{2.5}$ did not achieve attainment of the NAAQS.

The composition of PM in the USA can be very region specific. For example in the west of the country nitrate tends to dominate whereas in the east coast sulfate tends to dominate. However, the reduction in SO_2 emissions in the east of the country has led to reductions in the sulfate fraction.

3.3.4 Nitrogen Oxides. At the national level, the concentration of NO_2 has been steadily decreasing since the 1980s, see Figure 26. The annual average value permanently fell below the NAAQS by the late 1980s but many areas did not attain levels below the NAAQS until after 2000. The reduction in atmospheric concentrations follows the reduction in emissions, see Figure 20. Small decreases were seen in the 1980s, followed by larger

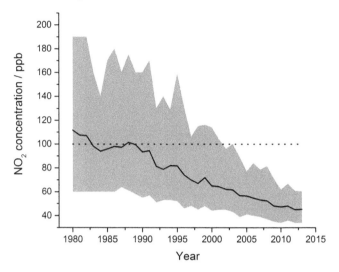

Figure 26   Annual atmospheric concentration of nitrogen dioxide (NO$_2$) in the USA, from 1980–2013. Data is generated from the measurements taken at 29 independent sites. The black line is the annual 98th percentile of daily maximum 1 hour average. The dashed line is the National Ambient Air Quality Standard. The grey area is banded by the 10th and 90th percentile values.
(Data source: http://www.epa.gov/airtrends/nitrogen.html).

decreases from 1990 onwards which coincided with the implementation of the CAAA that, in particular, targeted NO$_x$ emissions from power stations.

3.3.5   Sulfur Dioxide.   Large decreases in the national average of ~80% have been observed from 1980 to the present date, as shown in Figure 27. These decreases mirror the reduction in emissions shown in Figure 21. Large variations in SO$_2$ emissions occur across the country and as such large variations occur in the regional concentrations. In particular, locations which have not attained the NAAQS are clustered around the "Rust Belt" states of Pennsylvania, Ohio, Indiana and Illinois which have a large proportion of coal-powered power stations and hence SO$_2$ emissions. The large differences between certain regions are highlighted in Figure 28 which shows the differences between annual concentrations in the West and Central EPA defined regions. Whilst at the national level it took until 2004 to achieve the SO$_2$ NAAQS, it took the Central region which encompasses the "Rust Belt" until 2009 to achieve it, with many districts still significantly above the prescribed standards. In comparison, the West region has always been much under the NAAQS since before measurements started in the 2000s.

3.3.6   Ground-level Ozone.   Ground level ozone has been reducing steadily since the 1980s with a reduction of ~30% achieved, see Figure 29.

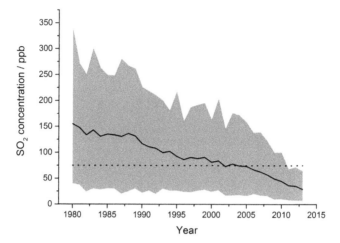

Figure 27   Annual atmospheric concentration of sulfur dioxide (SO₂) in the USA, from 1980–2013. Data is generated from the measurements taken at 47 independent sites. The black line is the annual 99th percentile of daily maximum 1 hour average. The dashed line is the National Ambient Air Quality Standard. The grey area is banded by the 10th and 90th percentile values.
(Data source: http://www.epa.gov/airtrends/sulfur.html).

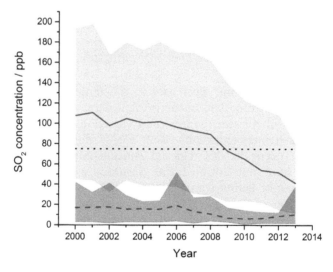

Figure 28   Annual atmospheric concentration of sulfur dioxide (SO₂) for the USA West and Central regions, from 2000–2013. Data from the West and Central regions are generated from measurements taken at 18 and 73 independent sites, respectively. The solid and dashed lines are the annual 99th percentile of daily maximum 1 hour average for the Central and West regions, respectively. The dashed line is the National Ambient Air Quality Standard. The light and dark grey areas are banded by the 10th and 90th percentile values for the Central and West regions, respectively.
(Data source: http://www.epa.gov/airtrends/sulfur.html).

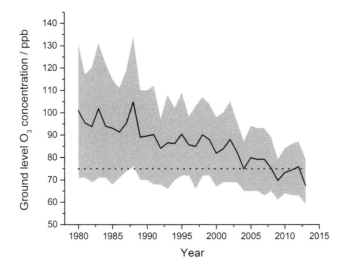

Figure 29   Annual atmospheric concentration of ground level O_3 in the USA, from 1980–2013. Data is generated from the measurements taken at 222 independent sites. The black line is the annual fourth highest non-overlapping 8 hour average. The dashed line is the National Ambient Air Quality Standard. The grey area is banded by the 10th and 90th percentile values.
(Data source: http://www.epa.gov/airtrends/ozone.html).

However, it is only very recently that the ozone NAAQS has been achieved, with four of the last five years achieving the NAAQS at the national level. Many regions and districts still do not attain NAAQS requirements. Since ozone is not emitted directly into the atmosphere, but produced through the complex reactions of VOCs, NO_x and sunlight,[3] the atmospheric concentrations cannot be directly compared to a direct emissions inventory. However, in general it can be stated that the reduction in NO_x and VOCs have resulted in significant reduction in O_3. Ozone production and loss at ground level is strongly linked to meteorological parameters and therefore to be able to understand the effect of pollutant control programs on ground level ozone, the effect of meteorological conditions throughout the lifecycle of ozone needs to be taken into account.[32]

3.4   Conclusions

In the years since 1970 and the formation of the EPA, all six of the EPA designated criteria pollutants have been significantly reduced. This has been through a variety of regulatory activities including improvements in road vehicles and power stations. In recent years all NAAQS are being achieved at the national average level. However, as recently as 2010, 123 million USA inhabitants, roughly one third of the country, lived in counties that exceeded the NAAQS in one or more categories, so much still remains to be achieved.[28]

4   Less Developed Countries: China as a Case Study

Air pollution in developing countries has become a growing concern for the general public and the government. This is particularly true for China after the widespread haze hovering above thousands of square kilometres in China with over 700 million population affected sometimes. In megacities such as Beijing, the air quality index sometimes was over 500, which was considered the "maximum" limit of the scale. In 2012 stricter air pollution monitoring of ozone and $PM_{2.5}$ were ordered to be gradually implemented so that by 2015 all but the smallest cities would be included. Huge government budget has been set up to clean up the air. This has already led to some progress in cutting the emissions of air pollutants and improving the ambient air quality in recent years.

4.1   Trends in Emissions since 1970

4.1.1   Particulate Matter (Including Black Carbon).   Figure 30 shows the emission trend in PM_{10} from the Emissions Database for Global Atmospheric Research (EDGAR 4.2). EDGAR is a joint project of the European Commission Joint Research Centre (JRC-IES, Ispra, Italy) and the Netherlands Environmental Assessment Agency (PBL) which stores global inventories of direct and indirect greenhouse gas emissions from anthropogenic sources including halocarbons and aerosols both on a per country and region basis as well as on a grid.[33] Total PM_{10} emissions in China increased steadily from 6100 kilotonnes in 1970 to 14 069 kilotonnes in 1996, followed by a substantial decrease to 10 956 kilotonnes in 2001; from 2001, the total PM_{10} emissions increased again and reached 14 935 kilotonnes in 2008 (see Figure 30). The total PM_{10} emissions from EDGAR 4.2 is significantly smaller than that from the Lei *et al.*[34] and Zhao *et al.*[35] inventories (see Figure 31). The trend in PM_{10} emissions from 1990 to 2005 is similar in EDGAR 4.2 and Lei *et al.*[34] However, from 2005 to 2008, PM_{10} emissions in EDGAR 4.2 increased by more than 16%, while that in Zhao *et al.*[35] had little change.

Based on EDGAR 4.2,[33] the residential and other sectors have always been a major contributor to the PM_{10} emissions and their contribution to total PM_{10} emissions have changed little since 1970; the second largest contributor to PM_{10} emissions is from public electricity and heat production; and PM_{10} emissions from the production processes have been increasing steadily since the 1990s (see Figure 30).

A $PM_{2.5}$ emission inventory was not included in EDGAR 4.2. A recent study by Lei *et al.*[34] calculated that the $PM_{2.5}$ emissions increased from 9280 kilotonnes in 1990 to 12 950 kilotonnes in 2005. A similar trend in emissions is also noted for Black Carbon (BC) by a Health Effects Institute (HEI) report.[36] Zhao *et al.*[35] reported little change in the $PM_{2.5}$ emissions, which are over 12 200 kilotonnes from 2005 to 2010.

The significant decrease in PM_{10} emissions from 1996 to 2001 is mainly due to the reduction in emissions from public electricity and heat

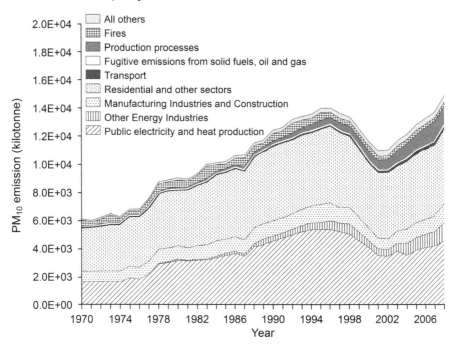

Figure 30 Trends in PM_{10} emissions in China (from EDGAR 4.2).[33] Note: Transport includes domestic aviation, road transportation, rail transportation, inland navigation, other transportation; production processes includes cement production, lime production, production of other minerals, production of chemicals, production of metals, production of pulp/paper/food/drink; fires includes Savanna burning, agricultural waste burning, forest fires, grassland fires, and fossil fuel fires; and all others include direct soil emissions, manure management, solvent and other product use: degrease and waste incineration.

production in EDGAR 4.2.[33] This decrease was also noted in another study by Lei *et al.*[34] but the exact time of the decrease is slightly different (see Figures 30 and 31). $PM_{2.5}$ and BC emissions also decreased in the late 1990s in China but increased again from early 21st century according to Lei *et al.*[34] From 2005 to 2010, the BC emissions remained relatively constant at 1600 kilotonnes according to Zhao *et al.*[35]

4.1.2 Sulfur Dioxide. Figure 32 shows the total SO_2 emissions in China from EDGAR 4.2.[33] Contrary to PM_{10} emissions, which only increased approximately two-fold from 1970 to 2008, SO_2 emissions increased nearly five-fold during the same period; from 1970, the total SO_2 emissions increased steadily until 1997 when it started to decrease; however, this decreasing trend did not continue for long and it started to increase again from 2001 (see Figure 32). The general trend from EDGAR 4.2,[33] Lei *et al.*,[34] Hao and Wang[37] and Ministry of Environmental Protection of China (MEP)[38] appears to be similar. From 2008, both Lei *et al.*[34] and

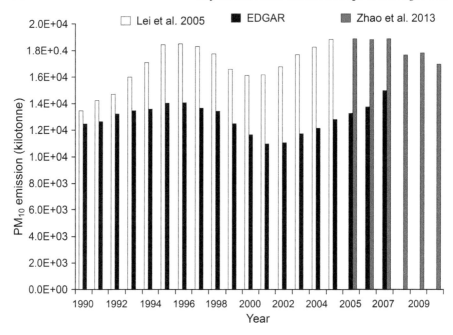

Figure 31   Trend in total PM_{10} emissions in China from different studies (EDGAR 4.2;[33] Lei *et al.*;[34] Zhao *et al.*[35]).

MEP[38] inventories showed a decrease in SO_2 emissions. There are some but not huge discrepancies in the SO_2 emissions particularly from 2004 to 2007 from EDGAR, Lei *et al.*,[34] Zhao *et al.*,[35] Ohara *et al.*[39] and the MEP[38] inventories. However, there is a major difference between EDGAR 4.2, Lei *et al.*[34] and Zhao *et al.*[35] in SO_2 emissions in 2005, with the former as high as 39 554 kilotonnes while the latter two inventories give figures of only around 30 000 kilotonnes. For the year 2000, the MEP,[38] Lei *et al.*,[34] Streets *et al.*,[40] EDGAR 4.2,[33] Hao and Wang[37] and Su *et al.*[41] all gave SO_2 emissions of approximately 20 000 kilotonnes.

There are also important differences in the contribution of different sources to SO_2 emissions from different inventories. EDGAR 4.2[33] indicated a steady increase in SO_2 emissions from public electricity and heat production from 1970 to 2008 (see Figure 32) while Lei *et al.*[34] calculated a steady and significant decrease in SO_2 emissions from this sector from 2004 to 2008 (see Figure 33). The decrease in SO_2 emissions in the late 1990s and early 2000s was mainly due to the decrease in emissions from manufacturing industries and construction in EDGAR 4.2 (see Figure 32), while Lei *et al.*[34] attributed a significant increase to industry (see Figure 33).

4.1.3   Oxides of Nitrogen.   Figure 34 shows the trend in emission of NO_x in China from EDGAR 4.2.[33] The nitrogen oxides (NO_x) increased from about 3000 kilotonnes in 1970 to over 20 000 kilotonnes in 2008; this increase is driven by an almost continuous increase in emissions from

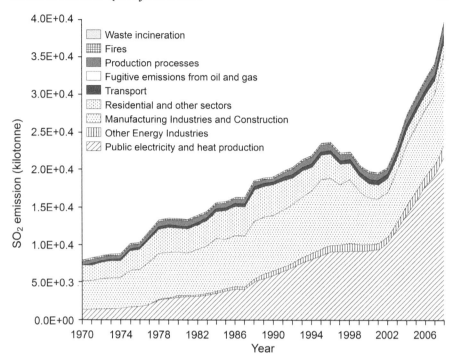

Figure 32   Trends in SO$_2$ emissions in China (from EDGAR 4.2).[33] Note: transport includes domestic aviation, road transportation, rail transportation, inland navigation, other transportation; production processes includes production of chemicals, production of metals, production of pulp/paper/food/drink; fires includes savanna burning, agricultural waste burning, forest fires, grassland fires, fossil fuel fires.

public electricity and heat production, manufacturing industries and construction, and transport (see Figure 34). There was a decrease in NO$_x$ emissions in the late 1990s and early 2000s, but the extent of decrease was much less significant than that in the emissions of PM$_{10}$ and SO$_2$ during the same period (see Figures 30–34). In other inventories such as Zhao *et al.*,[35] the NO$_x$ emissions did not even show such a decrease from the late 1990s to early 2000s. A recent comparison showed surprisingly close total NO$_x$ emissions estimates for China in literature sources.[35]

4.1.4 Carbon Monoxide. Figure 35 shows the total CO emissions in China from EDGAR 4.2.[33] The CO emissions increased from about 37 200 kilotonnes in 1970 to over 107 000 kilotonnes in 2008; this increase is mainly driven by the increase in emissions from residential and other sectors and by production processes since the mid-1990s; transport also contributed to the increases in CO emissions after the early 1990s but its contribution decreased after 2002 (until 2008); there was a decrease in CO emissions in the late 1990s and early 2000s (see Figure 35), but the degree

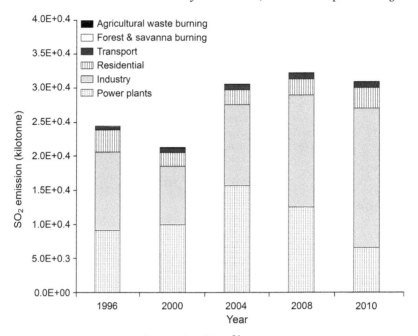

Figure 33 Trends in SO_2 emissions in China.[34]

of decrease was much less significant compared to the emissions of PM_{10} and SO_2 during the same period (see Figures 30–35).

4.1.5 Volatile Organic Compounds. Figure 36 shows the trend in non-methane hydrocarbon (NMHC) emissions in China from EDGAR 4.2.[33] NMHC emissions increased almost continuously from about 8000 kilotonnes in 1970 to 22 000 kilotonnes in 2008; the three largest contributors to the NMHC are public electricity and heat production, fugitive emissions from solid fuel, oil and gas, and solvent use; the increase in total NMHC emissions was mainly attributed to the latter two sources (see Figure 36). A similar trend and total emission amount was reported except for the substantial increase from 2003 to 2005[36] based on the The Frontier Research Center for Global Change 2007 inventory.

4.2 Ambient Air Monitoring – Development of the Networks

Routine PM_{10} monitoring in major cities in mainland China was not implemented until the late 1990s. Before that, only Total Suspended Particulates (TSP) were monitored. Beijing city has recently established a $PM_{2.5}$ network but the national $PM_{2.5}$ monitoring network will not be in place until 2015. Unlike the UK where black smoke has been monitored for over 50 years, no such data were found within China. The only data that date back to 1970s appear to be for visibility, although this is not a direct index for air quality.

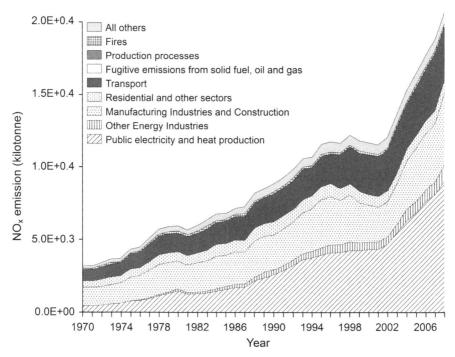

Figure 34 Trends in NO$_x$ emissions in China (from EDGAR 4.2).[33] Note transport includes domestic aviation, road transportation, rail transportation, inland navigation, other transportation; production processes includes production of chemicals, production of metals, production of pulp/paper/food/drink; fires includes savanna burning, agricultural waste burning, forest fires, grassland fires, fossil fuel fires; all others include direct soil emissions, manure management, manure in pasture/range/paddock and waste incineration.

The ranges of ambient SO$_2$ concentrations have been reported in the China Environment Yearbook by MEP since 1991 although the national average values were missing until 2002. The ranges of NO$_x$ were reported sometimes in the China Environment Yearbook from 1991. In reports, NO$_x$ has been replaced by NO$_2$ since 2002. Beijing is again leading the air quality monitoring in mainland China and established a SO$_2$, PM$_{10}$ and NO$_2$ monitoring network in 1998. It is unclear whether there is a NHMC network in China but comprehensive data on NMHC have not been reported yet.

4.3   Trends in Airborne Concentrations

4.3.1   Particulate Matter and Visibility.   Figure 37 shows the trend in PM$_{10}$ mass concentrations in China. The World Bank indicator dataset provides the longest coverage of national average PM$_{10}$ mass concentrations dating

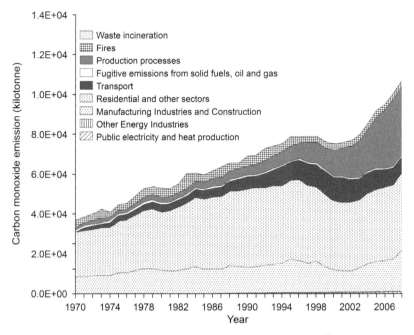

Figure 35   Trends in CO emissions in China (from EDGAR 4.2).[33] Note: production
processes include production of other minerals, production of chemicals,
production of metals, production of pulp/paper/food/drink; transport
includes domestic aviation, road transportation, rail transportation,
inland navigation, other transportation; fires includes savanna burning,
agricultural waste burning, forest fires, grassland fires, fossil fuel fires.

back to 1990. The national average PM_{10} mass concentration in China was
over 120 $\mu g\ m^{-3}$ in the early 1990s; it decreased substantially from 1991 to
1995 with little change until 2005, from when the PM_{10} mass concentrations started to decrease continuously. The China Environment Yearbook did not report an average PM_{10} mass concentration in key
environmental protection cities until 2002. The trend in PM_{10} mass concentrations from 2005 in the China Environment Yearbook was similar to
that published by the World Bank (see Figure 37).

Figure 38 shows the trend in PM_{10} mass concentrations in Beijing. The
annual average concentration of PM_{10} in urban Beijing decreased from
180 $\mu g\ m^{-3}$ in 1999 to 108 $\mu g\ m^{-3}$ in 2013. Especially after 2006, there was a
continuous decrease in PM_{10} pollution in Beijing. Qu *et al.*[43] found decreasing PM_{10} concentrations over 16 Chinese northern cities and 11 central
cities, but relatively constant PM_{10} concentrations over southern cities from
2000 to 2006. Long-term measurements of the $PM_{2.5}$ mass concentration in
China are rare in the literature. He *et al.*[44] reported that the $PM_{2.5}$ mass
concentration was over 100 $\mu g\ m^{-3}$ from 2001 to 2008 at an urban site in
Beijing. It fluctuated year by year but varied by less than 15% if using 2001 as
a basis.[44]

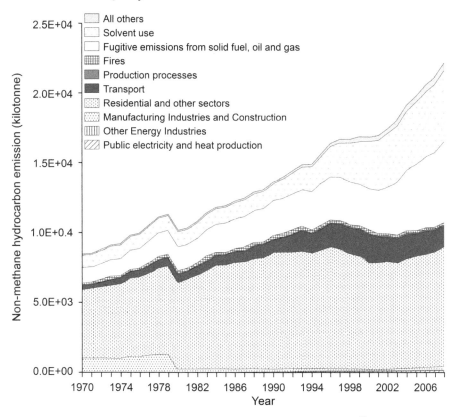

Figure 36 Trends in NMHC emissions in China (from EDGAR 4.2).[33] Note: transport includes domestic aviation, road transportation, rail transportation, inland navigation, other transportation; production processes include production of chemicals, production of metals, production of pulp/paper/food/drink; fires include Savanna burning, agricultural waste burning, forest fires, grassland fires, fossil fuel fires; solvent and other product use (paint, degrease, chemicals, other); and all others include solid waste disposal on land, waste incineration, and other waste handling.

Visibility is one of the indirect indicators of PM pollution and this is the only index in China that has a long-term dataset. In Guangzhou and Beijing, the visibility showed a decreasing trend from 1973 to 2007 while in Xiamen and Shantou, there was continuous decrease in visibility until the early 2000s (see Figure 39; Chang *et al.*[45]), after which the visibility remained relatively constant.[46] There was an increase in the annual average visibility in Beijing when the PM_{10} mass concentration substantially decreased, *i.e.*, from 1997 to 2007 (see Figures 38 and 39).

4.3.2 Sulfur Dioxide. Figure 37 shows the trend in average SO_2 mass concentrations in Chinese cities. The annual average SO_2 in Chinese cities increased slightly from 90 μg m^{-3} in 1990 to 98 μg m^{-3} in 1994 but

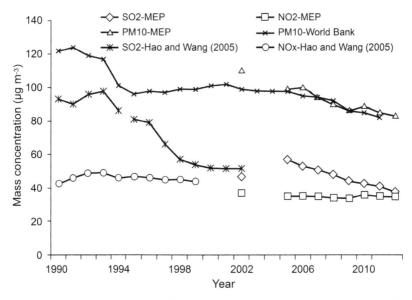

Figure 37 Trend in PM_{10}, SO_2 and NO_2 annual average mass concentration in key environmental protection cities in China (data from Hao and Wang and MEP China[5,37] and country average PM_{10} from World Bank).[42]

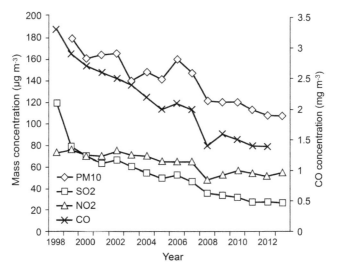

Figure 38 Trend in PM_{10}, SO_2, NO_2 and CO annual average mass concentration in Beijing (data from Beijing Environmental Protection Bureau).[47]

decreased sharply thereafter to 52 $\mu g\ m^{-3}$ in 2002 (see Figure 37). The data from China Environment Yearbook reported annual average mass concentrations of SO_2 in Chinese key environmental protection cities from 2002, but data were missing in 2003 and 2004; it showed that SO_2 decreased from 57 $\mu g\ m^{-3}$ in 2005 to 37 $\mu g\ m^{-3}$ 2012 (see Figure 37).

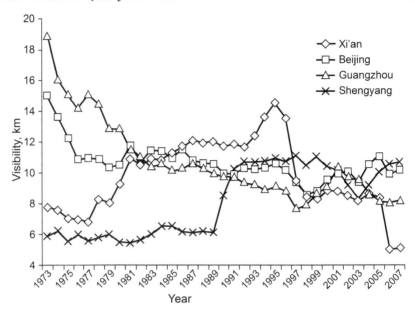

Figure 39 Mean annual visibility in four major cities in China from 1973 to 2007. (This figure is re-drawn based on data from Chang *et al.*[45] Original visibility data are provided by Dr Yu Song from Peking University).

Figure 38 shows the trend in SO_2 mass concentrations in Beijing from 1998–2013. The SO_2 mass concentration in Beijing was as high as 120 μg m^{-3} in 1998 but decreased to 27 μg m^{-3} in 2013 (see Figure 38). Data from 7 other major cities in China, in particular Chongqing city also reported an overall decreasing trend in SO_2 concentration from 1980 to 2005.[36]

4.3.3 Oxides of Nitrogen. Figure 37 shows the trend in NO_x or NO_2 mass concentration in Chinese cities. The average NO_x concentration in Chinese cities increased from 1990 to 1993 and then decreased slightly until 1999; since 2002, average NO_2 mass concentrations in key environment protection cities remained relatively constant according to China Environment Yearbook (see Figure 37).

The average NO_x mass concentration in major cities from 1990 to 1999 was 46 μg m^{-3},[37] whereas that in key environmental cities from 2005 to 2012 was only 35 μg m^{-3}.[38] Since the majority of NO_x emitted from vehicles is NO, it is expected that NO_2 concentrations will be appreciably lower than that of the NO_x in cities. Therefore, there appear to be discrepancies in the NO_x and NO_2 mass concentrations. One possibility is that average NO_x and NO_2 mass concentration was calculated based on different sets of cities. However, it is unlikely that this has led to as large a discrepancy as observed because of the number of cities used for averaging. The second possibility is a major increase in NO_x mass concentration after the year 2002 (the concentration of NO_x would be two or more times higher than that of NO_2). This is unlikely considering that the trends in NO_x from 1990 to 1999 and NO_2 from 2005 to 2012 were relatively consistent.[35] The third possibility may be related to the quality of the data reported.

Figure 38 includes the trend in NO_2 in Beijing. The NO_2 mass concentration was as high as 74 $\mu g\ m^{-3}$ in 1998; only a small decrease in NO_2 was observed from 1998 to 2013 and the NO_2 mass concentration has exceeded that of SO_2 since 2000; by 2013, the mass concentration of NO_2 was more than twice that of SO_2 in Beijing (see Figure 38).

4.3.4 Carbon Monoxide. The average CO mass concentration was not reported in the China Environment Yearbook. Long-term measurement data are rarely available. The annual reports from Beijing, Guangzhou and Dalian Environmental Protection Bureau showed that CO mass concentration in Beijing decreased substantially and almost continuously from 1998 to 2013; that in Guangzhou decreased from about 4 mg m^{-3} in 1989 to 1.6 mg m^{-3} in 2005; and that in Dalian decreased slightly from 1985 to 1998, but dropped substantially from nearly 2 mg m^{-3} to only 0.5 mg m^{-3} in 2001, and then remained relatively constant.[48] A long-term monitoring dataset in Hong Kong showed that the CO increased from 1994 to 2008 with a positive trend of 3.5 ppb year^{-1}.[49]

4.3.5 Ground-level Ozone. Ozone is not reported in the China Environment Yearbook. Wang *et al.*[49] reported an increasing trend of O_3 at a site in Hong Kong from 1994 to 2007 (see Figure 40). A similar increasing trend in surface ozone concentrations is also reported by Tang *et al.*[50] in Beijing. Xu *et al.*[51] showed that the annual average O_3 did not vary significantly from 1995–2006 at the Linan Regional Background Station in eastern China.

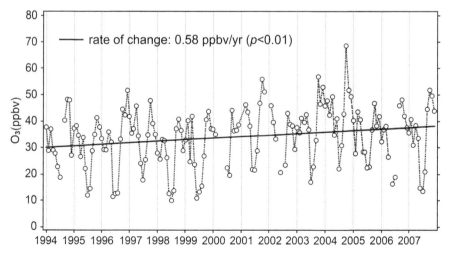

Figure 40   Trend in O_3 mass concentration in Hong Kong.[49]

4.4   Commentary upon Differences in Trends in Emissions and Air Quality

Most emission inventory studies have shown a significant temporal variation and an increasing trend in pollutant emissions from 1990 to 2008 (see Figures 30–38). However, most ground-level observation data show an opposite trend of reduction in ambient concentrations of major pollutants (except NO_2 which remained relatively constant but was only reported after 2002) over Chinese cities (see Figures 37 and 38). This raises a question over the accuracy of some of the emission inventories and/or the quality and representativeness of the ambient air quality data reported in the literature and the governmental reports:

(i)   Air quality control policy: Qu *et al.*[43] argued that the different trends in emissions and concentration of PM_{10} may partly be due to the shifting of pollutant emissions such as PM_{10} to more and more dispersed emissions sources. In addition, air pollution control policies to move industrial plants from urban to rural areas may also explain in part the decrease in the concentrations of PM^{34} and other pollutants over the cities. However, this cannot explain the decrease in national average PM_{10} mass concentration (see Figure 37) in China[42] if it represents accurately the national average concentration. Furthermore, the discrepancies may partially be explained by the shifting of ground- to high-level emissions, *i.e.* stack emissions from industry and heat and electricity production. However, although such an emission mode may lead to a decrease in ground-level pollutant concentrations, it will inevitably increase the overall burden of pollutants in the atmosphere. All these factors may contribute to the increase in haze frequency in recent years in China.

(ii) Emission inventories: the trend in emissions can only be accurate when total emissions from all the years considered are accurate. However, the emission inventories compiled more recently are more detailed and thus more likely to be accurate, while those earlier inventories for pre-2000 emissions may be less accurate. In a more recent study, Zhao *et al.*[35] compiled a new emission database for major air pollutants from 2005 to 2010 in China. They showed a decrease in total emissions of PM_{10}, $PM_{2.5}$, and SO_2 and a small increase in emissions of CO from 2005 to 2010. This contradicts the trend shown in EDGAR 4.2.[33] Only the trend in NO_x emissions in EDGAR 4.2 from 2005 onwards is consistent with Zhao *et al.*[35] The decrease in SO_2 emissions in Zhao *et al.*[35] seems to be consistent with the decrease in SO_2 concentration at a background site in the Yangtze River Delta region during the same period.[52]

(iii) Ground-level monitoring data: ground-level air quality monitoring data are from governmental reports. The methodology to calculate the city average and national average mass concentrations of air pollutants may not be consistent over the years. For example, the addition of suburban and rural monitoring data in a city would potentially lead to a decrease in the city average mass concentrations of air pollutants. But this would not represent the true trend in air pollution. In addition, the air quality monitoring data are not fully automatic but the methodologies for data reporting and processing were not available to the general public.

Acknowledgements

The authors are grateful to Dr Alison Loader (Ricardo-AEA) for provision of the raw data used in Figures 9 and 10 and to Dr Rangaswamy Mohanraj who analysed the data and prepared the graphs, to Dr. Yu Song (Peking University) and Dr. Litao Wang (Hebei Engineering University) for provision of raw data used in Figures 37 and 39.

References

1. J. S. Bower, J. Broughton and J. R. Stedman, A winter NO_2 smog episode in the U.K, *Atmos. Environ.*, 1994, **28**(3), 461–475.
2. P. D. E. Biggins and R. M. Harrison, The atmospheric chemistry of automotive lead, *Environ. Sci. Technol.*, 1979, **13**(5), 558–565.
3. R. L. Canfield, C. R. Henderson, D. A. Cory-Slechta, C. Cox, T. A. Jusko and B. P. Lanphear, Intellectual impairment in children with blood lead concentrations below 10 µg per deciliter, *N. Engl. J. Med.*, 2003, **348**(16), 1517.
4. S. Eggleston, M. P. Hackman, C. A. Heyes, J. G. Irwin, R. J. Timmis and M. L. Williams, Trends in urban air pollution in the United Kingdom during recent decades, *Atmos. Environ.*, 1992, **26B**(2), 227–239.
5. D. L. R. Bailey and P. Clayton, The measurement of suspended particle and total carbon concentrations in the atmosphere using standard smoke shade methods, *Atmos. Environ.*, 1982, **16**(11), 2683–2690.
6. http://naei.defra.gov.uk/overview/ap-overview.
7. Air Quality Expert Group, *Trends in primary nitrogen dioxide in the UK*, AQEG, Department for Environment, Food and Rural Affairs, London, 2007.
8. C. Holman, *Pollution: Causes, Effects and Control*, ed. R. M. Harrison, Royal Society of Chemistry, Cambridge, 5th edn, 2014, vol. 13, pp. 297–325.
9. J. M. Delgado-Saborit, C. Stark and R. M. Harrison, Carcinogenic potential, levels and sources of polycyclic aromatic hydrocarbon mixtures in indoor and outdoor environments and their implications for air quality standards, *Environ. Int.*, 2011, **37**(2), 383–392.
10. G. W. Campbell, J. R. Stedman and K. Stevenson, A survey of nitrogen dioxide concentrations in the United Kingdom using diffusion tubes, July-December 1991, *Atmos. Environ.*, 1994, **28**(3), 477–486.
11. A. J. Gair and S. A. Penkett, The effects of wind speed and turbulence on the performance of diffusion tube samplers, *Atmos. Environ.*, 1995, **19**(18), 2529–2533.

12. R. M. Harrison, J. Stedman and D. Derwent, Why are PM$_{10}$ concentrations in Europe not falling?, New directions, atmospheric science perspectives special series, *Atmos. Environ.*, 2008, **42**(3), 603–606.

13. A. Charron, R. M. Harrison, S. Moorcroft and J. Booker, Quantitative interpretation of divergence between PM$_{10}$ and PM$_{2.5}$ Mass measurements by TEOM and gravimetric (Partisol) instruments, *Atmos. Environ.*, 2004, **38**(3), 415–423.

14. A. M. Jones and R. M. Harrison, Temporal trends in sulphate concentrations at European sites and relationships to sulphur dioxide, *Atmos. Environ.*, 2011, **45**(4), 873–882.

15. Department for Environment, Food and Rural Affairs, *Air Pollution in the UK 2012*, Defra, London, 2013.

16. P. J. Lawther and R. E. Waller, Coal fires, industrial emissions and motor vehicles as sources of environmental carcinogens, *IARC Sci. Publ.*, 1976, **13**, 27–40.

17. D. J. T. Smith and R. M. Harrison, Concentrations, trends and vehicle source profile of polynuclear aromatic hydrocarbons in the U.K. atmosphere, *Atmos. Environ.*, 1996, **30**(14), 2513–2525.

18. M. S. Alam, J. M. Delgado-Saborit, C. Stark and R. M. Harrison, Using atmospheric measurements of PAH and quinone compounds at roadside and urban background sites to assess sources and reactivity, *Atmos. Environ.*, 2013, **77**, 24–35.

19. E. Paoletti, A. De Marco, D. C. S. Beddows, R. M. Harrison and W. J. Manning, Ozone levels in European and USA cities are increasing more than at rural sites, while peak values are decreasing, *Environ. Pollut.*, 2014, **192**, 295–299.

20. www.epa.gov/oust/regions/regmap.htm.

21. www.donorasmog.com.

22. R. P. Wayne, *Chemistry of Atmospheres*, Oxford, 3rd edn, 2000.

23. www.epa.gov/air/caa/amendments.html.

24. U.S. EPA, (2012) Our Nation's Air: status and trends through 2010, EPA-454/R-12-001.

25. J. L. Hand *et al.*, *Atmos. Environ.*, 2014, **94**, 671–679.

26. data.worldbank.org.

27. www.epa.gov/airmarkets.

28. www.epa.gov/ttn/chief/net/2011inventory.html.

29. R. C. Hudman *et al.*, *Geophys. Res. Lett.*, 2008, **35**(4), L04801.

30. D. D. Parish, *Atmos. Environ.*, 2006, **40**, 2288–2300.

31. G. C. Lough *et al.*, *Environ. Sci. Technol.*, 2005, **39**(3), 826–836.

32. L. Camalier *et al.*, *Atmos. Environ.*, 2007, **41**(33), 7127–7137.

33. http://edgar.jrc.ec.europa.eu/overview.php?v=42.

34. Y. Lei, Q. Zhang, K. B. He and D. G. Streets, Primary anthropogenic aerosol emission trends for China, 1995-2005, *Atmos. Chem. Phys.*, 2011, **11**, 931–954.

35. Y. Zhao, J. Zhang and C. P. Nielsen, The effects of recent control policies on trends in emissions of anthropogenic atmospheric pollutants and CO_2 in China, *Atmos. Chem. Phys.*, 2013, **13**, 487–508.
36. HEI, Special report 18: Outdoor air pollution and health in the developing countries of Asia: a comprehensive review, 2010.
37. J. Hao and L. Wang, Improving urban air quality in China: Beijing case study, *J. Air Waste Manage. Assoc.*, 2005, **55**, 1298.
38. http://jcs.mep.gov.cn/hjzl/zkgb/.
39. T. Ohara, H. Akimoto, J. Kurokawa, N. Horii, K. Yamaji, X. Yan and T. Hayasaka, An Asian emission inventory of anthropogenic emission sources for the period 1980-2020, *Atmos. Chem. Phys.*, 2007, 7, 4419.
40. D. G. Streets *et al.*, An inventory of gaseous and primary aerosol emissions in Asia in the year 2000, *J. Geo. Res.*, 2003, **108**(D21), 8809.
41. S. Su, B. Li, S. Cui and S. Tao, Sulfur dioxide emissions from combustion in China: from 1990 to 2007, *Environ. Sci. Technol.*, 2011, **45**, 8403.
42. http://www.worldbank.org.
43. W. J. Qu, R. Arimoto, X. Y. Zhang, C. H. Zhao, Y. Q. Wang, L. F. Sheng and G. Fu, Spatial distribution and interannual variation of surface PM_{10} concentrations over eighty-six Chinese cities, *Atmos. Chem. Phys.*, 2010, **10**, 5641.
44. K. He, F. Yang, F. Duan and Y. Ma, *Airborne Particles and Regional Complex Pollution*, China Science Press, 2011 (In Chinese).
45. D. Chang, Y. Song and B. Liu, Visibility trends in six megacities in China 1973-2007, *Atmos. Res.*, 2009, **94**, 161.
46. J. Deng, K. Du, K. Wang, C. S. Yuan and J. Zhao, Long-term atmospheric visibility trend in Southeast China, 1973e2010, *Atmos. Environ.*, 2012, **59**, 11.
47. http://www.bjepb.gov.cn/bjepb/323474/324034/324735/index.html.
48. Committee on Energy Futures and Air Pollution in Urban China and the United States, National Academy of Engineering and National Research Council in collaboration with Chinese Academy of Engineering and Chinese Academy of Sciences, *Energy Futures and Urban Air Pollution: Challenges for China and the United States*, 2007, http://www.nap.edu/catalog/12001.html.
49. T. Wang, X. L. Wei, A. J. Ding, C. N. Poon, K. S. Lam, Y. S. Li, L. Y. Chan and M. Anson, Increasing surface ozone concentrations in the background atmosphere of Southern China, 1994-2007, *Atmos. Chem. Phys.*, 2009, **9**, 3217.
50. G. Tang, Y. Li, Y. Wang, J. Xin and X. Ren, Surface Ocean trend details and intperetations in Beijing, 2001-2006, *Atmos. Chem. Phys.*, 2009, **9**, 8813.
51. X. Xu, W. Lin, T. Wang, P. Yan, J. Tang, Z. Meng and Y. Wang, Long-term trend of surface ozone at a regional background station in eastern China 1991-2006: enhanced variability, *Atmos. Chem. Phys.*, 2008, **8**, 2595.
52. H. X. Qi, W. L. Lin, X. B. Xu, X. M. Yu and Q. L. Ma, Significant downward trend of SO2 observed from 2005 to 2010 at a background station I the Yangtze Delta region, China, *Sci. China: Chem.*, 2012, **55**, 1451.

Mercury and Lead

ROBERT P. MASON

ABSTRACT

While both lead (Pb) and mercury (Hg) are not abundant in the Earth's crust, they are two trace elements that have largely impacted human health because of their enhancement in the biosphere due to human activity. Their global cycles have been altered by humans as both Pb and Hg have been used for thousands of years. Current inputs to the atmosphere for Hg are a factor of 5–6 times higher than pre-anthropogenic emissions. The inputs at the time of heightened use of Pb in gasoline (1960–1980's) were much larger relative to the pre-industrial flux. However, as emissions of Pb have been strongly curtailed due to regulation of their use in gasoline, paints and many other applications, the current inputs to the atmosphere are much lower, even given the heightened recent industrial activity in Asia and elsewhere. This is not the case for Hg, as emissions have not decreased over the last 20 years, although now the major emissions are from Asia, whereas previously they were from the western world. The impact of anthropogenic activity is recorded in various archives, and impacts have been documented by changing concentrations in the oceans and terrestrial environment. The success of various policy initiatives are highlighted as well as the impact of emissions and also regulation of these activities on human health.

1   Introduction

While both lead (Pb) and mercury (Hg) are not abundant in the Earth's crust (50 and 0.3 $\mu mol\,kg^{-1}$, respectively; Table 1),[1–4] they are two trace elements

Issues in Environmental Science and Technology No. 40
Still Only One Earth: Progress in the 40 Years Since the First UN Conference on the Environment
Edited by R.E. Hester and R.M. Harrison
© The Royal Society of Chemistry 2015
Published by the Royal Society of Chemistry, www.rsc.org

Table 1   Average concentrations of lead, mercury and methylmercury in various reservoirs and media, and freshwater partition coefficients (K_D) and bioaccumulation factors for predatory fish (BAF). Compiled from various sources.[1]

Reservoir	Lead (Pb)	Mercury (Hg)	Methylmercury (CH$_3$Hg)
Lithosphere (μmol kg^{-1})	50–100	0.3	–
Soil (μmol kg^{-1})	50	0.2	$1–5\times10^{-3}$
Freshwater (pM)^a	20–1000	0.5–20	0.1–5
Seawater (pM)^a	10–150	0.2–3	0.01–0.5
Remote aerosols (pmol m^{-3})	$10–10^4$	0.01–0.5	<0.05
Log $K_D{}^b$	5.6	5.3	4.5
Log BAF	2.3–4.0	3.0–3.8	6.3

^{a}0.4 micron filtered concentrations.
^{b}Median water column values for freshwater.

that have had a large impact on human health, and on wildlife, because of their enhancement in the biosphere due to human activity.[2,5–13] Both Pb and Hg have been mobilized through their intentional use (*e.g.* Pb in alkyl lead products in gasoline, in paint pigments, and in solder in cans and plumbing; Hg use in gold mining, in the chlor-alkali and other industry, in dentistry, in anti-bacterial applications and pesticides[2,14–17]). Both Pb and Hg have been used in batteries and as catalysts. Additionally, they have been added to the biosphere as they are trace constituents in coal and other petroleum products, especially for Hg,[6,18,19] and are in important metal ores (mainly sulfide ores), and can be released during waste incineration. Both have also been mobilized as a result of the purposeful mining of their primary and other metal ores.[20,21] Elevated concentrations of Pb are found in acid mine drainage throughout the world.

While the toxicological impact of Pb is mainly due to its inorganic form, the main form of Hg that is highly bioaccumulated in aquatic food chains, and drives most health advisories, is the methylated form (methylmercury, MeHg or CH$_3$Hg).[14,22–26] The Hg contamination incident in Minamata, Japan which began in the late 1950's helped highlight the health impacts of methylated Hg accumulation in fish. At the height of the contamination that resulted from industrial discharges into a small bay off Minamata, concentrations of MeHg in fish were as high as 20 mg kg^{-1}.[26] For comparison, current health advisory levels in the world range from 0.3–1 mg kg^{-1}.[27,28] The Minamata exposure was one of several major contamination incidents during the 1950–80's.[10,23] Other high-level exposure incidents highlighted the hazards of MeHg exposure and led to a reexamination of its uses and sources in the environment.

Additionally, while the human health and environmental impact of Pb has been substantially reduced through phasing out of its use and because of environmental regulation, this has occurred to a much lesser extent for Hg, although an international treaty has recently been signed (the Minamata Convention).[29] Additionally, two aspects of Hg biogeochemistry enhance the legacy of its impact: (1) its propensity to be widely distributed due to its volatility and relatively long residence time in the atmosphere;[19] and (2) the degree to which it is methylated and the enhanced toxicity and bio-accumulation of MeHg compared to inorganic Hg.[23,30]

While Hg is mostly released to the atmosphere in its reduced state as Hg^0, which is a gas in air at environmental concentrations, Pb is released to the environment as oxidized species (Pb^{II} compounds), and mostly attached to particles.[6–8] Some fraction of the Hg released to air is also in an ionic form, either as a gas (HgX_2, where $X = Br^-$, Cl^-, OH^-) or attached to aerosols. Direct discharges to water and the terrestrial environment are mostly the ionic forms of both metals. The metal ions form strong associations with particles and therefore are mostly transported in rivers with the solid phase.[1,31,32] Geogenically, Pb ores are mostly formed from hydrothermal solutions, and are sulfide ores (galena; PbS), but Pb is often found in association with other metals (copper (Cu), zinc (Zn) and silver (Ag)).[20] In the sedimentary rocks Pb is quite evenly distributed as Pb is also derived through the radioactive decay chain of uranium (U) and thorium (Th) with different decay pathways leading to the different Pb isotopes: ^{206}Pb is the stable product of the ^{238}U decay series, ^{207}Pb arises from decay of ^{235}U and its products and ^{208}Pb from the ^{232}Th decay series.[33] Thus, the isotopic ratio of Pb in the environment is altered due to the presence of either U or Th in the medium. It has been demonstrated that Pb from different locations has isotope ratios which are different enough to track the sources of the Pb in the ocean and other environments.[11] The short-lived Pb radioisotope (^{210}Pb; half-life ~ 22 years), which is often used in sediment archive core dating, is formed from decay of ^{222}Rn, which is a gas, and ultimately transforms to ^{206}Pb.

Lead is not highly solubilized during weathering, and is found in water mostly in the particulate phase. In solution, depending on pH, and in the absence of organic ligands, Pb is present either as the free ion or as carbonate and hydroxide complexes.[1,32] In seawater, chloride complexes are more important. Lead has one of the highest partition coefficients (distribution between dissolved and particulate fractions; K_D) of the trace metals ($>10^5$, typically), which is somewhat higher than that of inorganic Hg, which is also strongly particulate-bound (see Table 1). In aquatic waters, Hg is mostly present as a complex; chlorides dominate in seawater and hydroxides in freshwater in the absence of thiol-containing natural organic compounds to which Hg binds strongly. Some fraction of the Hg in environmental waters is Hg^0, present as a dissolved gas. Methylmercury is present as a +1 cation and can exist in solution as the free ion or as inorganic and organic complexes, and in most waters is <10% of the total Hg. In the ocean, in addition to MeHg, dimethylmercury (Me_2Hg or $(CH_3)_2Hg$) is also found. Methylmercury partitions to particles to a lesser degree than ionic Hg, but it is strongly bioaccumulated in the aquatic food chain due to its association with the proteins in biota (see Table 1). The high bioaccumulation of MeHg relative to inorganic Hg results in most of the Hg in higher food chain organisms (secondary consumers and predators) being MeHg (>80%) except for organisms that can demethylate MeHg (some marine mammals and birds).[34]

In the biosphere, anthropogenically-emitted Pb is strongly accumulated in the uppermost organic and alluvial layers of soils. The use of leaded gasoline since the early 1900's has been the main source of Pb to the atmosphere (see Figure 1).[6,11,25] In most countries, this source has been now eliminated, as discussed further below. Geochemically, Hg behaves similarly to other metals of its group (cadmium (Cd) and zinc (Zn)), and the precious metals. While Hg can exist in the environment as a sulfide ore (cinnabar or meta-cinnabar; HgS) or in its elemental form as a liquid, it is often a constituent in sulfide ore deposits, or is incorporated into gold (Au) and silver (Ag) deposits, due to its amalgamation with these metals.[10,16,35,36] High concentrations of Hg are known to exist in gas, oil, and coal deposits.[37] There are regions of the globe where Hg is naturally enriched in the upper mantle, primary in tectonically-active regions (spreading centers and subduction zones).[38] Mercury differs from Pb because of its ability to vaporize as Hg^0 under natural conditions from the terrestrial surface or the ocean.[39,40] The volatility of Hg facilitates its migration through the atmosphere and the differences in the volatility and solubility of the oxidized and reduced forms prolongs its cycling through the biosphere with its atmospheric residence time being estimated as 6 months to a year.[41,42] In contrast, Pb, like most metals that are primarily attached to aerosols in the atmosphere, has an atmospheric residence time of a few weeks,[5,43] and therefore is less globally distributed from its sources through the atmosphere than Hg.

2   The Anthropogenic Insult

Of the potentially toxic and bioaccumulative metals, it is probable that the global cycles of Hg and Pb have been affected most by humans. Both Pb and Hg have been used by humans for thousands of years. Bindler *et al.*[44] report evidence for Pb pollution in lake sediments around 3500 years ago in Sweden, and suggest about half of the cumulative burden of atmospherically deposited Pb was before industrialization (<1850). There is also similar evidence for contamination from pre-industrial Hg use in Sweden, in South America and elsewhere.[45,46] There was a historical peak in usage during Roman times as Pb was used in pipes and other applications, and in pewter.[25] The pre-anthropogenic inputs of Pb would have been primarily from windblown dust and volcanoes,[5,43,47] with both emission sources being of variable but similar magnitude[48] and are minor compared to the human-related inputs.[49] In contrast, for Hg, inputs from volcanoes are today an important contributor to atmospheric inputs (5–8%).[19,40,38]

Many studies in North America and Europe[50–54] have examined the long-term record of Pb pollution using sediment archives (see Figure 2) and also by examining the isotopic variations in the Pb isotopes as this is diagnostic of sources other than local ores and soils, as discussed further below. The sediment records overall display similar trends over time to that of the estimated Pb production (see Figure 2A) and reflect the increased usage during the Roman era, the subsequent decrease and the rapid increase

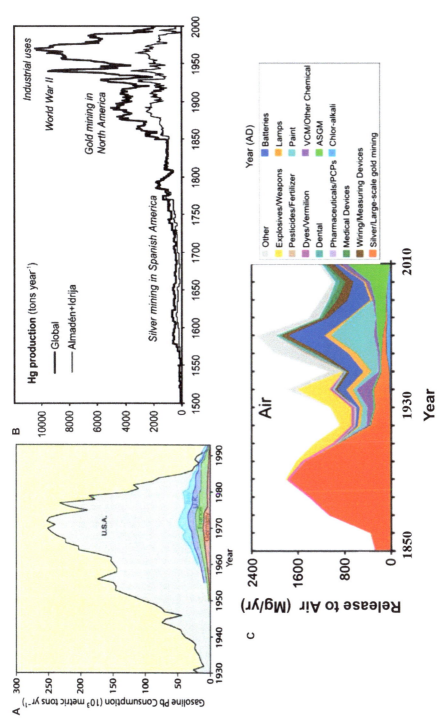

Figure 1 **A:** Historical usage of lead (Pb) in gasoline in the Western world, as depicted in Boyle *et al.*, *Oceanography*, 2014, **27**, 69–75 (2014),[11] (Reprinted with permission of the Oceanography Society); **B:** Estimated mercury (Hg) production since 1500 according to Hylander *et al.*[59] (Reprinted with permission of Elsevier from *Sci. Total Environ.*, 2003, **261**, 99–107); and **C:** Compilation of estimated emissions to air for Hg since 1850 taken from Horowitz *et al. Environ. Sci. Technol.*, 2014, **48**, 10242–10250.[65] (Copyright (2014) American Chemical Society).

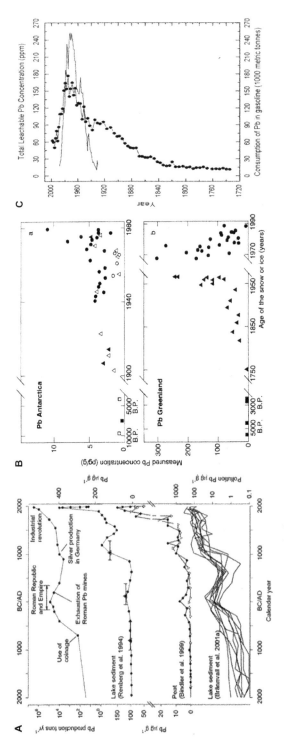

Figure 2 Historical records of lead (Pb) contamination. **A:** Estimated lead production, and lake sediment and peat core profiles compiled by Bindler *et al. J. Paleolimnol.*, 2008, **40**, 755–770.[44] (Reprinted with permission from Springer Science and Business Media); **B:** Ice core profiles for lead from both polar regions. (Reprinted with permission from Elsevier from Boutron *et al.*,[82] *Geochim Cosmochim. Acta*, 1994, **58**, 3217–3225); and **C:** Record of deposition of lead as recorded in the sediments of the Pettaquamscutt estuary, RI, USA (line with symbols) compared to USA leaded gasoline usage (no symbol). (Reprinted with permission from Elsevier from Lima *et al.*[83] from *Geochim Cosmochim Acta*, 2005, **69**, 1813–1824).

during industrialization. The global contamination by Pb is evident in the ice core records from both poles (see Figure 2B) which clearly show the increasing concentrations in the middle 1900's as the use of Pb in gasoline increased (see Figure 1A). The relationship between Pb use in gasoline in the USA and the Pb signal in a core from the permanently anoxic basin of the Pettaquamscutt estuary (see Figure 2C) further demonstrates the close link between usage and emissions of Pb and its deposition. This core also shows the decreasing concentrations and fluxes since the controls over Pb in gasoline and other emissions.

Mercury was also actively mined by the Romans and other civilizations since ancient times and used as a pigment (cinnabar), and in alchemy, and in medical and ritualistic uses in China and other parts of the world.[10,16] After the Roman period, there was a decrease in usage. An archival record from Antarctica provides evidence of Hg inputs during the early Chinese, Roman and South American (Incas, Mayan) civilizations.[55] Ancient use of cinnabar led to its extraction and use in South America, as documented by regional contamination in lake sediment cores.[56] Increased extraction and recovery of Hg^0 from cinnabar began with the use of Hg amalgamation as a method for extracting Ag by the Spanish in South America initially using the "patio" process.[10,15,57] The use of Hg in precious metal, mostly Au, mining subsequently spread to the later gold rushes in North America and other parts of the world[16,57-60] (see Figure 1).

A number of studies in South America have documented the local and regional contamination from early Hg and precious metal mining.[35,56-58,61-63] Also, in South American Hg mines, there was the potential for release of Hg during the processing of the cinnabar to recover Hg^0 for Ag mining.[64] For example, it has been estimated that the Hg released between 1560 and 1810 from cinnabar smelting at Huancavelica, the primary Hg source for Ag mining, was 17 000 Mg with 39 000 Mg being released during the associated Ag mining activities.[64] This estimate is comparable to other estimates for emissions during that period from Hg refining and Ag extraction.[66,67]

The input of Hg to the atmosphere through its use in Au and Ag mining and extraction in the last 400 years is a topic of recent debate, as there are various emission estimates based on analysis of mining and usage records (see Figure 1)[59,65-67] and the interpretation of historic archival records of these inputs[57,68-70] (see Figure 3). While many studies have examined the record of Hg deposition in sediment archives, there has also been examination of the historical record preserved in peat cores, ombrotrophic bogs, ice cores and mammalian tissues (*e.g.* eggshells, teeth, hair).[55,70-75] There is a difference of opinion in the literature about the accuracy of the record archived in the various media.[69,76] Recent evidence suggests that the suitability of bogs for estimating historic inputs is problematic[76] but that sediment cores reliably retain a Hg record, but perhaps not for Pb in all instances, once the metal is sequestered below the surface sediment layer mixed by bioturbation and physical mixing.[77]

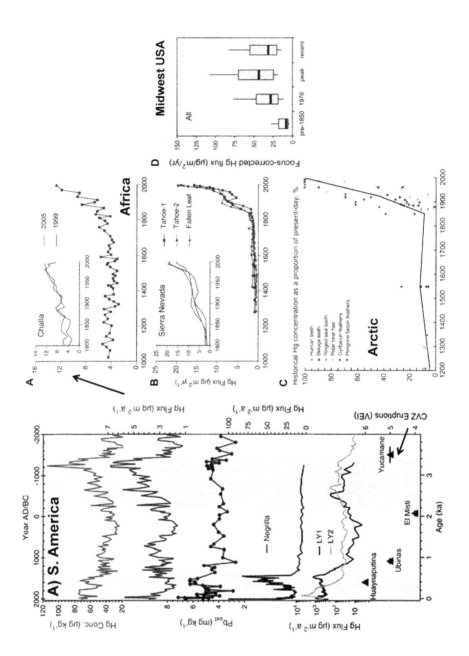

One issue with examining historic deposition is the dating of sediments older than those datable using ^{210}Pb ($\sim$150 years) and other recent age "markers" (*e.g.* ^{137}Cs peak) because of the need to make assumptions about sedimentation rate in the absence of other methods of dating the older sediments.[57,78] One study suggests that this could lead to an overestimation of pre-industrial inputs.[78] There is also the need to account for sediment focusing, watershed inputs and other factors.[57,79] Inputs from the watershed of legacy deposited Hg may confound the signal and lead to "smoothing" of the record.[69] The importance of regional deposition and regional factors (*e.g.* changes in precipitation with elevation; changes in the watershed) need to be properly evaluated in situations where there is the potential for such factors to compromise interpretation of the signal recorded in the sediments.[80,81] One study of lakes in the vicinity of (<50 km), and remote from (>150 km), urban areas showed that the deposition rate and the modern/pre-industrial flux ratio was related the lake's distance from the urban environment.[80] In such cases, these sediment cores will not truly represent the global input. In interpreting the pre-industrial inputs in South America, therefore, the relative proximity of some sites to mining areas suggests that these sites are reflecting regional rather than global inputs.[35,57,63] Clearly, the successful estimation of historical inputs requires a careful examination of multiple historical records.

The use of Hg amalgamation for Au extraction still occurs today, and is mostly carried out by small-scale artisanal miners on many continents.[28,84] Industrial mining mostly uses approaches based on Au extraction using cyanide and the subsequent Au recovery. The extent of emissions from mining amalgamation activities is hard to estimate and as noted above,[64] there are different views of the amount released to the atmosphere during Au and Ag recovery.[57,66] This is primarily a result of methods of processing the ores, and the extent to which the Hg is recovered after the amalgamation step. In many small scale artisanal activities currently occurring around the world, there is little attempt to recover the Hg because of its relatively low cost compared to that of Au (Au price : Hg cost ratio (g g^{-1}) is $\sim$1600), and

---

Figure 3 Historical profiles as recorded in archives. **A:** Historical mercury (Hg) flux, and labile lead (Pb), for various locations in South America taken from Beal *et al.*[61] compared with the lake cores from Huancavelica (LYI, LY2) and Laguna Negrilla taken from Cooke *et al.*[56,58] (Reprinted with permission of Wiley from *Global Biogeochem. Cycles*, 2014, **28**, 437–450); **B:** Sediment records of mercury flux in North America and Africa taken from Engstrom *et al.*[57] *Environ. Sci. Technol.*, 2014, **48**, 6533–6543. (Copyright (2014) American Chemical Society); **C:** Comparative concentrations (percent change) of mercury in various biotic records over time for the Arctic taken from Kirk *et al.*[86] (Reprinted with permission from Elsevier from *Environ. Res.*, 2012, **119**, 64–87); and **D:** Relative flux ratios for different time periods for lakes in the northwestern USA taken from Drevnick *et al.*[87] (Reprinted with permission from Elsevier from *Environ. Pollut.*, 2012, **161**, 252–260).

the need for retorting equipment for Hg recovery.[9,18,28,84] Thus, current emissions from these activities are high. This contrasts the earlier mining activities where Hg was more actively recovered. The recent AMAP/UNEP (Arctic Monitoring and Assessment Program/United Nations Environment Program) analysis[28] for the Minamata Convention suggests that current artisanal mining inputs to the atmosphere are now the largest single source, and are substantial in Central and South America, sub-Saharan Africa and in Southeastern Asia, according to the report. Other recent analyses do not conclude that artisanal mining is the dominant current source to the atmosphere, but all suggest that it is a major contributor.[18,65,66,85]

The long history of use of Pb and Hg by humans makes it difficult to estimate the pre-anthropogenic flux to the atmosphere and oceans. Furthermore, the natural component of Hg emissions to the atmosphere is harder to assess given that Hg once deposited can be re-emitted to the atmosphere if converted to Hg^0. Thus, the so-called "natural fluxes" contain both Hg derived from natural sources and anthropogenic Hg that has been deposited and then re-emitted.[88] As a result, most of the gas evasion from the ocean and from the terrestrial surface is previously deposited anthropogenic Hg (>75% based on the pre-anthropogenic estimates below). The geogenic Hg flux is relatively significant (see Table 2) compared to present day anthropogenic inputs, but if the pre-anthropogenic river flux (crustal erosion) was known it could be possible to construct a pre-anthropogenic Hg budget based on current relative fluxes between the surface reservoirs and the atmosphere (*e.g.* Driscoll *et al.*).[19]

The pre-anthropogenic river flux of metals can be estimated by using the current flux of an element whose river input has not been substantially altered by human activity, and crustal ratios to that element. Sodium (Na) is a metal whose global cycle has been little impacted by humans[31,89,90] and it

Table 2   Comparative emissions to air as estimated in various publications (Mmol year^{-1}). Data for mercury (Hg) are for 2000 and primarily taken from the summary in Pirrone and Mason.[85] Data for lead (Pb) are for the 1980–1990's and taken from various sources.[1,97]

Source	Hg flux (Mmol year^{-1})	Source	Pb flux (Gmol year^{-1})
Primary anthropogenic			
Fossil fuel combustion	7.1	Gasoline emissions	1.7
Metal production	2.0	Metal production	0.4
Other industrial emissions	2.0	Other sources	0.2
Other sources	1.5		
Artisanal gold mining	2.0		
Natural processes			
Ocean emissions	13.3	Geogenic	0.03
Terrestrial areal emissions	8.9	Dust/other sources	0.03
Biomass burning	3.4		
Geogenic	0.5		

is used here to estimate the fluxes, although its transport in rivers is mostly in the dissolved phase, while Pb and Hg are transported as particulate. Estimates were also made based on silicon (Si), which is transported mostly as particulate, acknowledging that the Si flux to the ocean has been substantially decreased by reservoir construction.[91–93] The Si-based estimates were of the same order as the estimates based on Na, but lower. Using this approach, the pre-anthropogenic river flux for Hg was estimated as 1.3 Mmol year^{-1} and that for Pb as 0.32 Gmol year^{-1}. These represent <10% of the current river flux estimates for each metal, which ranges from 14–27 Mmol year^{-1} for Hg with a large uncertainty around each estimate.[19,94,95] The current estimated river flux for Pb[1,3] is 5–10 Gmol year^{-1}. These river fluxes do not reflect the input to the open ocean as 80% or more of the Pb and Hg in rivers are deposited in the coastal zone.[3]

Using these estimates a pre-anthropogenic budget was developed for Hg with the following atmospheric inputs: geogenic inputs 1.9 Mmol year^{-1}; terrestrial emissions 2.0 Mmol year^{-1} and ocean evasion 2.9 Mmol year^{-1}; total emissions 6.8 Mmol year^{-1}. The geogenic inputs are balanced by Hg removal in the coastal zone (1 Mmol year^{-1}; 80% of the river flux) and deposition in the deep ocean (0.9 Mmol year^{-1}). This suggests that current inputs to the atmosphere for Hg (see Figure 4) are a factor of 5–6 times higher than pre-anthropogenic emissions, similar to the enrichment estimated by Sen *et al.*[89] This estimate is at the high end of some recent estimates (a factor of 2–5 increase in atmospheric inputs suggested by Engstrom *et al.*[57] and references therein), but lower than those of others,[42,95] but represents the total mobilization through both primary and secondary inputs of Hg to the atmosphere. As already noted, the natural flux for Pb is less than 10% of current inputs. Overall, it is abundantly clear that human activities have dramatically altered the cycling of both elements through the biosphere.

There have been many recent studies examining the extent of emissions of Hg and Pb to the atmosphere from both natural and anthropogenic sources. As an example, emission estimates for each metal and the dominant sources are compiled in Table 2. For Hg, primary anthropogenic industrial sources are coal and other fossil fuel combustion, ferrous and non-ferrous metal mining, cement and other manufacturing. For Pb, leaded gasoline was the major source in the latter part of the 20th century with inputs from non-ferrous metal mining being the second most important source (see Table 2). In Sweden, it is estimated that atmospheric Pb deposition rates during the 20th century were at least a hundred times greater than pre-anthropogenic rates, and are still elevated compared to these natural emissions even after the substantial recent reductions in emissions.[44] Pacyna *et al.*[96] documented a decrease of a factor of 10 in Pb in rain in Europe in 2000 compared to the maximum around 1975. The recent decadal scale variability in anthropogenic inputs for both Pb and Hg, which have changed due to the relative rate of industrialization around the world, and the incorporation, and then banning of, industrial processes and uses of each metal, raises an issue of

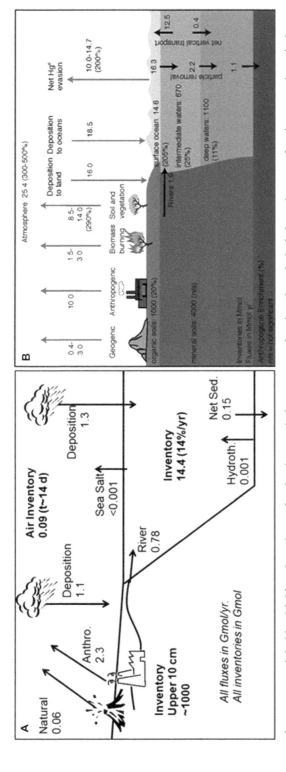

Figure 4   **A**: Global lead (Pb) cycle estimate for the time period 1980–1990 showing the relatively large impact of anthropogenic emissions. Figure adapted and redrawn from Mason, *Trace Metals Aquat. Syst.* (2013),[1] and compiled using information from various sources;[3,5,97] **B**: The global mercury (Hg) cycle modified from Mason *et al.*[131] (Reprinted with permission from Elsevier fom *Environ. Res.*, 2012, **119**, 101–117).

the timescale over which values are compared as the inputs to various parts of the globe have changed in recent decades, as shown in Figure 1.

Other sources of Pb are fossil fuel combustion and emissions from the non-ferrous metal industry and waste incineration (~ 50 Mmol year^{-1} in 1980).[8] Usage of Pb in gasoline peaked in the western world in the late 1970's but there are still some uses of Pb in gasoline. As an example of the changes over time, in the USA, Pb use in gasoline accounted for 83% of emissions to the atmosphere in 1970, with metal processing accounting for a further 11%, and fossil fuel combustion being 5% of the total of ~ 1 Gmol year^{-1}. In 1980, gasoline emissions still dominated (>80% of the total) but the total Pb emissions were a third of those a decade earlier, indicating that in addition to decreased use of leaded gasoline there were concomitant decreases in other emission sources. In 1990, emissions were <2% of the 1970 levels (~ 16 Mmol year^{-1}) with leaded gasoline, fossil fuel combustion and waste incineration emissions at 10–15% of the total, and metal refining now being the dominant emission source (50–60% of the total).[97] This is a substantial change in emissions over a very short geological timeframe and the impacts of these changes are still percolating through the biosphere.[11]

Studies have also documented the impact of human activities throughout the world since industrialization, such as the increased Pb content in sediment cores in North America,[98,99] Mexico,[100] Europe and Asia.[101–103] For example, the concentrations of Pb in sediment cores in Mexico increased, with a changing isotopic signature, around 1900 in concert with increased Pb production and export, and later the manufacture of alkyl Pb products. In central California, analysis of recent and archived lichen samples demonstrated Pb contamination since around 1890, likely due to Pb smelters in the region.[104] The isotopic signatures suggest that Pb in gasoline became the dominant source in the 1950's and concentrations peaked in the mid-1970. While there have been decreases in atmospheric concentrations related to the removal of Pb sources, there can be a strong lag in response due to the legacy of Pb in soils and their resuspension and transport, as discussed further below.[104] Many examples exist of investigators using the isotopic composition of Pb in media to identify the origins of the Pb as different ores mined in different parts of the world, and used by different countries for manufacturing, have different signatures.[11,83,105–112] These are discussed further in Section 3.2.

Mercury was used in many industrial processes that are now banned because of the demonstration of large-scale local contamination,[65,113] and many prior uses of Pb are now similarly curtailed, and these changes are reflected in sediment records (see Figures 2 and 3). Furthermore, the introduction of pollution control technology for cleaning stack gases of pollutants, and other modifications to factories, have resulted in a reduction in the emissions from such sources.[114,115] Evidence for contamination during the last century has been recorded in many locations in the world.[57,87,116–127] The later increase in contamination of Hg in Asia compared to North America and Europe is highlighted by data from lakes in

China[121] that show increases typically beginning in the mid-1900's and in many lakes, no evidence for a recent decrease in concentration, in contrast to the western countries. Studies in Canada[122] of multiple lakes has shown that the fluxes were inversely correlated with latitude and Hg flux ratios (1990's/pre-1850) ranged from <1 to 8. Overall, the results confirmed that anthropogenic emissions in mid-latitudes have been transported to the Arctic region.[86] Other sediment core studies in North America and Europe have reached similar conclusions for Hg[117,119] and for Pb.[107] Examination of Pb isotopes for sediment cores from Greenland led to the conclusion that inputs of Pb to the region during the 1970's, when inputs were maximal, were from both European and Russian sources.

3   The Global Biogeochemical Cycles of Mercury and Lead

The global budget for Pb is shown in Figure 4A.[3,6,8,82] The average residence time in the atmosphere of two weeks is comparable to, but somewhat lower than, that estimated for the other trace metals. Much of the Pb added to the atmosphere, especially since industrialization, was from automobile exhaust, and would have been removed by dry deposition faster than that added higher to the atmosphere from the stacks of combustion and other industrial sources. Overall, a large fraction of the Pb inputs to the ocean are atmospheric inputs, with river inputs being 40–60% of the total input. River inputs include anthropogenic contributions plus the transport of atmospheric Pb deposited to watersheds, and from background sources. It is clear that the inventory of Pb in the ocean has been substantially changed by anthropogenic inputs. Based on Figure 4, the estimated average rate of increase in the ocean is 14% year^{-1}, although it is relatively obvious that this change is not the same for all oceans (see Figure 5), and is changing over time. There is now relatively clear evidence that there has been an overall decrease in Pb concentration since the early 1980's for the North Atlantic Ocean (see Figure 5A), which can be mostly attributed to the decrease in deposition of Pb as a result of the phasing out of Pb in gasoline in North America and Europe.[11,128] However, while input from Pb use in gasoline has declined, anthropogenic sources of Pb to the atmosphere still continue due to inputs from fossil fuel burning, waste incineration and metal refining.[6,43]

 Hydrothermal inputs are insignificant for Pb and the burial in deep ocean sediments removes a small fraction of the estimated current input (<10%). The profiles for Pb in the ocean show enrichment in the surface that is reflective of the importance of atmospheric input (see Figure 5). As the residence time of Pb in the surface mixed layer of the ocean is short (~2 years), its concentration tracks that in the atmosphere.[11] However, the lifetime of Pb in the deeper waters is longer and therefore these waters show a lag in their response time to changes in atmospheric Pb levels (see Figure 3). The deep ocean circulation results in the "oldest" waters being in the deep Pacific Ocean, and because of the particle reactivity of Pb, this results in its deep North Pacific Ocean concentration being lower than that in the North

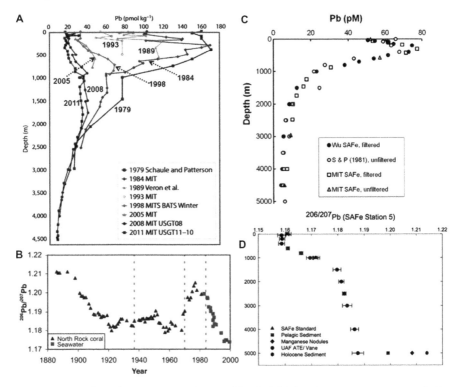

Figure 5 **A:** Changing concentrations of lead (Pb) in the upper 4500 m of the water column in the North Atlantic Ocean in the vicinity of Bermuda; **B:** Lead isotope ratios (^{206}Pb/^{207}Pb) with time as recorded in coral and in seawater in the vicinity of Bermuda. Both figures adapted from Boyle *et al.*[11] (Reprinted with permission of the Oceanography Society from *Oceanography*, 2014, **27**, 69–75); **C:** Vertical profile of lead at the SAFe station (30°N, 140°W) in the North Pacific compared to other measurements in the vicinity over time; and **D:** isotope ratios (^{206}Pb/^{207}Pb) for the North Pacific water column, compared to those of three deep ocean sediments (pelagic sediment, manganese nodules and Holocene sediment). Both figures taken from Wu *et al.*[129] (Reprinted with permission from Elsevier from *Geochim. Cosmochim. Acta*, 2010, **74**, 4629–4638).

Atlantic.[129] This contrasts the less particle reactive metals (*e.g.* Cd, Zn) and nutrients whose deep Pacific Ocean concentration is 3–7 times higher than that of the deep Atlantic.[130] The difference for Pb is also exacerbated by the anthropogenic signal that is present to a greater degree in the deep Atlantic compared to the Pacific.[11]

The global cycle of Hg (see Figure 4B) has received much attention lately because of the heightened concern about the health risks of consuming CH$_3$Hg-laden fish and marine organisms and the likely impact that anthropogenic emissions have had on the ocean Hg burden. A number of recent papers have discussed the global Hg cycle[8,7,40–42,132–135] and while there is relative agreement in the magnitude of the estimates for Hg emissions, there are also substantial differences in some aspects which reflect the complexity of the global Hg cycle and Hg biogeochemistry. For example, while most metals exist in the atmosphere only attached to particles, Hg can also exist in the gaseous phase, either as elemental Hg (Hg^0) or as ionic Hg species (*e.g.*, $HgCl_2$, $HgBr_2$; collectively termed "reactive gaseous Hg" (RGHg) based on the methodology of its measurement).[136-138] Thus, dry deposition of both gaseous and particulate species can occur, including the uptake of Hg^0 by vegetation,[139] and all forms of Hg can also be scavenged by wet deposition to differing degrees. Additionally, Hg can exist as a dissolved gas (mostly Hg^0) in ocean and freshwaters and therefore can be lost to the atmosphere *via* gas exchange.

The global budget shows the relative importance of anthropogenic sources to the atmosphere. The residence time of Hg in the atmosphere is substantially longer than that of the other metals. This reflects the complexity of its atmospheric chemistry and the fact that most (>90%) of the Hg in the atmosphere is Hg^0, which is removed from the atmosphere relatively inefficiently compared to that of particulate Hg or RGHg. Additionally, there is the potential for recycling of Hg between the surface reservoirs as deposited ionic Hg can be reduced and then re-emitted to the atmosphere as Hg^0. This flux is a very important part of the global Hg cycle (see Figure 4B) as it decreases the relative impact of anthropogenic emissions on ocean Hg concentrations.[40,42,94]

As with Pb, Hg concentrations in the ocean are not at steady state, and the ocean basins have been similarly differently impacted by recent anthropogenic activities.[42,94] A compilation of water column concentrations, though much more limited than that for Pb, show similar trends with decreasing concentrations in the North Atlantic and Mediterranean Sea, and increasing concentrations in the upper waters of the North Pacific Ocean (see Figure 6).[19,94,131] These changes correspond to the changing emissions profile globally. While the total global anthropogenic emissions have not changed substantially in the last 20 years, there has been a shift from 40–50% of emissions in 1990 being from North America and Europe to >50% of emissions now coming from Asia, with the North American/European contribution being now <20%.[7,19,66,140] For Hg, the dynamic nature of water movement and the biogeochemical cycles of Hg in the upper ocean further complicate the changes in concentration with time. The relative rate of change in the North Atlantic mirrors to some degree that of Pb whose concentration has decreased markedly in the last decades (see Figure 6).[11]

The changes in Pb emissions over time does not reflect the later industrialization in Asia to the same extent as found for Hg because of the relatively short time that leaded gasoline was used in Asia compared to the western world, and the much greater use of leaded gasoline in the USA compared to the rest of the world (see Figure 1). The global signal for Pb is also much more regional than that of Hg given its relative short residence time in the atmosphere. Overall, the mass budgets, while they may not be

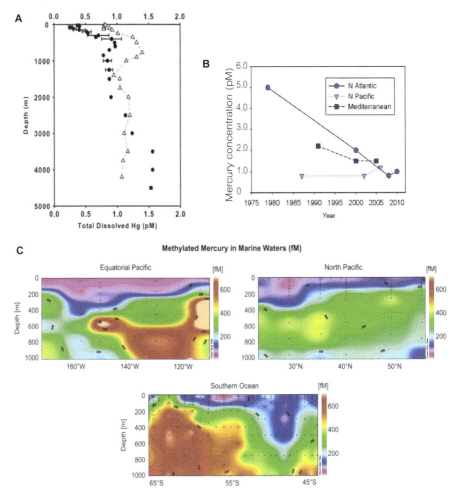

Figure 6   **A:** Vertical profiles for total mercury (Hg) in the North Atlantic near Bermuda (△) and in the North Pacific (●) (similar locations to the Pb data in Figure 5), taken from Lamborg *et al.*[170] *Limnol. Oceanograph. Methods*, 2012, **10**, 90–100. (Copyright (2012) by the Association for the Sciences of Limnology and Oceanography, Inc.); **B:** Changes in average measured concentrations over time for the upper ocean taken from Driscoll *et al.*[19] *Environ. Sci. Technol.*, 2013, **47**, 4967–4983. (Copyright (2013) American Chemical Society); and **C:** Distributions of methylated Hg in the upper waters of various oceans, modified from Mason *et al.*[131] (Used with permission of Elsevier from *Environ. Res.*, 2012, **119**, 101–117).

able to provide details of the cycling of metals at the Earth's surface, are still insightful as they provide an indication of the major processes involved. These budgets highlight areas where more research is needed in the global cycling of metals but highlight the extent that human activities have impacted these cycles.

3.1   Mercury Cycling in Aquatic Systems

As noted above, Hg^0 is the dominant volatile Hg species in the surface waters of aquatic systems. Given the relative surface area of the ocean compared to freshwater ecosystems, and the similar range in concentrations for surface waters, gaseous evasion flux from the ocean dominates over freshwater systems.[141] A number of box model and numerical modeling papers have recently focused on the estimation of the evasion of Hg^0 from the ocean to the atmosphere.[94,131,142,143] In surface waters, Hg^0 is produced as a net result of photochemical oxidation and reduction, as well as biologically-mediated reduction.[144–147] Overall, the net reduction and formation of Hg^0 in most surface waters is driven by the removal of Hg^0 to the atmosphere as surface water is mostly supersaturated. For the ocean, recent studies have focused on air–sea exchange, especially in Atlantic waters,[148–151] the Pacific,[152,153] coastal regions,[154–156] the Arctic[157] and the Mediterranean.[148,158,159] Photochemical processes dominate the reduction of Hg^{II} in most instances, and studies in freshwater and shallow coastal waters have demonstrated short-term changes in Hg^0 concentration and a diurnal cycle. The reduction of Hg^{II} appears to be enhanced in the presence of organic matter, and dissolved Hg speciation plays an important role in determining the rate and extent of this transformation.[147,160,161]

The distribution of Hg in ocean waters reflects the sources and cycling as well as its internal methylation, demethylation, oxidation and reduction. Given its global transport in the atmosphere, and the dominance of atmospheric deposition as a source to the ocean (see Figure 4B), and the strong re-evasion of deposited Hg, there is much less difference in concentration between ocean basins, both in the surface ocean and in deep waters (see Figure 6A). This is clearly a strong contrast to the distribution of Pb in the environment (see Figure 5) where the differences between ocean basins and between the surface and deep ocean are much greater. It likely also reflects the relatively larger timeframe for addition of Hg to the biosphere in the last 500 years. Modeling studies suggest that about half of the Hg in the deep ocean is from natural sources, with the rest having an anthropogenic source.[42] For the surface ocean, the anthropogenic component is much larger ($\sim 80\%$) with about half of that due to anthropogenic emissions since 1950. This reflects the relative short average residence time of Hg in the ocean (around 100 years based on Figure 4B) compared to the much longer timeframe of Hg emissions (see Figure 1).

The production and destruction of methylated Hg species, primarily CH_3Hg and $(CH_3)_2Hg$ in the ocean, is reflected in numerous profiles of Hg speciation in the ocean that demonstrate an enhancement in the concentration of methylated Hg at mid-depth, in regions where organic matter degradation is enhanced[131] (see Figure 6C). Note that in some studies, samples were acidified to preserve them for later analysis and because of the instability of $(CH_3)_2Hg$, this results in the quantification of total methylated Hg ($\sum(CH_3)_xHg$; $x = 1$–2). Profiles in the equatorial and North Pacific, North

Atlantic, Arctic and Southern Oceans, for example, show an enhancement in methylated Hg in the subsurface and in the region of the oxygen minimum zone.[131,162–167] While these profiles could be explained by various mechanisms, the most accepted notion is that the MeHg is being produced *in situ* during microbial organic matter degradation. This notion is reinforced by studies demonstrating water column methylation[166] and studies of the Hg isotopic composition of fish that shows differences between surface and subthermocline feeding fish that can only be explained by production of MeHg in the subsurface waters.[168] Additionally, Hg isotopic differences between coastal and open ocean fish further suggest that coastal waters are not the major source of MeHg to offshore fish.[169] While little is known about the microorganisms or processes whereby Hg is methylated in the ocean,[131] the fact that methylation occurs in upper ocean waters, and the dominance in many instances of $(CH_3)_2Hg$, suggests that the pathways may be different from those for freshwater and coastal environments, and in sediments.

The biogeochemical cycling of Hg in lakes has been the focus of recent research, driven by the elevated levels of CH_3Hg in freshwater fish and the associated human and wildlife health concerns.[34] Most of this work has been completed on temperate ecosystems in Europe and North America, but given the underlying principles, the relationships developed for these ecosystems are generally applicable. In lakes and freshwaters, the dominant process whereby Hg is methylated is biotic, with the major product being CH_3Hg with sulfate-reducing bacteria being the most important methylators,[171] although there is now increasing evidence for other microbes being important in specific locations.[172,173] The details of the methylation pathway have only recently been demonstrated for anaerobic organisms.[174] Additionally, biotic demethylation is not well understood except for systems with elevated Hg and CH_3Hg content, where the *mer* operons, a series of enzymes involved in Hg and CH_3Hg uptake and detoxification, are induced in some bacteria.[175] Abiotic photodemethylation has been also demonstrated in both freshwater and marine systems and appears to be enhanced by UV radiation in surface freshwaters, especially those with low Dissolved Organic Carbon (DOC) and/or Total Suspended Solids (TSS), *i.e.* low color and reflectance.

Evidence from lake studies support the importance of seasonal stratification in the buildup of CH_3Hg in bottom waters and in its supply to the food chain, and demonstrate the lack of $(CH_3)_2Hg$, in contrast to the oceans. Additionally, the importance of *in situ* production of methylated Hg, primarily in sediments, compared to external inputs is demonstrated in many studies.[171] One innovative approach to studying Hg cycling was the use of stable Hg isotopes that were deliberately added to a lake (Lake 658) in the Experimental Lakes Area of Canada (termed the METAALICUS project; http://wi.water.usgs.gov/mercury/metaalicus-project.html), aimed to simulate the impact of increased anthropogenic deposition, and also to assess its recovery after years of enhanced additions. The input was increased to correspond to the average Hg deposition rate of the eastern USA, or about 5 times the pre-addition level. Additionally, by adding different isotopes to the upland watershed, an associated wetland and the lake surface, it was possible to track the significance of the various pathways in contributing to the CH_3Hg accumulating in the food chain. Prior to the addition of the isotopes, the inputs of Hg to the lake were dominated by watershed inputs with atmospheric deposition being a relatively minor component.[176,177]

These studies showed that the system responded rapidly to the Hg isotope added directly to the lake surface with the documentation of enriched isotopic CH_3Hg in sediment trap material (sinking particulate) within weeks after the addition of the Hg^{II} isotope. Furthermore, evidence of its presence in the surface sediment and the zooplankton and benthos (organisms which live, on, in or near the seabed) was found within a month.[176] These results suggest that Hg input from the atmosphere is rapidly cycled through the system and is actively methylated and transported within the ecosystem and bioaccumulated into the food chain. However, a significant portion of the isotope ($\geq 20\%$) directly added was lost to the atmosphere through net Hg^{II} reduction and subsequent evasion,[176,178] as found in other studies.[88] Thus, re-emission is an important process in Hg cycling in lakes, as it is for the ocean. Sedimentation is the other major loss and is important especially for lakes with substantial terrestrial input as the particulate material supplied through runoff increases the relative sedimentation rate.

The METAALICUS project showed that the isotope added to the surface of the lake was methylated much more efficiently than the Hg already in the system, and especially compared with the Hg isotope added to the lake from the watershed and wetland.[176] This demonstrates that the bioavailability of Hg being added to a lake or aquatic system changes over time. This makes sense given that inorganic Hg added to the lake from the atmosphere through wet and dry deposition is relatively labile. Over time, however, this Hg will become incorporated into natural organic matter (NOM) and particulate material, bound to inorganic sulfides and form other strong associations that will reduce its availability to the methylating organisms, and thus decrease its bioavailability.

Finally, this project clearly demonstrated the preferential bioaccumulation of isotope added directly to the lake surface. This further confirms that while there is a yield of Hg and CH_3Hg from the watershed to a lake, this is not recently deposited Hg but that which has cycled through the watershed for an extended period, perhaps for many years to decades. Given this retention in the watershed, it is also apparent that any lake with a significant input of Hg from the watershed will respond more slowly to changes in atmospheric deposition. Recent attempts to examine such differences using models suggest that there would be a rapid initial decrease in concentration, that would be greatest for a seepage lake or river, within 5–10 years, and a slower decrease to steady state over decades to a century.[179]

One important way that human activity has exacerbated CH_3Hg levels in fish is through the impact of reservoir flooding/new impoundment construction. For example, a study in Finland of 18 newly formed reservoirs showed that fish concentrations increased compared to natural lakes by 30%

on average, and that these elevated concentrations remained for an extended period, with fish concentrations not returning to pre-inundation levels after 20–30 years.[180] Other studies have also found similar correlations, and similar timeframes of impact.[181,182] Impacts were also related to the extent of flooding that occurred. While flooding of soils with higher carbon levels lead to more CH_3Hg formation, the presence of higher levels of organic matter in the water tended to counter bioaccumulation.[182] While such effects are dramatically demonstrated with new reservoir formation, there is also evidence that seasonal draw-down and replenishing of reservoirs, and other water bodies, can also enhance the production of CH_3Hg within the system.

The large body of information available on Hg cycling and the production of MeHg in the environment has led to improved understanding and allowed the ability to begin to predict the impact of future changes in emissions on global cycling.[183] Inter-comparison of the measurements of Hg and modeling approaches[85] has significantly improved the quality of data available for managers and regulators so that there is the potential to make informed decisions about regulation of Hg emissions to environmental media. The evolving global policy framework is now providing a viable platform for future research directed at the remaining outstanding questions concerning Hg cycling and fate in the biosphere, and the role of humans in exacerbating its effects.

3.2 Lead in the Biosphere

The study of Pb in the ocean is couched in the realization, through the demonstration of Clair Patterson in the 1970's, that most prior measurements of Pb in the ocean were incorrect due to contamination during collection and handling.[184] A number of subsequent studies documented in detail the overall contamination of the ocean by anthropogenic Pb, the role of Pb released from gasoline, and the more recent decrease in ocean Pb as a result of the phasing out of Pb additives for gasoline.[11,128] Contamination of even the most remote polar regions has been shown[185] (see Figure 2). Isotopic changes in Pb in the ocean also reflect the anthropogenic signal.[105,106,129,186–188] However, while there have been substantial decreases in the concentrations of Pb in the North Atlantic, the Mediterranean[189] and other regions, there is still some Pb input to the atmosphere from other anthropogenic sources. There is also strong evidence that the concentrations of Pb in the North Pacific are not decreasing even given the phase-out of Pb in gasoline due to the large inputs of Pb into the atmosphere from other sources in Asia. Concentration gradients and isotopic signatures off the coast of Asia attest to the importance of these continental inputs.[190]

As noted above, much of the Pb in the ocean reflects the different source locations and spatial and temporal distribution of these inputs. For example, the concentrations in the upper waters of the North Atlantic in the vicinity of Bermuda have been decreasing in the last 30 years (see Figure 5A),[11,128] in response to the phasing out of Pb in gasoline in many countries. Concentrations of Pb have decreased from values above 150 pM in the late 1970's to concentrations around 50 pM today. This can be compared to the predicted pre-industrial Pb concentration of <5 pM for the deep waters of the North Pacific, and ∼14 pM for the Atlantic.[129] These differences in concentration reflect the differences in the natural source signal relative to the basin size as well as the importance of particulate scavenging in the deep ocean in removing Pb from the water column. The vertical profiles of Pb in the North Pacific[129] also reflect differences in the source signal as concentrations are higher in the Atlantic Ocean compared to the Pacific (see Figure 5). It can therefore be concluded that while the surface waters near Bermuda have decreased in concentration they are still substantially elevated above background and reflect the continual input of Pb from combustion and other industrial sources, and the legacy of prior inputs. The extent of Pb input to the Atlantic Ocean is a function of its circulation, as noted above, and there was evidence in the 1990's for pollution Pb (based on isotopic composition and concentration) in the South Atlantic as well,[191] and in the atmosphere.[192] One important conclusion of these studies is that the release of Pb from sinking particulate was not an important source to the deep ocean waters compared to that due to water mass transport. These authors[192] also showed from the isotopic signature that the sources of Pb to the different deep waters masses were different: with Antarctic Bottom Water containing a signal due to a mixture of natural and anthropogenic Southern Hemisphere sources while the signals in the North Atlantic Deep Water masses reflected the dominant sources (North America and Europe) during the time of their sinking from the surface.

The recent data from the North Pacific[129] show horizontal differences in concentration that likely reflect the differences in atmospheric inputs to the surface ocean. Highest surface concentrations are found in the mid-latitudes and this reflects the heightened input of Pb to these waters. Air masses from the Asian continent typically track through the mid-latitudes and therefore the concentration reflects this anthropogenic signal. The upper ocean waters are much higher in concentration than the deep waters, and concentrations in the surface waters and atmospheric aerosols vary by a factor of 3 with distance from Asia, and the waters also had the isotopic signature of Asian aerosols.[190] Also, the values obtained recently[129] are similar to previous measurements at this location, suggesting little change in the recent past.

As noted, there has been more study of Pb concentrations and isotopes in the Atlantic than the Pacific, with a focus on emissions in the western world.[11,105,186] However, there is the potential for emissions in Asia to impact both the North Pacific and the Indian Oceans. A detailed study in the Indian Ocean used the comparison of the Pb : Ca ratio in corals as well as the Pb isotope ratios to examine the signals from the regional inputs.[106] The emissions of Pb from its use in gasoline mostly peaked for various countries between mid-1970 and mid-1990, whereas the emissions from other uses such as coal combustion began to increase in the 1980's in most countries, except India and Australia where emissions began earlier. In many countries, emissions from coal were more comparable to the gasoline emissions, especially as coal use increased and leaded gasoline use decreased at the end of the 20th century. The relative concentrations of Pb (Pb : Ca) in the corals reflected these trends overall, but the isotopic signatures suggested that different regions of the Indian Ocean are being impacted differently because of the differences in the extent and timing of emissions. The signatures also suggested that long-range transport of Pb from outside the region was occurring.[11,106]

A recent global study of Pb levels and isotope signatures for marine sediments from around the world[111] showed that these reflected both the location (*e.g.* highest in harbors of large cities) and the Pb sources. The authors used the isotopic signatures to evaluate the extent to which gasoline Pb contributed to the Pb signal. The fraction of the Pb from gasoline ranged from 15 to 90% with higher fractions for USA locations than for countries such as New Zealand and South Africa. Similarly, a recent study in Mexico[110] that examined Pb isotopes and concentrations in various media concluded that historic use of gasoline was a substantial source for Pb in urban aerosols, in sewage effluent and street dust. Rural aerosol samples had little contribution from prior gasoline Pb use.

4   Policy Response and Impact

4.1   Lead

For Pb, human exposure can be due to elevated concentrations in soils, paint chips, air (aerosols, house dust) or water.[193,194] Exposure can occur within or outside the home. It has been long known that children less than 7 years old are most at risk due to the propensity to consume paint chips in homes that contain Pb, and to be potentially exposed through ingestion of contaminated soils. Exposure in children is also enhanced due to their higher intake per body weight and their higher relative uptake across the intestinal tract. Children are also more susceptible to neurological and development effects from Pb exposure. For Pb, it has been shown that nutritional deficiencies can exacerbate the impacts. The impact of Pb pollution on Pb levels in humans was dramatically demonstrated by examining levels in prehistoric skeletons[195] which showed levels 50–200 times lower than that of contemporary people living in remote regions. The related levels in current humans living in high Pb environments was substantially higher (>1000 times).

For Pb, there have been different global initiatives focused on its different uses. In term of the use of Pb in paints, at the 2002 World Summit on Sustainable Development (WSSD), it was decided to phase out its use, and this decision was reinforced at a later meeting of the International Conference on Chemicals Management (ICCM2) in 2009.[2,194] A global partnership to promote the phasing out of Pb in paints – the Global Alliance to Eliminate

Lead Paint (GAELP) – was formed to minimize children and occupational exposure. This initiative has resulted in a dramatic reduction of Pb use in paints. However, in 2013, the UNEP Governing Council further emphasized that issues related to Pb use still remained and mandated the continued activity of GAELP.[194]

There was much debate and controversy after the first studies relating blood Pb levels in children to Pb in gasoline were published.[196–200] For example, it was shown in New York City in the 1970's that the blood levels correlated with the levels of Pb in gasoline sold in the vicinity of the subjects, and the many other studies noted above also correlated levels of Pb in children's blood with the use of Pb in gasoline. There were also numerous studies that examined the impact of gasoline "sniffing" on exposure. However, as the evidence mounted, various countries reduced Pb levels in gasoline and/or began phasing out its use. This occurred at different rates in the western world and in developing countries. UNEP has played a role in phasing out leaded gasoline worldwide, another objective of the 2002 WSSD through which the global Partnership for Clean Vehicles and Fuels (PCVF) was established.[194] Most countries have implemented bans as a result of this initiative, or did so prior to the effort. In the USA, leaded gasoline use rapidly decreased when the use of catalytic converters was mandated in 1975 (see Figure 1), as Pb deactivated the catalysts in them. However, a complete ban was only enacted with the Clean Air Act in 1990, with complete phase-out in 1996. In Europe, Germany banned leaded gasoline in 1988 and full phase-out in all European countries was achieved in 2009. African nations completed phase-out in 2006 although leaded gasoline is still available in Algeria. New Zealand banned leaded gasoline in 1996, Australia in 2002, and other Asian nations over a similar time period, with full elimination by 2006.[194] A similar timeframe of elimination occurred in South America. Today only six countries still use leaded gasoline in motor vehicles, versus 82 countries when the PCVF was formed. The success of this action against Pb use in gasoline has been substantial given the difficulties in passing legislation earlier in the 20th century. There is still however some potential for exposure from leaded gasoline used in small aircraft. One study[201] found higher blood Pb levels in children living within 500 m of airports where such fuel was used. Recent research[194] suggests that the global effort to end the use of leaded gasoline translates into 1.2 million fewer deaths year^{-1} with economic benefits of US$2.4 trillion year^{-1}.

Numerous studies have cataloged the changes in Pb levels in children and adults in cities in response to implementation of the use of unleaded gasoline. In many instances there has been strong evidence of decreased levels after the phase-out (*e.g.* in the USA,[199,202,203] Sweden,[204] Spain,[205] Taiwan,[206] Canada[207] South Africa[208] and Argentina[209]) while the evidence is less clear in other countries due to the impact of additional Pb pollution sources (*e.g.* Uganda,[210] India,[211] Central and South America[212]).

Concentrations of Pb in the biosphere and in industrial products are now highly regulated, and acceptable levels of exposure exist in many regions of the world. For example, in the USA, child blood levels of 10 μg L^{-1} are considered the level for concern,[193] and the regulation for concentrations in air in the workplace (8 h exposure) is 50 μg m^{-3}. The blood level regulation of WHO is similar (www.who.int/ipcs/assessment/public_health/lead/en/). Permissible USA soil levels are <400 mg kg^{-1} (www.atsdr.cdc.gov/csem/csem.asp?csem=7&po=8). A study by the USGS of ∼4840 randomly collected soil samples in the USA found that the median Pb concentration was 17.8 mg kg^{-1} (mean 22.2 mg kg^{-1}) with a range in values from the detection limit to 2200 mg kg^{-1}.[213,214] For Hg, the same study found a median of 0.03 mg kg^{-1} (mean 0.04 mg kg^{-1}) and a range up to 4.6 mg kg^{-1}.[214] A major issue in terms of exposure is the "bioaccessability" of the Pb in soils which is typically assessed using laboratory assays. For example, a study in Scotland[215] showed that the bioavailable Pb fraction was higher for the stomach (50% on average), likely due to the low pH, while the overall accessibility for the stomach and intestine simulation was about half the amount. The bioaccessability depended on the speciation of the Pb in the soil and it was determined, using isotope ratios that the original source of the Pb was less important than its resultant speciation. A study in Canada confirmed the importance of Pb speciation and concluded that Pb carbonate phases were most bioavailable.[216]

The impacts of regulation have been documented in numerous studies, and some examples are detailed below. In terms of soil concentrations in the USA, for example, there is some evidence to support decreasing concentrations over time within city centers, but the overall range in concentrations has not changed substantially.[214] This study found a correlation between soil concentration and population. Studies of dust and atmospheric aerosols have shown that there is also a long lag in response due to the resuspension of soil dust and legacy contamination within houses. For example, little change in aerosol Pb since phasing out of leaded gasoline in Xiamen[218] and in Beijing, China has been found.[219] In Mumbai, India, there appeared to be a decrease in concentration of Pb in aerosols in suburban and residential areas.[220] Rasmussen *et al.*[216] showed in an extensive study of house dust in Canada that Pb was higher in older homes in the city center. These studies all suggest that there have been some decreases over time but that the legacy of contamination is still high in urban environments. Besides urban locations, other areas where legacy Pb contamination is still important are regions surrounding Pb mining operations.[97,221] As Pb mining can be of sulfide ores, there is also the potential for enhanced Pb release due to acid mine drainage which would increase the solubility of Pb, both through oxidation of the ore or due to reaction with Fe under acidic conditions (Fe^{3+} and PbS reacts to form Fe^{2+} and Pb^{2+}).[20]

The impact of regulation on exposure of children has also been documented in many studies. One example focused on examining relationships between concentrations in environmental media and blood levels in adults and children in Europe over time (see Figure 7).[217] Overall, the study showed a decrease in blood Pb levels (B-Pb) over time, and that these changes were

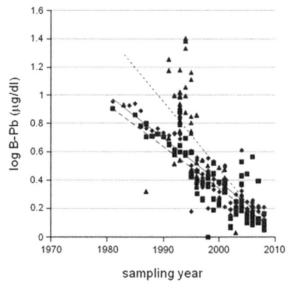

sampling year

♦ Adult men   ■ Adult women   ▲ Primary school children

Figure 7   Median blood lead levels (B-Pb) for different subsets of the European population as a function of time.
Taken from Bierkens.[217] (Used with permission from Elsevier from *Sci. Total Environ.*, 2011, **409**, 5101–5110).

relatively well correlated with temporal changes in air concentration and dietary intake. The study found that concentrations in air in 2007 had decreased by a factor of 7 from a mean of 0.3 µg m^{-3} in 1990, although there were still high levels and more than two orders of magnitude range in concentrations, and that the Pb levels of adult, but not children, correlated with the changes in air concentration. Similar relationships were found for dietary intake. The study concluded however that the relationships were insufficient to be used in future predictions of exposure.

A study in the USA[202] that examined different population groups and compared blood levels in children over time found similar decreases. Results showed a temporal shift in the number of children with elevated blood levels in all populations. In 1988–91, 8.6% of the children had Pb blood levels >10 µg dL^{-1} while the fraction was 1.4% in 1999–2004. The changes were most significant for the non-Hispanic black population although their levels were still higher than the other population groups. The major risk factors identified were living in older housing, age, and poverty level and being black.

One potential impact of higher historical exposure was the potential influence of Pb exposure on crime, and related societal issues. Two papers[222,223] in the literature represent the case that violent crime could be related to the exposure of city children to high Pb levels in their early years of development. The correlation in the USA with Pb use in gasoline shows very similar trends with a 22 year offset. The other study focused on a number of Western countries and reached very similar conclusions.

Another route of Pb exposure that has been examined is the potential for the use of Pb shot in hunting to cause wildlife exposure.[224] However, there is little detailed information about the extent of this problem.[221] One specific case that has been highlighted is the impact of Pb in ammunition on the recovery of the endangered Californian condor.[225] The issue is likely to be more important in areas such as shooting ranges where the amount of Pb may be elevated because of high usage. Another additional pathway of exposure within homes is where elevated radon (Rn) is found, typically in the basement.[226] Radon diffuses out of bedrock where it is formed by radioactive decay of radium (^{226}Ra). Given its short half-life (3.8 days), Rn will decay and its products being particle reactive will form radioactive dust particles containing ^{214}Pb, ^{210}Pb and other radioactive decay products. The final decay product is ^{206}Pb. Thus there is the potential for buildup of radioactivity and Pb in badly ventilated areas.

Overall, based on the examples above and from the dramatic reduction in Pb in some environmental reservoirs as the result of elimination of Pb in products and its emissions from anthropogenic sources, it can be concluded that environmental action and policy has had a very positive effect on Pb exposure. It is evident that the exposure of humans will further decrease over time although there will be a legacy effect due to the persistence of Pb in surface soils and the potential for its reintroduction into the atmosphere, and its leaching into waters. Given the large legacy of Pb in soils, the rate of decrease will be slower in urbanized areas, and in the vicinity of mining sites and other localized "hotspots", where there is still the potential for significant future exposure.

4.2 Mercury

Inorganic Hg poisoning was documented very early;[10,23] for example, it was noted in Roman times that mine workers in Hg mines did not live long. Mercury compounds have been used in medicine throughout history, even until recently, and as preservatives in medicine (*e.g.* phenylHg compounds, thimerosal/methiolate and merbromin/mercurochrome). The impacts of Hg use in the felt industry on the workers is captured in the stories of Lewis Carroll (the term "mad as a hatter" reflects the impact of inorganic Hg exposure). Methylated Hg compounds were used in agriculture as antifungal agents in seed grain until the 1970's when the practice was discontinued because of a number of mass poisonings, the most notable being in Iraq, and the exposure of both humans and certain wildlife species from their use.[23] Mercury is still used in products that are a source for exposure.[227] The most well-known and dramatic example of MeHg exposure *via* seafood was in Minamata Bay,[26] where MeHg and other forms of Hg were released from industry into the bay where most residents obtained their fish. Acute poisoning was evident and traced to the discharge from the factory. Fish Hg in excess of $20 \, \text{mg kg}^{-1}$ was found. These can be compared to the current regulatory levels around the world which are all $\leq 1 \, \text{mg kg}^{-1}$. After identification of the contamination in Minamata and removal of the inputs, levels in sediments decreased rapidly, and were further decreased after remediation of the most contaminated sediments in 1984. Since then, levels have continued to decrease as older sediments are buried at depth.

Levels in excess of $5 \, \text{mg kg}^{-1}$ have also been found in freshwater fish living downstream of paper pulp, chlor-alkali and other industrial plants that have used Hg in their chemical processes, and which historically discharged their contaminated effluents into rivers.[37] Such activities are curtailed in many countries and many of these industries do not use Hg anymore, and therefore the inputs to the environment have decreased locally in these regions of point source contamination. Most intentional use of Hg is being eliminated around the world. One ongoing potential impact is the use of Hg in dental amalgams, which can release Hg into the body by various processes. Recent evaluations in the USA and Canada, for example, suggest that there is a potential exposure of the population from this use, and that the exposure level is related to the number of amalgams.[228,229] The use of Hg in amalgams is another incidence where there has been substantial controversy about the health impacts over time.

However, one ongoing activity that is a large local source of contamination with substantial human health effects is related to artisanal gold mining in many regions of the world, as noted in recent reviews.[28,230,231] Health impacts primarily from inorganic Hg exposure are well documented in these compilations of studies in many countries in South America, Africa and Asia which conclusively show that levels in hair and urine, the primary biomarkers used in evaluating effects, are above WHO and other environmental health guidelines, and were substantially greater than levels in control populations examined.[231] Additionally, it is not only the mine workers that are exposed, but also the general population living in the mining regions. Most common effects reported are neurological (*e.g.* tremor, ataxia, memory, vision). Kidney dysfunction is also reported.[230] Additionally, there is the potential for exposure to MeHg if fish downstream from mining locations are consumed.[230]

In the last decades national guidelines for exposure (critical dose, acceptable levels) have been introduced in many countries.[227] In 2013 the Minamata Convention[29] was signed which is a legally-binding instrument to enable reduction of Hg releases to the biosphere. Prior to the Convention there was the establishment of a number of UNEP global partnerships which evaluated various aspects of Hg sources and cycling. These partnerships are ongoing and consist of activities to reduce Hg emissions from: artisanal and small-scale gold mining; coal combustion; the chlor-alkali industry; the cement industry; products; and waste management. Additional partnerships are examining air transport and fate of mercury, and Hg supply and storage. These partnerships form a framework for the Minamata Convention. The major highlights of the Minamata Convention on Mercury include a ban on new mercury mines, the phase-out of existing ones, control measures on air emissions, and the international regulation of the informal sector for artisanal and small-scale gold mining.[29] As of August 2014, 102 countries have become signatories to the convention. The convention has supported and led to a number of publications that have highlighted the concerns of Hg in the environment in specific parts of the world. This initiative is likely to promote research and regulation throughout the world and enhance activities to reduce Hg emissions to the biosphere in both developing and developed countries.

While local and regional exposure from inorganic Hg and MeHg are an important concern, the much larger impact on human health is from the exposure to MeHg accumulated in aquatic organisms, and particularly in seafood, as this is the main source of exposure for most humans.[232–234] Substantial wildlife impacts due to MeHg are focused on birds and mammals that consumed aquatic organisms. While the human health impact has been well-documented, global regulation has lagged behind this knowledge. Humans, and fish-consuming wildlife and birds are all potentially at risk from exposure at current environmental levels. For some human populations consumption of marine mammals is also a significant source of MeHg exposure.[86] The sections of the population most at risk are the young children and babies exposed *in utero*, and for this reason many advisories target women who may become or are pregnant.[227] The rate of elimination of MeHg from the human body is relatively slow so that there is the potential for exposure even if fish consumption ceased upon becoming pregnant. The early stages of neurological development are when the exposure is greatest. Fish and seafood MeHg concentrations vary greatly and so the exposure is related to both the amount consumed, and the type of fish as predatory (piscivorous) and long-lived fish have higher concentrations than invertebrates (*e.g.* shrimp) and primary consumers (planktivores). For example, a recent compilation of data on concentration shows that values range from ≤ 0.1 μg g^{-1} wet weight (*e.g.* shrimp, squid and pollock) to >1 μg g^{-1} wet weight (*e.g.* tilefish, swordfish, shark), with high variability in concentration for the larger fish species. Thus, large tuna and other sportfish (*e.g.* marlin) can also have levels above 1 μg g^{-1} wet weight. Canned tuna are typically smaller species and have concentrations of <0.5 μg g^{-1} wet weight.

The variation in concentration of more than 100 invertebrates and top predators indicates that there is the possibility to consume seafood in larger quantities without substantial exposure. This is important as seafood is an important source of omega-3 (*n*-3 polyunsaturated) fatty acids, and has other nutritional benefits, and for much of the population fish consumption is the main source of these supplements, and there is therefore the potential for loss of the health benefits by not consuming fish.[233,235] The fatty acids have been shown to be important in brain and visual system development in infants and to reduce the risks of certain forms of heart disease in adults.[235] They also range significantly in concentration in marine and freshwater fish from <2 mg g^{-1} (*e.g.* crabs, crayfish, flatfish, haddock, some tuna species) to

>10 mg g^{-1} (bluefin tuna, salmon, king mackerel). In many instances, fish high in Hg are relatively low in these fatty acids (*e.g.* shark, swordfish) so the benefits likely do not outweigh the risks of their consumption. Additionally, many large predatory species are overfished or endangered, another reason to limit their consumption.

In an analysis of USA fish consumption in 2007, it was concluded that while tuna, shrimp, pollock, salmon and cod were the top five consumed seafood, the intake of MeHg was dominated by fresh/frozen and canned tuna ($\sim$40% of the total).[232] Shrimp and pollock, while being important in terms of amount consumed, were relatively small contributors to the MeHg exposure with both contributing <10% each to the total intake. Swordfish, although consumed in relatively small amounts was responsible for 8% of the intake because of its high MeHg concentration. This assessment focused on the average USA consumer. Other studies have shown strong regional differences due to local and subsistence fisherman consuming higher quantities of local fish, and less tuna and other fish from the global fish supply. Studies in the USA have suggested that based on the USA reference dose, a fraction of the population (>90th percentile in terms of fish consumption) are exposed to MeHg at levels above this value.[232,236]

Health effect advisory levels differ across the globe depending on the type of guidelines issued, and the studies that are used to define the reference dose. Three major studies have occurred, of which two are ongoing: in the Faroe Islands; in the Seychelles; and the New Zealand study. The Faroe islanders also consume marine mammals. For example, in the USA, the EPA advisory reference dose is 0.1 µg MeHg kg^{-1} body weight day^{-1} for women of child-bearing age, which is often translated into one meal a week of fish with 0.3 µg MeHg g^{-1} wet weight. Other countries and organizations have a range in values, up to 1.6 µg kg^{-1} day^{-1}, and the difference for the estimates reflect differences in interpretation of the amount of exposure that causes effects, and the uncertainty factor associated with the calculation.[227,235] The EPA calculation has a safety factor of 10 included in the calculation, which is the highest value used. Overall, while it may be possible to reduce exposure to MeHg from choosing to consume specific species of fish,[233] overall reduction of inputs of Hg to the biosphere is the more obvious long-term strategy to reduce human exposure.[237]

5   Trends in the Future

The future trends for Hg and Pb in the ocean and terrestrial environment can be estimated based on past inputs and the likely scenarios for emissions to the atmosphere in the future. For Pb, levels in surface waters have been changing in line with regional emissions, decreasing in the North Atlantic since $\sim$1975 but increasing the western Pacific, although such increases are not yet evident in the middle of the North Pacific Ocean.[11,106,128] Recent economic development and a later removal of leaded gasoline in Asian countries have driven this trend and also led to higher concentrations in the

Indian Ocean. There has been a substantial change in the use and recycling of Pb in the last decades. Currently, while supply/mining of Pb has not changed dramatically in the last decade, the increased demand for Pb has been mostly met by recycling, which accounted for 45% of global usage in 2003.[238] In that year, Pb usage was mostly for batteries ($\sim$ 86% for all battery types) and for sheeting, alloys and cable sheathing, and for ammunition. Most of these usages are ones where Pb can be further recycled. As noted above, Pb emissions from mining and refining are now the major global contributors and these emissions are now being more carefully monitored and regulated in many parts of the world.[8] For example, decreases in Pb in atmospheric aerosols and rain continue in Europe and Pb emissions were estimated to have decreased by a factor of two in Europe between 2005 and 2010.[239] The decreases in Hg emissions in Europe are of similar magnitude.

Many international, national and regional/inter-country initiatives exist to monitor and regulate Pb releases to the environment.[238] As mentioned above, legacy Pb contamination of soils and dust in urban areas and in houses is an ongoing issue but there is evidence for a slow recovery which will overall decrease exposure to individuals from these sources in the future. Thus, there is the expectation of a decreasing level of exposure of humans from Pb in the future.

The same conclusion cannot be as easily made for Hg although the new and ongoing international, national and regional treaties are leading to more regulation and monitoring. The potential impacts of future emissions and climate change have been simulated[42,95,179] and it has been concluded that there would be a slow decrease in concentrations in the biosphere even if Hg emissions from anthropogenic sources were completely removed ("zero" future Hg emissions after 2015; an unlikely scenario). Under this scenario, surface ocean concentrations are predicted to only decrease by 40–60% by 2100.[95] However, another evaluation suggests only modest decreases in Hg emissions over time.[240] Under many future emission scenarios, the rate of change would increase, not decrease, and one projection, for example, would lead to a doubling in fish concentration in the North Pacific in 2050.[237] Because of the active recycling of Hg through the biosphere and the circulation in the ocean, the response time for MeHg in the ocean to changing anthropogenic emissions is slow given that its formation is within the ocean water column. Most future model evaluations similarly conclude that the rate of response of the biosphere to changing Hg emissions is in the order of decades because of the active recycling of Hg between the surface reservoirs (ocean and terrestrial environment) and the atmosphere, and the long-range transport of Hg in the gas phase.[42,95,179,237]

The impacts of climate change on Hg cycling and on net formation and bioaccumulation of MeHg are many and varied.[167] Increased warming of the biosphere will tend to increase evasion to the atmosphere and prolong the cycling of Hg in the biosphere. The predicted increased oxidative capacity of the atmosphere will enhance oxidation and deposition of Hg. In the ocean, increased primary productivity and organic matter degradation in the subsurface may enhance MeHg production which will be further enhanced by increased stratification of the water column and decreased oxygen content in the subsurface waters. In coastal waters the impact of eutrophication may be a decrease in net MeHg accumulation in fish.[241] In the terrestrial realm, increased biomass burning will increase the input of Hg to the atmosphere. One region where climate impacts are likely to be substantial is the Arctic as ice melting and other changes are predicted to substantially change MeHg levels.[242,243] Overall, there is much less certainty and understanding of the future exposure of humans to MeHg because of the uncertainties in future emissions and the impact of climate on the global Hg cycle.

References

1. R. P. Mason, *Trace Metals in Aquatic Systems*, John Wiley-Blackwell, Chichester, 2013.
2. UNEP, *Environmental Risks and Challenges of Anthropogenic Metals Flows and Cycles*, United Nations Environmental Programme, 2013.
3. R. Chester, *Marine Geochemistry*, Blackwell Science, Malden, 2003.
4. J. Drever, *The Geochemistry of Natural Waters*, Prentice Hall, Upper Saddle River, 1997.
5. R. Duce, P. S. Liss, J. T. Merrill *et al.*, *Global Biogeochem. Cycles*, 1991, **5**, 193–259.
6. J. Nriagu and J. M. Pacyna, *Nature*, 1988, **333**, 134–139.
7. E. G. Pacyna, J. M. Pacyna, F. Steenhuisen and S. Wilson, *Atmos. Environ.*, 2006, **40**, 4048–4063.
8. J. N. Rauch and J. M. Pacyna, *Global Biogeochem. Cycles*, 2009, **23**, DOI: 10.1029/2008gb003376.
9. L. D. de Lacerda and W. Salomons, *Mercury from Gold and Silver Mining: A Chemical Time bomb?*, Springer-Verlag, Berlin, 1998.
10. L. J. Goldwater, *Mercury: A History of Quicksilver*, York Press, Baltimore, Maryland, 1972.
11. E. A. Boyle, J.-M. Lee, Y. Echegoyen, A. Noble, S. Moos, G. Carrasco, N. Zhao, R. Kayser, J. Zhang, T. Gamo, H. Obata and K. Norisuye, *Oceanography*, 2014, **27**, 69–75.
12. Y. Erel and C. C. Patterson, *Geochim. Cosmochim. Acta*, 1994, **58**, 3289–3296.
13. K. J. R. Rosman, W. Chisholm, C. F. Boutron, J. P. Candelone and C. C. Patterson, *Geophys. Res. Lett.*, 1994, **21**, 2669–2672.
14. J. O. Nriagu, *The Biogeochemistry of Lead in the Environment*, Elsevier/North-Holland Biomedical Press, Amsterdam, 1978.
15. J. Nriagu, *Nature*, 1993, **363**, 589.
16. *Mercury: Sources, Measurements, Cycles and Effects*, ed. M. B. Parsons and J. B. Percival, Mineralogical Association of Canada, Halifax, Canada, 2005.
17. R. Fuge, in *Fundamentals of Medical Geology*, ed. O. Seleinus, Elsevier, Amsterdam, 2005, pp. 43–59.

18. N. Pirrone, S. Cinnirella, X. Feng, R. Finkelman, H. Friedli, J. Leaner, R. Mason, A. Mukherjee, G. Stracher, D. Streets and K. Telmer, *Atmos. Chem. Phys.*, 2010, **10**, 5951–5964.
19. C. T. Driscoll, R. P. Mason, H. M. Chan, D. J. Jacob and N. Pirrone, *Environ. Sci. Technol.*, 2013, **47**, 4967–4983.
20. D. Blowes, C. J. Ptacek, J. L. Jambor and C. G. Weisener, in *Environmental Geochemistry*, ed. B. Lollar, Elsevier, Amsterdam, 2005, vol. 9, pp. 149–203.
21. L. D. Hylander and R. B. Herbert, *Environ. Sci. Technol.*, 2008, **42**, 5971–5977.
22. D. A. Wright and P. Welbourn, *Environmental Toxicology*, Cambridge University Press, Cambridge, 2002.
23. T. W. Clarkson and L. Magos, *Crit. Rev. Toxicol.*, 2006, **36**, 609–662.
24. H. Passow, A. Rothstein and T. Clarkson, *Pharmacol. Rev.*, 1961, **13**, 185–224.
25. J. Nriagu, in *The Biogeochemistry of Lead in the Environment*, ed. J. Nriagu, Elsevier, Amsterdam, 1978, vol. Part A, pp. 1–14.
26. A. Kudo, Y. Fujikawa, S. Miyahara, J. Zheng, H. Takigami, M. Sugahara and T. Muramatsu, *Water Sci. Technol.*, 1998, **38**, 187–193.
27. USEPA, Listing of fish and wildlife advisories. CD-ROM. EPA-823-C-97-005, 1996.
28. AMAP/UNEP, *Technical Background Report for the Global Mercury Assessment 2013*, Arctic Monitoring and Assessment Programme, Oslo, Norway/UNEP Chemicals Branch, Geneva, Switzerland, 2013.
29. UNEP, *The Minamata Convention*, http://www.mercuryconvention.org/, Accessed August 7, 2014.
30. W. F. Fitzgerald and T. W. Clarkson, *Environ. Health Perspect.*, 1991, **96**, 159–166.
31. J. Gaillardet, J. Viers and B. Dupre, in *Treatise on Geochemistry*, ed. H. Holland and K. K. Turekian, Elsevier, Amsterdam, 2004, vol. 5, pp. 225–272.
32. W. Stumm and J. J. Morgan, *Aquatic Chemistry*, John Wiley and Sons, New York, 1996.
33. J. D. Blum and Y. Erel, in *Treatise on Geochemistry*, ed. J. I. Drever, Elsevier, Amsterdam, 2004, vol. 5.
34. J. G. Wiener, D. P. Krabbenhoft, G. H. Heinz and A. M. Scheuhammer, in *Handbook of Ecotoxicology*, ed. D. J. Hoffman, B. A. Rattner, G. A. Burton and J. J. Cairns, CRC Press, Boca Raton, 2003, pp. 409–463.
35. C. A. Cooke, P. H. Balcom, C. Kerfoot, M. B. Abbott and A. P. Wolfe, *Ambio*, 2011, **40**, 18–25.
36. R. Fuge, N. Pearce and W. Perkins, *Nature*, 1992, **357**, 369.
37. R. Ebinghaus, R. R. Turner, L. D. de Lacerda, O. Vasiliev and W. Salomons, *Mercury Contaminated Sites*, Springer, Berlin, 1999.
38. W. F. Fitzgerald and C. H. Lamborg, in *Environmental Geochemistry*, ed. B. Lollar, *Treatise of Geochemistry*, ed. H. Holland and K. K. Turekian, Elsevier, Amsterdam, 2005, vol. 9, pp. 107–147.

39. W. F. Fitzgerald, C. H. Lamborg and C. R. Hammerschmidt, *Chem. Rev.*, 2007, **107**, 641–662.

40. R. P. Mason, W. F. Fitzgerald and F. M. M. Morel, *Geochim. Cosmochim. Acta*, 1994, **58**, 3191–3198.

41. N. E. Selin, D. J. Jacob, R. M. Yantosca, S. Strode, L. Jaegle and E. M. Sunderland, *Global Biogeochem. Cycles*, vol. 22: Article # GB 2011, 2008.

42. H. M. Amos, D. J. Jacob, D. G. Streets and E. M. Sunderland, *Global Biogeochem. Cycles*, 2013, **27**, 410–421.

43. T. Berg, E. Steinnes, in *Biogeochemsitry, Availability and Transport of Metals in the Environment*, ed. A. Sigel, H. Sigel and R. K. O. Sigel, Metal Ions in Biological Systems, Taylor&Francis, Boca Raton, 2005, vol. 44.

44. R. Bindler, I. Renberg and J. Klaminder, *J. Paleolimnol.*, 2008, **40**, 755–770.

45. Y. M. Hermanns and H. Biester, *J. Paleolimnol.*, 2013, **49**, 547–561.

46. Y. M. Hermanns and H. Biester, *Sci. Total Environ.*, 2013, **445**, 126–135.

47. J. O. Nriagu, *Nature*, 1989, **338**, 47–49.

48. T. Hinkley, *Geophys. Res. Lett.*, 2007, **34**, DOI: 10.1029/2006gl028736.

49. S. J. Eisenreich, N. A. Metzer, N. R. Urban and J. A. Robbins, *Environ. Sci. Technol.*, 1986, **20**, 171–174.

50. J. G. Farmer, L. J. Eades, A. B. Mackenzie, A. Kirika and T. E. BaileyWatts, *Environ. Sci. Technol.*, 1996, **30**, 3080–3083.

51. M. L. Brannvall, R. Bindler, O. Emteryd, M. Nilsson and I. Renberg, *Water, Air, Soil Pollut.*, 1997, **100**, 243–252.

52. C. E. Dunlap, E. Steinnes and A. R. Flegal, *Earth Planet. Sci. Lett.*, 1999, **167**, 81–88.

53. H. Miller, I. W. Croudace, J. M. Bull, C. J. Cotterill, J. K. Dix and R. N. Taylor, *Environ. Sci. Technol.*, 2014, **48**, 7254–7263.

54. D. Petit, J. P. Mennessier and L. Lamberts, *Atmos. Environ.*, 1984, **18**, 1189–1193.

55. L. G. Sun, X. B. Yin, X. D. Liu, R. B. Zhu, Z. Q. Xie and Y. H. Wang, *Sci. Total Environ.*, 2006, **368**, 236–247.

56. C. A. Cooke, P. H. Balcom, H. Biester and A. P. Wolfe, *Proc. Natl. Acad. Sci. U. S. A.*, 2009, **106**, 8830–8834.

57. D. R. Engstrom, W. F. Fitzgerald, C. A. Cooke, C. H. Lamborg, P. E. Drevnick, E. B. Swain, S. J. Balogh and P. H. Balcom, *Environ. Sci. Technol.*, 2014, **48**, 6533–6543.

58. C. A. Cooke, H. Hintelmann, J. J. Ague, R. Burger, H. Biester, J. P. Sachs and D. R. Engstrom, *Environ. Sci. Technol.*, 2013, **47**, 4181–4188.

59. L. D. Hylander and M. Meili, *Sci. Total Environ.*, 2003, **304**, 13–27.

60. R. Bindler, R. L. Yu, S. Hansson, N. Classen and J. Karlsson, *Environ. Sci. Technol.*, 2012, **46**, 7984–7991.

61. S. A. Beal, M. A. Kelly, J. S. Stroup, B. P. Jackson, T. V. Lowell and P. M. Tapia, *Global Biogeochem. Cycles*, 2014, **28**, 437–450.

62. S. A. Beal, B. P. Jackson, M. A. Kelly, J. S. Stroup and J. D. Landis, *Environ. Sci. Technol.*, 2013, **47**, 12715–12720.

63. C. A. Cooke, A. P. Wolfe and W. O. Hobbs, *Geology*, 2009, **37**, 1019–1022.

64. N. A. Robins and N. A. Hagan, *Environ. Health Perspect.*, 2012, **120**, 627–631.
65. H. M. Horowitz, D. J. Jacob, H. M. Amos, D. G. Streets and E. M. Sunderland, *Environ. Sci. Technol.*, 2014, **48**, 10242–10250.
66. D. G. Streets, M. K. Devane, Z. Lu, T. C. Bond, E. M. Sunderland and D. J. Jacob, *Environ. Sci. Technol.*, 2011, **45**, 10485–10491.
67. J. O. Nriagu, *Sci. Total Environ.*, 1994, **149**, 167–181.
68. R. Hudson, S. Gherini, W. Fitzgerald and D. Porcella, *Water, Air, Soil Pollut.*, 1995, **80**, 265–272.
69. M. E. Goodsite, P. M. Outridge, J. H. Christensen, A. Dastoor, D. Muir, O. Travnikov and S. Wilson, *Sci. Total Environ.*, 2013, **452**, 196–207.
70. P. F. Schuster, D. P. Krabbenhoft, D. L. Naftz, L. D. Cecil, M. L. Olson, J. F. Dewild, D. D. Susong, J. R. Green and M. L. Abbott, *Environ. Sci. Technol.*, 2002, **36**, 2303–2310.
71. P. M. Outridge, K. A. Hobson, R. McNeely and A. Dyke, *Arctic*, 2002, **55**, 123–132.
72. G. M. Vandal, W. F. Fitzgerald, C. F. Boutron and J. P. Candelone, *Nature*, 1993, **362**, 621–623.
73. J. Benoit, W. F. Fitzgerald and A. W. H. Damman, in *Mercury as a Global Pollutant: Towards Integration and Synthesis*, ed. C. J. Watras and J. W. Huckabee, Lewis, 1994.
74. N. Givelet, F. Roos-Barraclough, M. E. Goodsite, A. K. Cheburkin and W. Shotyk, *Environ. Sci. Technol.*, 2004, **38**, 4964–4972.
75. F. Roos-Barraclough, N. Givelet, A. K. Cheburkin, W. Shotyk and S. A. Norton, *Environ. Sci. Technol.*, 2006, **40**, 3188–3194.
76. H. Biester, R. Bindler, A. Martinez-Cortizas and D. R. Engstrom, *Environ. Sci. Technol.*, 2007, **41**, 4851–4860.
77. J. B. Percival and P. M. Outridge, *Sci. Total Environ.*, 2013, **454**, 307–318.
78. C. A. Cooke, W. O. Hobbs, N. Michelutti and A. P. Wolfe, *Environ. Sci. Technol.*, 2010, **44**, 1998–2003.
79. P. C. Van Metre and C. C. Fuller, *Environ. Sci. Technol.*, 2009, **43**, 26–32.
80. P. C. Van Metre, *Environ. Pollut.*, 2012, **162**, 209–215.
81. D. R. Engstrom, S. J. Balogh and E. B. Swain, *Limnol. Oceanogr.*, 2007, **52**, 2467–2483.
82. C. F. Boutron, J. P. Candelone and S. M. Hong, *Geochim. Cosmochim. Acta*, 1994, **58**, 3217–3225.
83. A. L. Lima, B. A. Bergquist, E. A. Boyle, M. K. Reuer, F. O. Dudas, C. M. Reddy and T. I. Eglinton, *Geochim. Cosmochim. Acta*, 2005, **69**, 1813–1824.
84. K. H. Telmer and M. M. Veiga, in *Mercury Fate and Transport in the global Atmosphere*, ed. N. Pirrone and R. P. Mason, Springer, Dordrecht, 2009, pp. 131–172.
85. N. Pirrone and R. P. Mason, *Mercury Fate and Transport in the Global Atmosphere*, Springer, 2009.
86. J. L. Kirk, I. Lehnherr, M. Andersson, B. M. Braune, L. Chan, A. P. Dastoor, D. Durnford, A. L. Gleason, L. L. Loseto, A. Steffen and V. L. St Louis, *Environ. Res.*, 2012, **119**, 64–87.

87. P. E. Drevnick, D. R. Engstrom, C. T. Driscoll, E. B. Swain, S. J. Balogh, N. C. Kamman, D. T. Long, D. G. C. Muir, M. J. Parsons, K. R. Rolfhus and R. Rossmann, *Environ. Pollut.*, 2012, **161**, 252–260.

88. R. P. Mason, in *Mercury Fate and Transport in the Global Atmsophere: Emissions, Measurements and Models*, ed. N. Pirrone and Mason, R., Springer, Norwell, MA, USA, 2009, ch. 17, pp. 173–191.

89. I. S. Sen and B. Peucker-Ehrenbrink, *Environ. Sci. Technol.*, 2012, **46**, 8601–8609.

90. M. Meybeck, in *Treatise on Geochemistry*, ed. J. J. Drever, Elsevier, Amsterdam, 2004, vol. 5.

91. G. G. Laruelle, V. Roubeix, A. Sferratore, B. Brodherr, D. Ciuffa, D. J. Conley, H. H. Durr, J. Garnier, C. Lancelot, Q. L. T. Phuong, J. D. Meunier, M. Meybeck, P. Michalopoulos, B. Moriceau, S. N. Longphuirt, S. Loucaides, L. Papush, M. Presti, O. Ragueneau, P. Regnier, L. Saccone, C. P. Slomp, C. Spiteri and P. Van Cappellen, *Global Biogeochem. Cycles*, 2009, **23**, DOI: 10.1029/2008gb003267.

92. C. J. Vorosmarty, M. Meybeck, B. Fekete, K. Sharma, P. Green and J. P. M. Syvitski, *Global and Planetary Change*, 2003, **39**, 169–190.

93. J. Milliman and R. H. Meade, *J. Geology*, 1983, **91**, 1–21.

94. E. M. Sunderland and R. P. Mason, *Global Biogeochem. Cycles*, 2007, **21**, DOI: 10.1029/2006GB002876.

95. H. M. Amos, D. J. Jacob, D. Kocman, H. M. Horowitz, Y. Zhang, S. Dutkiewicz, M. Horvat, E. S. Corbitt, D. P. Krabbenhoff and E. M. Sunderland, *Environ. Sci. Technol.*, 2014, published online; 10.1021/es502134t.

96. E. G. Pacyna, J. M. Pacyna, J. Fudala, E. Strzelecka-Jastrzab, S. Hlawiczka, D. Panasiuk, S. Nitter, T. Pregger, H. Pfeiffer and R. Friedrich, *Atmos. Environ.*, 2007, **41**, 8557–8566.

97. E. Callender, in *Environmental Geochemistry*, ed. B. Lollar, Elsevier, Amsterdam, 2005, vol. 9, pp. 67–105.

98. S. N. Chillrud, S. Hemming, E. L. Shuster, H. J. Simpson, R. F. Bopp, J. M. Ross, D. C. Pederson, D. A. Chaky, L. R. Tolley and F. Estabrooks, *Chem. Geol.*, 2003, **199**, 53–70.

99. T. M. Church, C. K. Sornmerfield, D. J. Velinsky, D. Point, C. Benoit, D. Amouroux, D. Plaa and O. F. X. Donard, *Mar. Chem.*, 2006, **102**, 72–95.

100. M. F. Soto-Jimenez, S. A. Hibdon, C. W. Rankin, J. Aggarawl, A. C. Ruiz-Fernandez, F. Paez-Osuna and A. R. Flegal, *Environ. Sci. Technol.*, 2006, **40**, 764–770.

101. T. Hosono, C. C. Su, K. Okamura and M. Taniguchi, *J. Geochem. Explor.*, 2010, **107**, 1–8.

102. Z. Q. Zhao, C. Q. Liu, W. Zhang and Q. L. Wang, *Appl. Geochem.*, 2011, **26**, S267–S270.

103. A. R. Flegal, C. Gallon, P. M. Ganguli and C. H. Conaway, *Crit. Rev. Environ. Sci. Technol.*, 2013, **43**, 1869–1944.

104. A. R. Flegal, C. Gallon, S. Hibdon, Z. E. Kuspa and L. F. Laporte, *Environ. Sci. Technol.*, 2010, **44**, 5613–5618.

105. A. E. Kelly, M. K. Reuer, N. F. Goodkin and E. A. Boyle, *Earth Planet. Sci. Lett.*, 2009, **283**, 93–100.
106. J.-M. Lee, E. A. Boyle, I. S. Nurhati, M. Pfeiffer, A. J. Meltzner and B. Suwargadi, *Earth Planet. Sci. Lett.*, 2014, **398**, 37–47.
107. R. Bindler, I. Renberg, N. J. Anderson, P. G. Appleby, O. Emteryd and J. Boyle, *Atmos. Environ.*, 2001, **35**, 4675–4685.
108. C. Reimann, B. Flem, K. Fabian, M. Birke, A. Ladenberger, P. Negrel, A. Demetriades, J. Hoogewerff and G. P. Team, *Appl. Geochem.*, 2012, **27**, 532–542.
109. A. J. Veron, T. M. Church, C. C. Patterson and A. R. Flegal, *Geochim. Cosmochim. Acta*, 1994, **58**, 3199–3206.
110. M. F. Soto-Jimenez and A. R. Flegal, *J. Geochem. Explor.*, 2009, **101**, 209–217.
111. M. M. Larsen, J. S. Blusztajn, O. Andersen and I. Dahllof, *J. Environ. Monit.*, 2012, **14**, 2893–2901.
112. Y. Hirao, H. Mabuchi, E. Fukuda, H. Tanaka, T. Imamura, H. Todoroki, K. Kimura and E. Matsumoto, *Geochem. J.*, 1986, **20**, 1–15.
113. E. B. Swain, P. M. Jakus, G. Rice, F. Lupi, P. A. Maxson, J. M. Pacyna, A. Penn, S. J. Spiegel and M. M. Veiga, *Ambio*, 2007, **36**, 45–61.
114. USEPA, *Mercury study report to Congress*. EPA 452-R-97-004, 1997c.
115. E. B. Swain, D. R. Engstrom, M. E. Brigham, T. A. Henning and P. L. Brezonik, *Science*, 1992, **257**, 784–787.
116. H. D. Yang, D. R. Engstrom and N. L. Rose, *Environ. Sci. Technol.*, 2010, **44**, 6570–6575.
117. W. F. Fitzgerald, D. R. Engstrom, C. H. Lamborg, C. M. Tseng, P. H. Balcom and C. R. Hammerschmidt, *Environ. Sci. Technol.*, 2005, **39**, 557–568.
118. N. C. Kamman and D. R. Engstrom, *Atmos. Environ.*, 2002, **36**, 1599–1609.
119. C. H. Lamborg, W. F. Fitzgerald, A. W. H. Damman, J. M. Benoit, P. H. Balcom and D. R. Engstrom, *Global Biogeochem. Cycles*, 2002, **16**, Article # 1104.
120. D. R. Engstrom and E. B. Swain, *Environ. Sci. Technol.*, 1997, **31**, 960–967.
121. H. D. Yang, R. W. Battarbee, S. D. Turner, N. L. Rose, R. G. Derwent, G. J. Wu and R. Q. Yang, *Environ. Sci. Technol.*, 2010, **44**, 2918–2924.
122. D. C. G. Muir, X. Wang, F. Yang, N. Nguyen, T. A. Jackson, M. S. Evans, M. Douglas, G. Kock, S. Lamoureux, R. Pienitz, J. P. Smol, W. F. Vincent and A. Dastoor, *Environ. Sci. Technol.*, 2009, **43**, 4802–4809.
123. J. C. Varekamp, M. R. B. ten Brink, E. L. Mecray and B. Kreulen, *J. Coastal Res.*, 2000, **16**, 613–626.
124. V. J. A. Phillips, V. L. St Louis, C. A. Cooke, R. D. Vinebrooke and W. O. Hobbs, *Environ. Sci. Technol.*, 2011, **45**, 2042–2047.
125. M. A. Mast, D. J. Manthorne and D. A. Roth, *Atmos. Environ.*, 2010, **44**, 2577–2586.
126. C. H. Conaway, P. W. Swarzenski and A. S. Cohen, *Appl. Geochem.*, 2012, **27**, 352–359.

127. E. Perry, S. A. Norton, N. C. Kamman, P. M. Lorey and C. T. Driscoll, *Ecotoxicology*, 2005, **14**, 85–99.
128. J. F. Wu and E. A. Boyle, *Geochim. Cosmochim. Acta*, 1997, **61**, 3279–3283.
129. J. F. Wu, R. Rember, M. B. Jin, E. A. Boyle and A. R. Flegal, *Geochim. Cosmochim. Acta*, 2010, **74**, 4629–4638.
130. W. S. Broecker and T.-H. Peng, *Tracers in the Sea*, Eldigio Press, NY, 1982.
131. R. P. Mason, A. L. Choi, W. F. Fitzgerald, C. R. Hammerschmidt, C. H. Lamborg, A. L. Soerensen and E. M. Sunderland, *Environ. Res.*, 2012, **119**, 101–117.
132. R. Mason and G. Sheu, *Global Biogeochem. Cycles*, 2002, **16**, Article # 1093.
133. N. E. Selin, *Annu. Rev. Environ. Resour.*, 2009, **34**, 43–64.
134. T. Bergan, L. Gallardo and H. Rodhe, *Atmos. Environ.*, 1999, **33**, 1575–1585.
135. C. H. Lamborg, W. F. Fitzgerald, J. O'Donnell and T. Torgersen, *Geochim. Cosmochim. Acta*, 2002, **66**, 1105–1118.
136. M. S. Landis, R. K. Stevens, F. Schaedlich and E. M. Prestbo, *Environ. Sci. Technol.*, 2002, **36**, 3000–3009.
137. C.-J. Lin and S. O. Penhoken, *Atmos. Environ.*, 1999, **33**, 2067–2079.
138. A. J. Hynes, D. L. Donohoue, M. E. Goodsite and I. M. Hedgecock, in *Mercury Fate and Transport in the Global Atmosphere*, ed. N. Pirrone and R. P. Mason, Springer, 2009.
139. A. W. Rea, S. E. Lindberg, T. Scherbatskoy and G. J. Keeler, *Water, Air, Soil Pollut.*, 2008, **133**, 49–67.
140. N. Pirrone, G. J. Keeler and J. O. Nriagu, *Atmos. Environ.*, 1996, **30**, 3379.
141. R. P. Mason, in *Dynamics of Mercury Pollution on Regional and Global Scales*, ed. N. Pirrone and K. R. Mahaffey, Springer, New York, 2005, pp. 213–239.
142. A. Soerensen, E. Sunderland, C. Holmes, D. Jacob, R. Yantosca, H. Skov, J. Christensen, S. Strode and R. Mason, *Environ. Sci. Technol.*, 2010, **44**, 8574–8580.
143. S. A. Strode, L. Jaegle, N. E. Selin, D. J. Jacob, R. J. Park, R. M. Yantosca, R. P. Mason and F. Slemr, *Global Biogeochem. Cycles*, 2007, **21**, Article # 1017.
144. J. D. Lalonde, M. Amyot, J. Orvoine, F. M. M. Morel, J. C. Auclair and P. A. Ariya, *Environ. Sci. Technol.*, 2004, **38**, 508–514.
145. M. Amyot, G. A. Gill and F. M. M. Morel, *Environ. Sci. Technol.*, 1997, **31**, 3606–3611.
146. L. M. Whalin and R. P. Mason, *Anal. Chim. Acta*, 2006, **558**, 211–221.
147. K. Gardfeldt and M. Jonsson, *J. Phys. Chem. A*, 2003, **107**, 4478–4482.
148. K. Gardfeldt, J. Sommar, R. Ferrara, C. Ceccarini, E. Lanzillotta, J. Munthe, I. Wangberg, O. Lindqvist, N. Pirrone, F. Sprovieri, E. Pesenti and D. Stromberg, *Atmos. Environ.*, 2003, **37**(Suppl. 1), S73–S84.
149. A. L. Soerensen, R. P. Mason, P. H. Balcom and E. M. Sunderland, *Environ. Sci. Technol.*, 2013, **47**, 7757–7765.
150. R. P. Mason, N. M. Lawson and G. R. Sheu, *Deep Sea Res., Part II*, 2001, **48**, 2829–2853.

151. J. Kuss, C. Zulicke, C. Pohl and B. Schneider, *Global Biogeochem. Cycles*, 2011, **25**, DOI: 10.1029/2010GB003998.
152. J. P. Kim and W. F. Fitzgerald, *Science*, 1986, 1131–1133.
153. R. P. Mason and W. F. Fitzgerald, *Deep Sea Res., Part I*, 1993, **40**, 1897–1924.
154. W. Baeyens and M. Leermakers, *Mar. Chem.*, 1998, **60**, 257–266.
155. L. Whalin, E.-H. Kim and R. P. Mason, *Mar. Chem.*, 2007, **107**, 278–294.
156. M. E. Andersson, K. Gardfeldt, I. Wangberg, F. Sprovieri, N. Pirrone and O. Lindqvist, *Mar. Chem.*, 2007, **104**, 214–226.
157. M. E. Andersson, J. Sommar, K. Gardfeldt and O. Lindqvist, *Mar. Chem.*, 2008, **110**, 190–194.
158. R. Ferrara, C. Ceccarini, E. Lanzillotta, K. Gardfeldt, J. Sommar, M. Horvat, M. Logar, V. Fajon and J. Kotnik, *Atmos. Environ.*, 2003, 37(Suppl. 1), S85–S92.
159. E. Lanzillotta, C. Ceccarini and R. Ferrara, *Sci. Total Environ.*, 2002, **300**, 179–187.
160. K. Gardfeldt, J. Sommar, D. Stromberg and X. B. Feng, *Atmos. Environ.*, 2001, **35**, 3039–3047.
161. M. Amyot, G. Mierle and D. J. McQueen, *Geochem. Cosmochem. Acta*, 1997, **61**, 975.
162. R. P. Mason and W. F. Fitzgerald, *Nature*, 1990, **347**, 457–459.
163. R. Mason, K. Rolfhus and W. Fitzgerald, *Mar. Chem.*, 1998, **61**, 37–53.
164. D. Cossa, L. E. Heimburger, D. Lannuzel, S. R. Rintoul, E. C. V. Butler, A. R. Bowie, B. Averty, R. J. Watson and T. Remenyi, *Geochim. Cosmochim. Acta*, 2011, **75**, 4037–4052.
165. E. M. Sunderland, D. P. Krabbenhoft, J. W. Moreau, S. A. Strode and W. M. Landing, *Global Biogeochem. Cycles*, 2009, **23**, DOI: 10.1029/2008gb003425.
166. I. Lehnherr, V. L. St Louis, H. Hintelmann and J. L. Kirk, *Nat. Geosci.*, 2011, **4**, 298–302.
167. C. Lamborg, K. Bowman, C. Hammerschmidt, C. Gilmour, K. Munson, N. Selin and C.-M. Tseng, *Oceanography*, 2014, **27**, 76–87.
168. J. D. Blum, B. N. Popp, J. C. Drazen, C. A. Choy and M. W. Johnson, *Nat. Geosci.*, 2013, **6**, 879–884.
169. D. B. Senn, E. J. Chesney, J. D. Blum, M. S. Bank, A. Maage and J. P. Shine, *Environ. Sci. Technol.*, 2010, **44**, 1630–1637.
170. C. H. Lamborg, C. R. Hammerschmidt, G. A. Gill, R. P. Mason and S. Gichuki, *Limnol. Oceanogr.: Methods*, 2012, **10**, 90–100.
171. J. M. Benoit, C. C. Gilmour, A. Heyes, R. P. Mason and C. L. Miller, in *Biogeochemistry of Environmentally Important Trace Elements*, 2003, vol. 835, pp. 262–297.
172. E. Kerin, C. Gilmour, E. Roden, M. Suzuki, J. Coates and R. Mason, *Appl. Environ. Microbiol.*, 2006, **72**, 7919–7921.
173. C. C. Gilmour, M. Podar, A. L. Bullock, A. M. Graham, S. D. Brown, A. C. Somenahally, A. Johs, R. A. Hurt, Jr., K. L. Bailey and D. A. Elias, *Environ. Sci. Technol.*, 2013, **47**, 11810–11820.
174. J. M. Parks, A. Johs, M. Podar, R. Bridou, R. A. Hurt, Jr., S. D. Smith, S. J. Tomanicek, Y. Qian, S. D. Brown, C. C. Brandt, A. V. Palumbo,

J. C. Smith, J. D. Wall, D. A. Elias and L. Liang, *Science*, 2013, **339**, 1332–1335.

175. T. Barkay, S. B. Miller and A. O. Summers, *FEMS Microbiol. Rev.*, 2003, **27**, 355–384.
176. R. C. Harris, J. W. M. Rudd, M. Almyot, C. L. Babiarz, K. G. Beaty, P. J. Blanchfield, R. A. Bodaly, B. A. Branfireun, C. C. Gilmour, J. A. Graydon, A. Heyes, H. Hintelmann, J. P. Hurley, C. A. Kelly, D. P. Krabbenhoft, S. E. Lindberg, R. P. Mason, M. J. Paterson, C. L. Podemski, A. Robinson, K. A. Sandilands, G. R. Southworth, V. L. S. Louis and M. T. Tate, *Proc. Natl. Acad. Sci. U. S. A.*, 2007, **104**, 16586–16591.
177. H. Hintelmann, R. Harris, A. Heyes, J. P. Hurley, C. A. Kelly, D. P. Krabbenhoft, S. Lindberg, J. W. M. Rudd, K. J. Scott and V. L. St Louis, *Environ. Sci. Technol.*, 2002, **36**, 5034–5040.
178. A. J. Poulain, D. M. Orihel, M. Amyot, M. J. Paterson, H. Hintelmann and G. R. Southworth, *Chemosphere*, 2006, **65**, 2199–2207.
179. N. Selin, E. Sunderland, C. Knightes and R. Mason, *Environ. Health Perspect.*, 2010, **118**, 137–143.
180. P. Porvari, *Sci. Total Environ.*, 2005, **213**, 279–290.
181. R. Bodaly, V. St Louis, M. Paterson, R. Fudge, B. Hall, D. Rosenberg and J. Rudd, in *Metal Ions in Biological Systems*, ed. A. Sigel and H. Sigel, Marcel Dekker, Inc., New York, 1997, vol. 34, pp. 259–287.
182. B. D. Hall, V. L. St Louis, K. R. Rolfhus, R. A. Bodaly, K. G. Beaty, M. J. Paterson and K. A. Cherewyk, *Ecosystems*, 2005, **8**, 248–266.
183. R. Harris, D. P. Krabbenhoft, R. P. Mason, M. W. Murray, R. Reash and T. Saltman, *Ecosystem Responses to Mercury Contamination*, CRC Press, Boca Raton, 2007.
184. C. C. Patterson, Settle, in *Accuracy in Trace Analysis: Sampling, Sample Handling and Analysis*, ed. P. LaFleur, US National Bureau of Standards, 1976, vol. Special Publication #42, pp. 321–351.
185. A. R. Flegal, H. Maring and S. Niemeyer, *Nature*, 1993, **365**, 242–244.
186. D. Weiss, E. A. Boyle, J. F. Wu, V. Chavagnac, A. Michel and M. K. Reuer, *J. Geophys. Res.: Oceans*, 2003, **108**, Article # 3306.
187. L. Alleman, A. Veron, T. Church, A. Flegal and B. Hamelin, *Geophys. Res. Lett.*, 1999, **26**, 1477–1480.
188. A. J. Veron, T. M. Church and A. R. Flegal, *Environ. Res.*, 1998, **78**, 104–111.
189. A. Annibaldi, C. Truzzi, S. Illuminati and G. Scarponi, *Mar. Chem.*, 2009, **113**, 238–249.
190. C. Gallon, M. A. Ranville, C. H. Conaway, W. M. Landing, C. S. Buck, P. L. Morton and A. R. Flegal, *Environ. Sci. Technol.*, 2011, **45**, 9874–9882.
191. L. Y. Alleman, T. M. Church, P. Ganguli, A. J. Veron, B. Hamelin and A. R. Flegal, *Deep Sea Res., Part II*, 2001, **48**, 2855–2876.
192. L. Y. Alleman, T. M. Church, A. J. Veron, G. Kim, B. Hamelin and A. R. Flegal, *Deep Sea Res., Part II*, 2001, **48**, 2811–2827.

193. USEPA, *Lead Exposure*, www.epa.gov/superfund/lead/health.htm#lead, Accessed July 31, 2014.

194. UNEP, *Lead and Cadmium*, http://www.unep.org/chemicalsandwaste/ LeadCadmium/tabid/29372/Default.aspx, Accessed August 8, 2014.

195. A. R. Flegal and D. R. Smith, *N. Engl. J. Med.*, 1992, **326**, 1293–1294.

196. W. L. Yu, S. A. Vislay and R. E. Edwards, *Clin. Pediatr.*, 1977, **16**, 791–794.

197. I. H. Billick, A. S. Curran and D. R. Shier, *Environ. Health Perspect.*, 1980, **34**, 213–217.

198. P. Lindeberg, *Ambio*, 1981, **10**, 351–352.

199. R. R. Jones, *Atmos. Environ.*, 1983, **17**, 2367–2370.

200. P. C. Elwood and J. E. J. Gallacher, *J. Epidemiol. Community Health*, 1984, **38**, 315–318.

201. M. L. Miranda, R. Anthopolos and D. Hastings, *Environ. Health Perspect.*, 2011, **119**, 1513–1516.

202. R. L. Jones, D. M. Homa, P. A. Meyer, D. J. Brody, K. L. Caldwell, J. L. Pirkle and M. J. Brown, *Pediatrics*, 2009, **123**, e376–e385.

203. S. J. Eisenreich, N. A. Metzer, N. R. Urban and J. A. Robbins, *Environ. Sci. Technol.*, 1986, **20**, 171–174.

204. U. Stromberg, T. Lundh and S. Skerfving, *Environ. Res.*, 2008, **107**, 332–335.

205. M. Schuhmacher, M. Belles, A. Rico, J. L. Domingo and J. Corbella, *Sci. Total Environ.*, 1996, **184**, 203–209.

206. W. T. Wu, P. J. Tsai, Y. H. Yang, C. Y. Yang, K. F. Cheng and T. N. Wu, *Sci. Total Environ.*, 2011, **409**, 863–867.

207. S. T. Wang, S. Pizzolato, H. P. Demshar and L. F. Smith, *Clin. Chem.*, 1997, **43**, 1251–1252.

208. A. Mathee, H. Rollin, Y. von Schirnding, J. Levin and I. Naik, *Environ. Res.*, 2006, **100**, 319–322.

209. S. A. Martinez, L. Simonella, C. Hansen, S. Rivolta, L. M. Cancela and M. B. Virgolini, *Hum. Exp. Toxicol.*, 2013, **32**, 449–463.

210. L. K. Graber, D. Asher, N. Anandaraja, R. F. Bopp, K. Merrill, M. R. Cullen, S. Luboga and L. Trasande, *Environ. Health Perspect.*, 2010, **118**, 884–889.

211. V. Kalra, J. K. Sahu, P. Bedi and R. M. Pandey, *Indian J. Pediatr.*, 2013, **80**, 636–640.

212. J. Finkelman, *Environ. Health Perspect.*, 1996, **104**, 10 11.

213. C. Reimann, D. B. Smith, L. G. Woodruff and B. Flem, *Appl. Geochem.*, 2011, **26**, 1623–1631.

214. D. B. Smith, W. F. Cannon, L. G. Woodruff, F. Solano, J. E. Kilburn and D. L. Fey, *Geochemical and Mineralogical Data for Soils of the Conterminous United States*, US Geological Survey, 2013.

215. J. G. Farmer, A. Broadway, M. R. Cave, J. Wragg, F. M. Fordyce, M. C. Graham, B. T. Ngwenya and R. J. F. Bewley, *Sci. Total Environ.*, 2011, **409**, 4958–4965.

216. P. E. Rasmussen, C. Levesque, M. Chenier, H. D. Gardner, H. Jones-Otazo and S. Petrovic, *Sci. Total Environ.*, 2013, **443**, 520–529.

217. J. Bierkens, R. Smolders, M. Van Holderbeke and C. Cornelis, *Sci. Total Environ.*, 2011, **409**, 5101–5110.
218. L. Zhu, J. Tang, B. Lee, Y. Zhang and F. Zhang, *Mar. Pollut. Bull.*, 2010, **60**, 1946–1955.
219. Y. Sun, G. S. Zhuang, W. J. Zhang, Y. Wang and Y. H. Zhuang, *Atmos. Environ.*, 2006, **40**, 2973–2985.
220. R. M. Tripathi, R. Raghunath, A. V. Kumar, V. N. Sastry and S. Sadasivan, *Sci. Total Environ.*, 2001, **267**, 101–108.
221. E. M. Mager, in *Homeostatis and Toxicology of Non-Essential Metals*, ed. C. M. Wood, A. P. Farrell and C. J. Brauner, Elsevier, Amsterdam, 2012, vol. 31B.
222. R. Nevin, *Environ. Res.*, 2007, **104**, 315–336.
223. H. W. Mielke and S. Zahran, *Environ. Int.*, 2012, **43**, 48–55.
224. M. A. Pokras and M. R. Kneeland, *Ecohealth*, 2008, **5**, 379–385.
225. M. E. Finkelstein, D. F. Doak, D. George, J. Burnett, J. Brandt, M. Church, J. Grantham and D. R. Smith, *Proc. Natl. Acad. Sci. U. S. A.*, 2012, **109**, 11449–11454.
226. J. D. Appleton, in *Medical Geology*, ed. O. Selinus, Elsevier, Amsterdam, 2004, pp. 227–262.
227. W. McKelvey and E. Oken, in *Mercury in the Environment*, ed. M. Bank, University of California Press, Berkeley, California, 2012, pp. 267–287.
228. G. M. Richardson, *Human and Ecological Risk Assessment*, 2014, **20**, 433–447.
229. G. M. Richardson, R. Wilson, D. Allard, C. Purtill, S. Douma and J. Graviere, *Sci. Total Environ.*, 2011, **409**, 4257–4268.
230. H. Gibb and K. G. O'Leary, *Environ. Health Perspect.*, 2014, **122**, 667–672.
231. A. K. B. Kristensen, J. F. Thomsen and S. Mikkelsen, *Int. Arch. Occup. Environ. Health*, 2014, **87**, 579–590.
232. E. M. Sunderland, *Environ. Health Perspect.*, 2007, **115**, 235–242.
233. E. Oken, A. L. Choi, M. R. Karagas, K. Marien, C. M. Rheinberger, R. Schoeny, E. Sunderland and S. Korrick, *Environ. Health Perspect.*, 2012, **120**, 790–798.
234. D. Mergler, H. A. Anderson, L. H. M. Chan, K. R. Mahaffey, M. Murray, M. Sakamoto and A. H. Stern, *Ambio*, 2007, **36**, 3–11.
235. K. R. Mahaffey, E. M. Sunderland, H. M. Chan, A. L. Choi, P. Grandjean, K. Marien, E. Oken, M. Sakamoto, R. Schoeny, P. Weihe, C.-H. Yan and A. Yasutake, *Nutr. Rev.*, 2011, **69**, 493–508.
236. K. Mahaffey, *JAMA*, 1998, **280**, 737–738.
237. E. M. Sunderland and N. E. Selin, *Environ. Health*, 2013, **12**, DOI: 10.1186/1476-069X-12-2.
238. UNEP, Technical Report for the 26th GC/GMEF of UNEP, 2011.
239. J. M. Pacyna, E. G. Pacyna and W. Aas, *Atmos. Environ.*, 2009, **43**, 117–127.
240. N. E. Selin, *Environ. Toxicol. Chem.*, 2014, **33**, 1202–1210.
241. C. T. Driscoll, C. Y. Chen, C. R. Hammerschmidt, R. P. Mason, C. C. Gilmour, E. M. Sunderland, B. K. Greenfield, K. L. Buckman and C. H. Lamborg, *Environ. Res.*, 2012, **119**, 118–131.

242. T. A. Douglas, L. L. Loseto, R. W. Macdonald, P. Outridge, A. Dommergue, A. Poulain, M. Amyot, T. Barkay, T. Berg, J. Chetelat, P. Constant, M. Evans, C. Ferrari, N. Gantner, M. S. Johnson, J. Kirk, N. Kroer, C. Larose, D. Lean, T. G. Nielsen, L. Poissant, S. Rognerud, H. Skov, S. Sorensen, F. Wang, S. Wilson and C. M. Zdanowicz, *Environ. Chem.*, 2012, **9**, 321–355.

243. J. A. Fisher, D. J. Jacob, A. L. Soerensen, H. M. Amos, E. S. Corbitt, D. G. Streets, Q. Wang, R. M. Yantosca and E. M. Sunderland, *Global Biogeochem. Cycles*, 2013, **27**, 1226–1235.

Persistent Organic Pollutants

MOHAMED ABOU-ELWAFA ABDALLAH

ABSTRACT

This chapter provides an overview of Persistent Organic Pollutants (POPs), which are widely considered as a global concern. The structure and role of the Stockholm Convention on POPs is discussed. The specific criteria for listing certain chemicals as POPs are explained. Both "legacy" and "new" POPs are classified according to their major applications, human exposure pathways and the required actions under the Stockholm Convention to minimize their risk to human and wildlife. Historical and current sources of POPs are discussed in line with their emissions to various environmental compartments. Causes for concern and potential adverse effects arising from exposure of human and wildlife to current levels of POPs are assessed. Different pathways of POPs transport and global distribution are evaluated and the impact of legislation on reducing levels of POPs in the global environment over the past 40 years is investigated. The expected influence of climate change on the behaviour, distribution and concentrations of POPs is explained and solutions for management of the problem are briefly discussed.

1 Introduction

Persistent organic pollutants (POPs) are synthetic organic chemicals which resist photolytic, biological and chemical degradation, and bioaccumulate in fatty tissues through aquatic and terrestrial food chains. They are also semi-volatile, enabling them to undergo long-range atmospheric transport to

Issues in Environmental Science and Technology No. 40
Still Only One Earth: Progress in the 40 Years Since the First UN Conference on the Environment
Edited by R.E. Hester and R.M. Harrison
© The Royal Society of Chemistry 2015
Published by the Royal Society of Chemistry, www.rsc.org

regions where they have never been used or produced, posing a risk of causing adverse effects to human health and the environment.[1]

Most of these chemicals proved beneficial in pest and disease control, crop production, and industry leading to their large scale industrial production following World War II. POPs include several chemicals which generally fall within two main categories:

1. Intentionally produced chemicals currently or previously applied in agriculture, disease control, manufacturing or industrial processes. Examples include PCBs (which were widely applied as hydraulic and heat exchange fluids in electrical transformers and large capacitors) and DDT (which is still used for vector-control of malaria in some parts of the world).
2. Unintentionally produced chemicals, such as dioxins, that result from some industrial processes and from combustion of municipal and medical waste.

Over the last few decades, the uncontrolled releases of POPs, combined with their physico-chemical properties, have resulted in their wide distribution all over the world. They have been detected on every continent, at sites representing every major climatic zone including remote regions such as the open oceans, the deserts, the Arctic and the Antarctic, where no significant local sources exist and the only reasonable explanation for their presence is long-range transport from other parts of the globe.

The ubiquity of POPs has led to extensive contamination of environmental media and living organisms, and resulted in prolonged exposure of many species, including humans, for periods of time that span generations to various levels of different POPs. Exposure to POPs, either acute or chronic, can be associated with a wide range of adverse health effects, including allergies, endocrine disruption, neurotoxicity, developmental toxicity, reproductive disorders, disruption of the immune system, cancer and death.[2] This has resulted in a growing international concern, on the scientific, public and governmental levels, over the environmental and health effects of POPs.

2 The Stockholm Convention on POPs

In response to the substantially growing global concern over POPs, the international community adopted the United Nations Environment Programme (UNEP) Stockholm Convention on POPs on 22nd May 2001. More than 90 countries signed the convention and Canada was the first to ratify it. The convention entered into force on 17th May 2004 after ratification of the 50th party. Currently, there are 179 participants to this treaty which aims "to protect human health and the environment from persistent organic pollutants which have become a prominent global problem." UNEP is the leading international environmental entity that supports the agenda and implementation of environmental sustainability for the United Nations. The COP, or the Conference of the Parties of the Stockholm Convention, governs the POPs convention, with its members being the convention's parties.[3]

Under the convention's treaty, the parties have initially agreed to reduce or eliminate the production, use, and/or release of 12 POPs, nicknamed "the dirty dozen", which have been recognized as causing adverse effects to humans and the ecosystem (see Table 1).

The control measures that must be taken by the convention parties are classified by the Annex in which chemical substances are listed:

- *Annex A:* parties must take measures to eliminate the production and use of the chemicals listed under Annex A. Specific exemptions for use or production are listed in the Annex and apply only to Parties that register for them.
- *Annex B:* parties must take measures to restrict the production and use of the chemicals listed under Annex B in light of any applicable acceptable purposes and/or specific exemptions listed in the Annex.
- *Annex C:* parties must take measures to reduce the unintentional release of chemicals listed under Annex C with the goal of continuous minimization and, where feasible, ultimate elimination.

The convention allows its parties to submit proposals for the addition of new POPs to the COP. The submitted proposals are reviewed by the POPs Review Committee (POPRC) comprising 31 government-designated experts in areas of chemical assessment or management from all UN regions. Based on the recommendations of the POPRC, and after the approval of the COP, 11 new chemicals have been added to the POPs list since 2009, which are currently known as the "new POPs" (see Table 2).

Furthermore, stockpiles and wastes containing POPs must be managed and disposed of in a safe, efficient, and environmentally sound manner, taking into account international norms, standards, and guidelines. Also, to ensure implementation of various articles of the convention, provisions are made for information exchange and increasing public awareness.[3]

Although the Stockholm Convention provides obligations on elimination and restriction of chemicals listed in Annex A and B. The implementation of these obligations may be subject to specific exemptions. Parties that have notified the Secretariat for registration of a specific exemption are allowed to continue to use or produce a chemical for a particular purpose.

3 POPs Criteria

Among thousands of synthetic chemicals, relatively few substances possess the necessary properties to make them POPs. For a certain chemical to be listed as POP, the Stockholm Convention specified the following criteria:

(a) *Persistence:* the ability to resist physical, chemical or biological degradation in various environmental media. Evidence should be

Table 1 The "dirty dozen".

POP	Category	CAS no.	Use/source	Major pathway of human exposure	Action[a]
Aldrin	Pesticide	309-00-2	Pesticide to kill termites, grasshoppers, corn rootworm, and other insect pests	Diet (dairy products and meat)	Annex A
Chlordane	Pesticide	57-74-9	Termiticide, broad-spectrum insecticide	Air	Annex A
DDT	Pesticide	50-29-3	Insecticide used primarily on cotton, and insects that carry malaria and typhus	Diet	Annex B
Dieldrin	Pesticide	60-57-1	To control termites, textile pests and insect-borne diseases	Diet	Annex A
Endrin	Pesticide	72-20-8	Insecticide and rodenticide	Diet	Annex A
Heptachlor	Pesticide	76-44-8	To kill termites, cotton insects, grasshoppers and malaria-carrying mosquitoes	Diet	Annex A
Hexachlorobenzene	Pesticide, industrial chemical, by-product	118-74-1	Fungicide, used to make fireworks, ammunition, synthetic rubber, unintentionally produced during combustion and the manufacture of certain chemicals	Diet	Annex A, Annex C
Mirex	Pesticide	2385-85-5	Insecticide, termiticide and flame retardant	Diet (meat and fish)	Annex A
Toxaphene	Pesticide	8001-35-2	Insecticide used to control pests on crops and livestock.	Diet	Annex A
Polychlorinated biphenyls (PCBs)	Industrial chemical, by-product	1336-36-3	Heat exchange fluids, in electric transformers and capacitors, by-product of combustion	Diet	Annex A, Annex C
Polychlorinated dibenzo-*p*-dioxins (PCDD)	By-product	several	Incomplete combustion, manufacture of chlorinated chemicals	Diet (particularly of animal origin)	Annex C
Polychlorinated dibenzofurans (PCDF)	By-product	several	Incomplete combustion, manufacture of chlorinated chemicals	Diet (particularly of animal origin)	Annex C

[a]Annex A substances: slated for "elimination"; Annex B substances: slated for "restriction" for which there is a specified "acceptable purpose"; and Annex C substances: continuing minimization and, where feasible, ultimate elimination of the total releases derived from anthropogenic sources.

Table 2 The "new POPs".

POP	Category	CAS no.	Use/source	Major pathway of human exposure	Action[a]
Chlordecone	Pesticide	143-50-0	Agricultural pesticide	Diet	Annex A
Hexabromobiphenyl	Industrial chemical	36355-01-8	Flame retardant	Diet, dust	Annex A
α- and β-hexachlorocyclohexane	Pesticide, by-product	319-84-6 319-85-7	Insecticide	Diet	Annex A
Lindane (γ-hexachlorocyclohexane)	Pesticide	58-89-9	Insecticide for seed, soil, tree and wood treatment, foliar applications and against ectoparasites in animals and human	Diet	Annex A
Tetra- and penta-bromodiphenyl ether	Industrial chemical	5436-43-1 60348-60-9	Flame retardant	Diet, dust	Annex A
Hexa- and hepta-bromodiphenyl ether	Industrial chemical	several	Flame retardant	Diet, dust	Annex A
Perfluorooctane sulfonic acid (PFOS), and perfluorooctane sulfonyl fluoride (PFOS-F)	Industrial chemical	1763-23-1	Electric and electronic parts, firefighting foam, photo imaging, hydraulic fluids and textiles	Diet, dust	Annex B
Pentachlorobenzene (PeCB)	Pesticide, industrial chemical, by-product	608-93-5	Used in PCB products, dyestuff carriers, fungicide and flame retardant. Produced unintentionally during combustion and industrial processes. Impurity in solvents or pesticides	Diet	Annex A, Annex C
α-, β- Endosulfan and Endosulfan sulfate	Pesticide	959-98-8 33213-65-9 1031-07-8	Broad-spectrum insecticide used to control crop pests, tsetse flies and ectoparasites of cattle and as a wood preservative	Diet	Annex A
Hexabromocyclododecane (HBCD)	Industrial chemical	several	Flame retardant	Dust, Diet	Annex A

[a] Annex A substances: slated for "elimination"; Annex B substances: slated for "restriction" for which there is a specified "acceptable purpose"; and Annex C substances: continuing minimization and, where feasible, ultimate elimination of the total releases derived from anthropogenic sources.

provided that the substance's half-life in water is greater than 2 months, or that its half-life in soil or sediment is greater than 6 months. Alternatively, evidence that the substance is otherwise sufficiently persistent to be of concern within the scope of the Convention should be presented.

(b) *Bioaccumulation:* the ability of a chemical to accumulate in living tissues to levels higher than those in the surrounding environment. To meet the Convention criteria, the following evidence should be provided:

    i. The bio-concentration factor (BCF)[†] or bio-accumulation factor (BAF)[‡] in aquatic species for the substance is greater than 5000 or in absence of BCF/BAF data, the $\log K_{ow}$[§] is greater than 5; or

    ii. A substance with a BCF or BAF in aquatic species significantly lower than 5000 presents other reasons for concern, such as high toxicity/eco-toxicity; or

    iii. Monitoring data in biota indicating that the bio-accumulation potential of the substance is sufficient to be of concern within the scope of the Convention.

(c) *Toxicity:* there is *hitherto* no established criterion for toxicity in international negotiations due to the multidimensional and multifaceted nature of this parameter. Therefore, the Convention requires:

    i. Evidence of adverse effects to human health or to the environment that justifies consideration of the chemical within the scope of this convention; or

    ii. Toxicity or eco-toxicity data, compared where possible with available detected or predicted levels of a substance, indicating a potential for damage to human health or the environment. The assessment of damage should include a consideration of toxicological interactions among substances.

(d) *Long-range transport*: this can be established by one of the following evidence:

    i. Measured levels of the chemical in locations distant from the sources of its release that are of potential concern; or

    ii. Monitoring data showing that long-range environmental transport of the chemical, with the potential for transfer to a receiving environment, may have occurred by way of air, water or migratory species; or

---

[†]Bioconcentration factor (BCF) is the concentration of a chemical in or on an organism or specified tissues thereof divided by the concentration of the chemical in the surrounding medium.

[‡]Bioaccumulation Factor (BAF) is the concentration of a substance in an organism divided by the concentration of the chemical in the surrounding medium measured in an intact ecosystem (takes into account accumulation through ingested food, as well as concentration from the surrounding medium).

[§]$\log K_{ow}$ is the logarithm of the partition coefficient, *i.e.* the ratio of the chemical's solubility in *n*-octanol and water at equilibrium.

iii. Environmental fate properties and/or model results that demonstrate that the chemical has a potential for long-range environmental transport through air, water or migratory species, and the potential for transfer to a receiving environment in locations distant from the sources of its release. For a chemical migrating significantly through the air, its half-life in air should be greater than two days.

In applying the screening criteria for identification of POPs, other factors, such as dispersion mechanisms, patterns of use, influences on marine environment and tropical climates should be taken into account, as well as the need to conserve biodiversity and protect endangered species.[4]

4   Sources and Applications of POPs

Source apportionment aims to identify and rank the major sources of pollutants to the environment. This allows identification of major release pathways, and hence the development and application of emission-control strategies.

The most common approach to develop a source inventory is to determine an emission factor for a specific source activity (*e.g.* 10 µg per tonne of waste burnt), then multiply this by an activity factor – *i.e.* the extent to which the activity is practiced (*e.g.* 3 million tonnes of waste burnt per year). In this way, an estimate of annual pollutant emissions for a specific source can be derived – in the case illustrated, annual emissions would amount to 30 g. However, the accuracy of the derived source inventory varies widely according to the quality of input data and nature of the emission source. For example, quantifying pesticide releases using well-documented industry production figures is more reliable than estimating emission factors for unintentional releases of dioxins from municipal waste incineration. This section provides a summary of the main applications of POPs and the sources of their current and past release to the environment.

4.1   Pesticide POPs

4.1.1   Organochlorine Pesticides.   For organochlorine pesticide POPs, the major source is deliberate application to crops and soils. The nine pesticide POPs (see Table 1) were introduced into commercial use after World War II and dramatically changed modern pest control. These compounds were effective against a wide range of pests for long periods of time. Ironically, the desired prolonged effect by the manufacturers at the time has led to the high environmental persistence of the banned pesticide POPs. Data on the total amount entering the environment and regional/global usage patterns of legacy agrochemical POPs is highly uncertain and often poorly known.[5] Although the use of many POP pesticides has been banned or restricted in the industrialized countries since the 1970s, their presence still remains of concern due to their environmental persistence (estimated soil half-lives up to 12 years). The problem is particularly severe in many developing countries and countries with economies in transition, because of stockpiles and uncontrolled dumpsites. According to the Food and Agriculture Organization (FAO) estimates, there are ∼500 000 tonnes of obsolete pesticide stocks in developing countries. Some of these stockpiles are held in the open for prolonged periods of time, leading to the uncontrolled release of these chemicals to air and contamination of land and water resources. In Africa, up to 120 000 tonnes of obsolete pesticide POPs were estimated and are present in virtually every country on the continent.[6] Stockpiles in Asia are currently reported at 6000 tonnes, a figure that excludes China, where the problem of pesticide waste is believed to be widespread. In the Middle East and Latin America together, approximately 10 000 tonnes have been declared.[7] Stockpiles of POP pesticides were also reported from Eastern European countries. For example, it was reported that approximately 3000 tonnes of obsolete pesticides are stored in warehouses in Moldova. Studies have shown conclusively that these materials have contaminated the sites and surrounding soils and nearby surface waters. Another 4000 tonnes, including 650 tonnes of DDT and 1300 tonnes of HCH are stored in one warehouse situated in the south of the country.[8]

While the use of cyclodiene pesticides (aldrin, dieldrin, endrin, chlordane, heptachlor, and mirex) was banned in the United States by the EPA during the 1970s, they continued to be used under restricted conditions as termiticides well into the 1980s. Aldrin, dieldrin, and endrin are chemically similar in structure, the latter two being stereoisomers of each other (see Figure 1).[9] Furthermore, Aldrin was reported to rapidly transform into dieldrin in air and soil.[10] Endrin was initially applied as a general

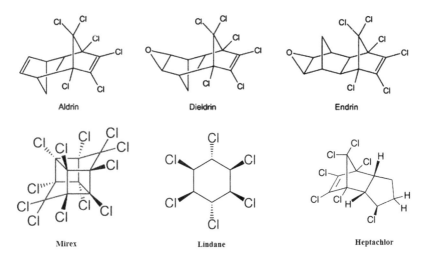

Figure 1   Chemical structures of some pesticide POPs.

insecticide, mainly on cotton and tobacco. It also proved useful as an avicide (bird-killing agent) and rodenticide, exploiting its high toxicity to vertebrates. Aldrin and dieldrin were most frequently used to control soil pests affecting corn and citrus crops, as well as termites in buildings. Dieldrin was also applied for indoor house spraying to control mosquitoes that carry malaria. The acute toxicity of dieldrin (either in its own formulations or in aldrin commercial products) to non-target wildlife led to increasing environmental concerns, particularly after numerous mass bird deaths were associated with its use as a seed treatment.[11] However, long before regulatory actions to restrict these insecticides were instigated, technical problems had emerged, owing to the development of resistance in target pests and secondary pest upsets caused by the elimination of natural predators. The development of other pesticides (*e.g.* organophosphates, pyrethroids and carbamates), which were considered more effective and less environmentally destructive also reduced demand for the cyclodiene insecticides.[9]

The production and usage of chlordane, heptachlor and mirex reached its peak during the 1950s and 1960s. Chlordane was manufactured and used as a complex mixture (dominated by *cis*- and *trans*-chlordane, heptachlor and nonachlor), often referred to as "technical chlordane". It was directly applied to soil to create a long-acting (effective for 25 years) chemical barrier against subterranean termites. Heptachlor was applied to both soil and seeds for pest control and to wood for termite protection. The last registered use for heptachlor in the United States, banned in 1988, was as insecticide in cable boxes to prevent nest-building by fire ants in electrical equipment.[12] According to a UNEP report issued on December 2000, both chlordane and heptachlor were still used in several African, Asian, and Eastern European countries for termite control.[13] Mirex was produced in lower volumes and applied mainly against fire ants in the southeast United. It has also been used to combat leaf cutters in South America and harvester termites in South Africa. Until 1991, mirex was applied as an additive flame-retardant in plastics, rubber, paint paper and electrical goods.[12]

4.1.2 The Exceptional Case of DDT. DDT (dichlorodiphenyltrichloroethane) was the first organochlorine insecticide developed, and is probably the most famous and controversial pesticide ever made. An estimated 2 million tonnes have been produced and applied worldwide since 1940. DDT exhibited several advantages over other synthetic organochlorine pesticides including relatively low mammalian toxicity, effectiveness against a broad range of pests, and low cost of production. In addition to its principal agricultural use on several crops (mainly cotton) from 1945–1975, DDT was also applied for the control of louse-borne typhus during World War II. Moreover, DDT played a major role in the malaria eradication programs of the 1950s and 1960s. Although the eradication of malaria proved elusive in several tropical countries, DDT contributed to saving countless lives in many malarious countries.[9]

The use of DDT in the industrial world peaked in the early 1960s but declined afterwards due to several technical problems including mounting insect resistance and secondary pest growth. Moreover, there was mounting evidence of the environmental impacts from DDT and its persistent metabolites, DDE (dichlorodiphenyldichloroethylene) and DDD (dichlorodiphenyldichloroethane), see Figure 2.

These impacts were particularly severe on bird populations through eggshell thinning and chick mortality in raptors (*e.g.* bald eagles) and oceanic birds (*e.g.* pelicans) – a phenomenon brought to public attention by Rachel Carson in her 1962 book, *Silent Spring*. This was followed by a plethora of scientific data confirming that high levels of DDE (a metabolite of DDT) in certain birds of prey caused their eggshells to thin so dramatically they could not produce live offspring.[14] One species particularly sensitive to DDE was the bald eagle. Public concern about the eagles' decline, together with other potential long-term adverse effects of DDT exposure on humans and wildlife, prompted the United States Environmental Protection Agency (US EPA) to cancel the registration of DDT in 1972. The bald eagle has since experienced one of the most dramatic species recoveries in modern history.[9] However, DDT production and usage for disease vector control (mostly malaria) continues in 25 developing countries, primarily because of its low cost and persistence. While other approaches such as early detection and treatment, or pyrethroid-treated bednets, may prove cost-effective and sustainable in the long term, DDT's lower cost remains an advantage, at least in some countries. Therefore, the Stockholm Convention strives for a balance between the public health benefits of DDT for malaria control, the availability and cost of alternatives, and the impacts of DDT on ecosystems and human health. While working towards reducing and ultimately eliminating the use

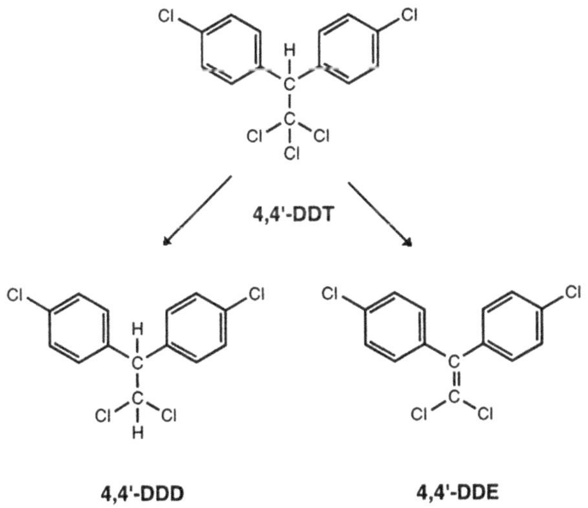

Figure 2   Chemical structure of DDT and its major metabolites.

of DDT, parties may use DDT only for disease vector control in accordance with World Health Organization recommendations.[13]

4.2 Byproduct POPs

*4.2.1 Polychlorinated Dibenzo-*p*-dioxins and Polychlorinated Dibenzofurans (PCDD/Fs).* Polychlorinated dibenzo-*p*-dioxins and polychlorinated dibenzofurans (PCDD/Fs) have never been intentionally produced – other than on a laboratory scale. The main sources of these compounds in the environment are principally as by-products of anthropogenic activities, in particular, the manufacture and use of organochlorine chemicals (*e.g.* PCBs; polychlorinated biphenyls) and combustion processes (such as waste incineration and steel manufacture). PCDD/Fs can also be released from natural sources, notably forest fires that may arguably be of significance in some countries.[15]

   The mechanism of PCDD/Fs formation during combustion activities is still not completely understood. It appears that PCDD/Fs formation occurs *via de novo* synthesis – *i.e.* formation from the basic chemical "building blocks" of carbon, chlorine, oxygen and hydrogen. Consequently, PCDD/F formation can occur from the combustion of any "fuel", providing sources of these four elements are present, when the required high temperature is reached. The rate of formation is enhanced if levels of "precursor" compounds (*e.g.* chlorophenols and chlorobenzenes) are higher in the "fuel". Laboratory experiments and studies on working waste incinerators revealed that PCDD/F formation occurs in post-combustion zones (*i.e.* oxygen-rich regions), such as electrostatic precipitators, where temperatures are in the range 250 °C–350 °C. In these regions, a series of reactions, catalyzed by the presence of metal chlorides, occur on the surface of fly ash particles and, as a result, PCDD/Fs are formed.[16] The presence of PCDDs in chlorophenols and products produced *via* chlorophenol intermediates (such as chlorophenoxy acetic acid derivatives; one of the principal constituents of "Agent Orange" – a defoliant used in the Vietnam War, and heavily contaminated with 2,3,7,8-TCDD – was the *n*-butyl ester of 2,4,5-trichlorophenoxyacetic acid) is more easily explained. PCDD formation occurs *via* the spontaneous reaction of two chlorophenol molecules (or chlorophenate ions), which yields PCDD(s), the exact identity(ies) of which are dependent on the chlorination pattern of the reactants, and the possibility of Smiles rearrangement products.[17]

   Dioxins have been central to a number of environmental controversies in the past few decades. In 1976, an explosion at a trichlorophenol herbicide production plant in Seveso, Italy, led to widespread environmental contamination. This was evidenced by local livestock and wildlife mortality, very high human exposures and clinical illness (*e.g.* chloracne, a severe and prolonged acne-form condition) and evacuation of the surrounding region.[18] Residential dioxin exposure and evacuation also occurred at Times Beach, Missouri, following the spraying of dioxin contaminated waste oil for dust control in the early 1970s.[9] Furthermore, burning and high-temperature treatment of PCB mixtures can lead to the production of PCDFs. This occurred during the Yusho and Yu-Cheng poisoning incidents in Japan and Taiwan, respectively, where the cooking of rice oil contaminated with PCBs resulted in the production of PCDFs. The presence of both PCBs and PCDFs in the cooked food caused chloracne and other toxic effects in adults, fetal mortality and developmental defects in their offspring.[19]

However, the existence of other sources of dioxins is evidenced by the failure of source inventories of atmospheric emissions of PCDD/Fs to account for substantial proportions of deposition to the earth's surface. The exact identity and contribution of the various sources of dioxins is subject to uncertainty, but essentially focus on combustion activities and inadvertent contamination of other chlorinated chemicals (*e.g.* pentachlorophenol and chloranil).[20]

4.3 Industrial Chemicals

4.3.1 Polychlorinated Biphenyls (PCBs). Polychlorinated biphenyls (PCBs) have been widely used since 1930 as dielectrics in transformers and large capacitors, heat exchange fluids, paint additives, in carbonless copy paper, and in plastics. The major advantages of PCBs for such a variety of industrial applications included their chemical inertness, resistance to heat, non-flammability, low vapour pressure and high dielectric constant. The total amount of PCBs produced globally was estimated at 1.5 million tonnes.[21]

By the late 1970s, the use and marketing of PCBs in the European Community and North America were heavily restricted. However, the use of existing stocks may have continued until the mid-1980s. Soils were reported to be the main environmental sink of PCBs. The principal loss mechanism from soils is *via* volatilization. Until relatively recently, such volatilization was thought to represent the principal source of PCBs to the contemporary atmosphere. However, an increasing body of research shows that emissions of PCBs from remaining applications (such as fluorescent light ballasts and building sealants) has led to substantial contamination of indoor air and dust, and *via* ventilation of contaminated indoor air is driving outdoor contamination. The evidence that indoor air ventilation is maintaining present-day outdoor air concentrations rather than volatilization from soil consists of three strands. Firstly, studies in both the late 1990s and the mid-2000s showed concentrations of ΣPCB in indoor air in the UK West Midlands were on average 30 times higher than those in outdoor air.[22,23] Secondly, examination of spatial variation in PCB concentrations in outdoor air along an urban–rural transect across the West Midlands reveals a clear "urban pulse", whereby concentrations in the city centre (where there is a greater density of contaminated indoor environments) exceed substantially those in rural and suburban locations.[24] The final and most conclusive strand of evidence involves exploitation of the chiral properties of some PCBs. Nineteen PCBs containing three or four chlorine atoms exist as stable diastereomers (*i.e.* each compound exists in two forms, known as "enantiomers", which are mirror images that differ only in the way they rotate the plane of polarized light and may biodegrade at different rates). In commercial PCB formulations, these 19 PCBs are present as racemates – *i.e.* concentrations of the two enantiomers are equal, such that the enantiomer fraction (EF) is racemic *i.e.* $= 0.5$ (EF $=$ concentration of the $(+)$ enantiomer divided by the sum of the concentrations of both enantiomers). By comparison, due to enantiomeric differences in resistance to biodegradative processes, the EFs of chiral PCBs found in soils – and which are preserved upon volatilization – deviate from 0.5, in theory varying between close to zero to *ca.* 1. Hence knowledge of EFs of chiral PCBs in samples of outdoor air, indoor air, and soil provides an indicator of the extent to which the contemporary ambient atmospheric burden is due to volatilization from soil, and how much arises from the ventilation of indoor air. Measurements in the West Midlands of the EF of the most abundant chiral PCB (PCB 95) have demonstrated that the racemic signature observed in almost all outdoor air samples is significantly different from the non-racemic value observed in soil, but matches the racemic values observed in indoor air.[24]

Although PCBs are principally addressed as intentionally produced industrial chemicals, they can also be produced in small amounts during incineration processes involving chlorinated organic chemicals and are therefore included in the unintentional "byproduct" category under the Stockholm Convention (see Table 1). Because of the magnitude of past use of PCBs and the continuing economic importance of previously manufactured PCB-containing equipment, parties to the Stockholm Convention must are requested to identify, label, and remove PCB-containing equipment from use by 2025. During this period, the Stockholm Convention mandates a series of measures to reduce exposures and risk from further releases of PCBs to the environment.[9,13]

4.3.2 Brominated Flame Retardants.

4.3.2.1 Polybrominated Diphenyl Ethers (PBDEs)
Polybrominated diphenyl ethers (PBDEs) do not occur naturally in the environment, but are produced synthetically as mixtures (see Table 3) applied as additives (*i.e.* not covalently bonded to the polymer matrix) to flame-retard a wide range of consumer products as follows:[25]

- Penta-BDE mixture: used in polyurethane foam products such as furniture and upholstery in domestic furnishing, and in the automotive and aviation industries.
- Octa-BDE mixture: used in plastic products, such as housings for computers, automobile trim, telephone handsets and kitchen appliance casings.
- Deca-BDE mixture: used in the high impact polystyrene (HIPS) housing for electrical and electronic equipment, and in various applications in the textile, automotive, and aviation industries.

Table 3   Composition of commercial PBDE formulations.

Component	% Contribution (w/w)			Main congener
	Penta-BDE	Octa-BDE	Deca-BDE	
Tribromodiphenyl ethers	0.2			BDE-28
Tetrabromodiphenyl ethers	36.1			BDE-47
Pentabromodiphenyl ethers	55.1			BDE-99
Hexabromodiphenyl ethers	8.6	5.5		BDE-153
Heptabromodiphenyl ethers		42.3		BDE-183
Octabromodiphenyl ethers		36.1	0.1	BDE-197
Nonabromodiphenyl ethers		13.9	2.5	BDE-207
Decabromodiphenyl ether		2.1	97.4	BDE-209

All PBDE commercial mixtures were high production volume chemicals with estimated global production in 2001 of 7500, 3790 and 56 100 tonnes of penta-BDE, octa-BDE and deca-BDE, respectively.[26] Information is limited on how PBDEs incorporated in plastics and foams are released from products to the environment. However, several studies have shown that PBDEs can be released during its production, application to consumer products, emissions from waste disposal/recycling and in small amounts from landfills (by leaching). PBDEs can also be released to the air during the life-span of flame retarded products.[27] In this regard, the greater the vapour pressure, and the lower the octanol–air partition coefficient, the more likely the BDE congener will volatilize from the plastic product. Another release mechanism reported recently is the flaking off, crumbling, and/or general abrasion of PBDE-containing particles/fibres from treated polymers. Although such a loss mechanism may be significant, there is currently no method to quantify this avenue of loss, or compare it to other emission mechanisms.[28,29]

Sewage treatment operations can also result in environmental release of PBDEs. Both sewage sludge (from the treatment process) as well as the effluent from sewage treatment plants (STP) near PBDE point sources contained appreciable amounts of PBDEs.[30]

A limited life cycle analysis can provide some information on the magnitude of PBDEs releases to the environment. Following the releases from PBDE production processes, volatilization from products, and waste disposal, the Toxics Release Inventory (TRI) reports a total release of 32.2 tonnes of deca-BDE to the air, land, and water of the United States in 2007. According to the TRI, total environmental releases peaked in 1999, with a release of 53.9 tonnes, dropped in 2003 to 36.3 tonnes, followed by an increase in 2004 to 44.8 tonnes, and then declined in 2005, 2006, and 2007.[26]

For comparison, a mathematical model was used to estimate the emissions of octa-BDE to the European environment in 2003. Total EU emissions estimated from the use of plastics were 7–15 tonnes year^{-1} to air, 0.2–0.9 tonnes year^{-1} to waste water treatment plants, 7–14 tonnes year^{-1} direct to surface water and 20–42 tonnes year^{-1} to urban/industrial soil. These total emissions were dominated by estimates over the service life of polymers and from waste disposal.[31]

The application of sewage sludge for soil amendment resulted in 13.16 tonnes year^{-1} of PBDEs being applied to agricultural land in the United States in 2009. Estimates of life cycle analysis reported that 1.19 tonnes year^{-1} of PBDEs are discharged annually into surface waters from sewage treatment plant effluent.[25]

4.3.2.2 Hexabromocyclododecane (HBCD)

Hexabromocyclododecane (HBCD) is an additive flame retardant applied mainly in the building industry where it is incorporated (typically at <3% by weight) into extruded (XPS) or expanded polystyrene (EPS) foam materials. Secondary applications include upholstered furniture, automotive interior textiles, electric and electronic equipment. With global production of 16 700 tonnes in 2001 and an EU consumption of 11 000 tonnes in 2007, HBCD is the most widely used cycloaliphatic additive brominated flame retardant.[32]

Like PBDEs, HBCD can be released to the environment *via* a number of different pathways including: emission during production, the manufacture of flame-retarded products, by leaching from consumer products during their life-span, or following product disposal/recycle.[33]

Very little is known as to what extent end-products containing HBCD are landfilled, incinerated, left in the environment or recycled. An environmental assessment study of a HBCD production site (5000 tonnes year^{-1}) was carried out in 2001. Results showed releases to air of 3400 kg year^{-1} and to wastewater from the onsite STP of 2000 kg year^{-1}. The annual release of HBCD in solid waste including sludge from the onsite STP was estimated to 1136 tonnes year^{-1}.[32]

A life cycle study of HBCD emissions from its various sources to the European environment in 2007 reported total emissions of ~2 tonnes year^{-1}. These total emissions were dominated by estimates over industrial use (textiles and insulation boards) and formulation of EPS and HIPS.[32]

4.3.3 Perfluorooctane Sulfonate (PFOS).

The perfluorooctane sulfonate (PFOS) anion is not a substance produced as such. It is, or was, commercially available in the form of salts, derivatives and polymers. The term PFOS-related substance is used to refer to any or all of the substances which contain the PFOS moiety (defined as the $C_8F_{17}SO_2$ group) and may break down in the environment to give PFOS. There are currently a number of lists of PFOS-related substances ranging from 48–183 substance, which are extensively applied in a wide range of applications (see Figure 3).[1]

The total historical worldwide production of POSF was estimated to be 96 000 tonnes (or 122 500 tonnes, including unusable wastes) between 1970–2002. Global releases of PFOS were estimated at 45 250 tonnes to air and water between 1970–2012 from direct (manufacture, use, and consumer products) and indirect (intermediates and/or impurities) sources. Measured oceanic data suggests 235–1770 tonnes of PFOS currently reside in ocean

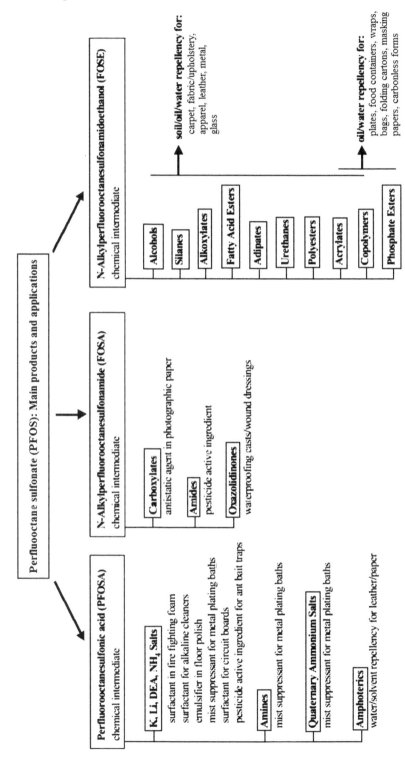

Figure 3 Major applications of PFOS-related substances.

surface waters. While large uncertainties surround emissions from indirect sources, Estimates indicate that direct emissions from PFOS-derived products are the major source to the environment, primarily through losses from stain repellent treated carpets, waterproof apparel, and aqueous fire-fighting foams.[34]

4.4   Hexachlorobenzene (HCB)

Hexachlorobenzene (HCB) is included under all three general categories of POPs: as a pesticide, an industrial chemical, and an unintended byproduct (see Table 1). HCB was first introduced in 1945 as antifungal for seed and soil treatment. The harmful effects of HCB were sadly demonstrated in Turkey during the late 1950s, when $\sim$3500 people ingested bread inadvertently made from HCB-treated grains. Exposed adults suffered from in a metabolic defect of blocked hemoglobin synthesis (porphyria cutanea tarda), which causes light-sensitive skin lesions, colored urine and in some cases, death. All children born to porphyric mothers during this epidemic died, with an estimated 1000–2000 children dying from related skin lesions, exacerbated by the HCB transfer in breast milk.[35]

HCB was also produced for various industrial applications including pyrotechnic coloring in military ordnance, synthetic rubber production, and as a chemical intermediate in dye manufacture and organic chemical synthesis.[9] While the use of HCB as pesticide is banned, an exemption is available under the Stockholm Convention for the use of HCB as an intermediate for synthesis of chlorophenols in China, with an estimated annual production of 3000–4000 tonnes. HCB is also produced as an unintended byproduct during the manufacture of chlorinated solvents, as an impurity of certain pesticides (*e.g.* chlorothalonil, DCPA, and PCP) and during incineration practices. Estimates of global emissions of HCB to the atmosphere from different categories of sources were reported as follows: pesticides application – 6500 kg year^{-1}; manufacturing – 9500 kg year^{-1}; combustion – 7000 kg year^{-1}, includes 500 kg from biomass burning. This adds up to total current HCB emissions of approximately 23 000 kg year^{-1} with an estimated range 12 000–92 000 kg year^{-1}. A substantial portion of HCB measured in the atmosphere was suggested to come from volatilization of "old" HCB on the soil from past contamination.[36]

5   Toxicity/Adverse Effects of POPs

Bearing in mind their environmental persistence and ubiquity, the main cause for concern raised by POPs is their ability to cause adverse effects to human health and wildlife. It should be noted that a detailed review of the toxicology and ecotoxicology of POPs is beyond the scope of this chapter. Therefore, only a brief summary of the main adverse effects of each class of compounds will be presented. A general consideration in relation to the toxic effects of POPs is the difficulties associated with assessing the effects in humans, in the understandable absence of clinical data due to various technical and ethical issues. Alternatively, evidence for adverse effects in humans is reliant on epidemiological surveys and extrapolations from animal studies, with all the problems inherent in such indirect measurements of human toxicity. This is further complicated by the fact that POPs rarely exist as single compounds in the environment and individual field studies are frequently insufficient to provide compelling evidence of cause and effect in their own right. A final general point is the tendency of POPs to partition into fatty tissues. As a result, their adverse effects are compounded by their bioaccumulation in species at the top of food chains (*e.g.* marine mammals, birds of prey and humans) which enables them to achieve toxicologically relevant concentrations even though direct exposure may appear limited and acute exposure may appear of little relevance to chronic toxicity.[15]

5.1 Ecotoxicity

Several studies have associated POPs with significant environmental impact in a wide range of species and at virtually every trophic level.[37] Although the acute effects of POPs intoxication have been well documented, adverse effects associated with chronic low-level exposure in the environment is of particular concern. This may be attributed to the long half-life of POPs in biological organisms, which allows accumulation of seemingly small concentrations *via* continuous exposure over extended periods of time.[38] For organochlorine pesticides and PCBs, experimental evidence suggests that such prolonged low-level exposures may be associated with chronic non-lethal effects including dermal effects, immunotoxicity, impairment of reproductive performance and carcinogenicity.[39,40] Following association of immunotoxicity with exposure to different POPs by several authors, detailed studies revealed immune dysfunction as a plausible cause for increased mortality among marine mammals. Furthermore, consumption of diets contaminated with persistent organic pollutants in seals was found to cause vitamin and thyroid deficiencies and concomitant susceptibility to microbial infections and reproductive disorders.[41] Investigators have also reported that immune deficiency was induced in a variety of wildlife species by a number of prevalent POPs, including TCDDs, PCBs, chlordane, HCB, toxaphene and DDT. Chronic exposure to POPs was significantly ($p < 0.05$) correlated with population declines in a number of marine mammals including the common seal, the harbour porpoise, bottlenosed dolphins and beluga whales.[12] More importantly, a direct cause and effect relationship was established in mink and otters, from the Great Lakes basin, between PCB exposure and immune dysfunction, reproductive failure, increased kit mortality, deformations and adult mortality.[42] Reproductive impairment was also shown for seals in the Baltic Sea and the Dutch Wadden Sea and for beluga whales in the St. Lawrence Seaway, Canada. Similarly, environmental levels of PCBs and dioxins were significantly correlated with reduced viability of larvae in several species of fish. Moreover, several papers have documented how organochlorines, particularly DDE, a stable metabolite of DDT, can reduce egg-shell thickness in birds of prey resulting in high off-spring mortality. It is re-assuring to see how, as POP levels have declined in the past few years following legislation restricting their production and usage, populations of affected species have increased again. Examples include harbour seals in the Southeast North Sea, white-tailed eagles in the Baltic and bald-eagles in the Great Lakes.[43]

In addition to reproductive problems, thyroid dysfunction and variations of thyroid hormone levels were reported in polar bears from Svalbard upon exposure to high levels of PCBs and HCB. Incidence of thyroid lesions associated with increased body burdens of POPs was documented in coho, pink and Chinook salmon sampled in the Great Lakes over the last two decades. Disruption of thyroid functions was also reported by several authors in various mammal and bird species upon exposure to high levels of PFOS, PBDEs and HBCD.[44]

Several studies have reported associations between exposure to various POPs and cancer. DDT, PCDD/Fs, toxaphene, PFOS and PBDEs are probably the most obvious examples.Supporting this is the observation that wildlife, including stranded carcasses of St. Lawrence beluga whales, with reported high incidence of tumours have contained significantly elevated concentrations of PCBs, mirex, chlordane and toxaphene.[45]

It is worth noting that the health effects of PCDD/Fs have been the subject of a huge research and review effort, costing in excess of $1 billion, by the US EPA and various European government agencies. These compounds present particular challenges because they occur in mixtures of up to 210 possible congeners. Only 17 individual congeners chlorinated at the 2, 3, 7 and 8 positions (see Figure 4) are of toxicological significance and can act collectively on a range of biological end-points. This structure-dependent toxicity is due to the fact that the toxic effects of PCDD/Fs are mediated by initial interaction with the arylhydrocarbon (Ah) receptor. The toxic effects

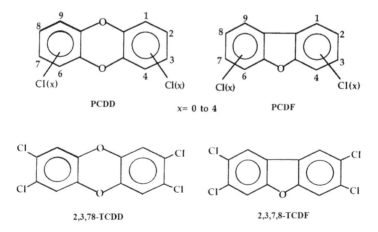

Figure 4 Chemical structures of biologically-relevant PCDD/Fs.

resulting from such compound–receptor interactions are additive, so that toxicity equivalent factors and concentration data should be combined to determine the total toxicity equivalent (ΣTEQ) loading present in the exposed tissue/target organism. This receptor-mediated theory of toxicity also lends a rationale to the considerable species-specific variations in the toxicity of 2,3,7,8-TCDD (guinea-pigs have an LD_{50} of 0.6 µg kg^{-1} body weight (bw), *c.f.* hamsters which have an LD_{50} of 3500 µg kg^{-1} bw), as the precise nature of the Ah-receptor varies widely between species.[15]

5.2 Human Toxicity

As with wildlife species, humans encounter a broad range of environmental exposures to various mixtures of chemicals. Therefore, our current understanding of the toxicological effects of POPs is still limited. The weight of scientific evidence suggests that some POPs have the potential to cause significant adverse effects to human health, at the local level, as well as at regional and global levels, through long-range transport. Initially, POPs emerged to public concern due to both acute and chronic adverse effects associated with occupational and accidental localised high-level exposure. Earliest reports of POPs exposure related to human health impact include an episode of HCB poisoning of food in south-east Turkey, resulting in the death of 90% of those affected and causing an array of adverse health effects including liver cirrhosis, porphyria and urinary, arthritic and neurological disorders.[35] In another acute incident in Italy in 1976, release of 2,3,7,8-TCDD to the environment resulted in a proportional increase of chloracne and an increased leukaemia and thyroid cancer-related mortality.[18] More recently, several studies have reported on dioxin-related health effects, especially for non-carcinogenic end-points, such as immunotoxicity, reproductive diseases and neurotoxicity. Experimental studies on laboratory animals and cell cultures associated exposure to high levels of POPs with a wide range of adverse effects including immune dysfunction, neurological deficits, reproductive anomalies, behavioural abnormalities, and carcinogenesis. Scientific evidence on the relationship between chronic exposure to sub-lethal concentrations of POPs (such as that which would occur as a result of long-range transport) and adverse health effects was more difficult to establish, but gives cause for serious concern. For example, dietary intake of PCBs and PCDD/Fs in fatty fish (salmon and herring) was linked to significant reduction of lymphocytes in Swedish adults,[46] while infants with high organochlorine dietary intake *via* breast milk may experience rates of infection ~10–15 times higher than those with lower intake levels.[47] It should be noted that while several studies have indicated that developing foetus and neonates may be vulnerable to POPs exposure *via* transplacental and lactational transfer of maternal burdens, most authors concluded that the benefits of breast feeding outweigh the risks.

High dietary intake of PCBs by the residents of the Canadian arctic was suggested to place this population at special risk for reproductive and developmental effects. Pre-natal exposure to TCDD/Fs, PCBs and PBDEs was associated with cryptorchidism in male offspring,[45] while high levels of PBDEs in serum were associated with reduced fecundability in American women.[48] PBDEs were also found to interfere with sexual development in rats causing delayed onset of puberty and decreased follicle formation. In addition, PBDEs and HBCD were shown to have thyroid disrupting properties *via* competitive binding to thyroid hormone receptors. However, the most critical toxicological end-point for PBDEs and HBCD appears to be developmental neurotoxicity. Of particular interest are the effects of neonatal exposure in mice, which has been demonstrated to adversely affect learning and memory functions in adult animals. If replicated in humans, such effects are of potential concern.[15]

Associations have been suggested between human exposure to certain chlorinated organic contaminants and cancer. The levels of DDE and PCBs were higher in serum of breast cancer patients than for control subjects, although statistical significance was achieved only for DDE. Women with high blood levels of HCH and PCBs were reported to be at higher risk of breast cancer. Furthermore, deca-BDE and HBCD were reported to induce tumours in rats and mice *via* a non-mutagenic mechanism.[15]

It should be noted that the risk of occupational and accidental high-level exposure is greatest in developing countries, where the use of POPs in tropical agriculture has resulted in a large number of deaths and injuries. For example, a study in the Philippines showed that in 1990 endosulfan became the number one cause of pesticide-related acute poisoning among rice farmers and mango sprayers. Furthermore, occupation, bystander and near-field exposure to toxic chemicals is often difficult to minimize in developing countries. This may be attributed to poor or non-existent training, lack of safety equipment, and substandard working conditions. In addition, the lack of ambient environment monitoring programmes and inconsistencies in medical monitoring, diagnosis, reporting and treatment records contribute to a lack of epidemiological data on POPs exposure from developing countries.[12]

5.3 Exposure Pathways and Combined Adverse Effects

In both humans and wildlife, establishing a direct link between exposure to POPs and health effects is substantially difficult due to the wide diversity of contaminants and different exposure routes. While each POP can pose serious environmental and human health risks on its own, the cumulative adverse effects that a combination of these chemicals may have is potentially of even greater concern.

Wildlife species, some of which are utilized by countries or jurisdictions as "indicators" of risk to humans, are exposed to POPs *via* their natural environment and diet. Laboratory and field studies indicate a variety of adverse effects on wildlife, including immunotoxicity, dermal aberrations, impairment of reproductive performance, deformations, hormonal deficiencies, cancer, increased mortality and overall population declines. Toxic implications of human exposure to various POPs include allergies, hypersensitivity, nervous system damage, reproductive and immune dysfunction, neurobehavioural and developmental disorders, endocrine disruption and cancer.[2]

Since these contaminants accumulate in fatty tissues, the first exposure in humans may occur in the foetus, when a percentage of the maternal "body burden" of accumulated xenobiotics is transported across the placenta. Human post-natal exposure to POPs may start *via* breast-feeding and continue through diet, occupation, natural and indoor environments. Traditionally, human exposure to POPs *via* diet exceeded considerably that received *via* other exposure pathways including inhalation, ingestion and dermal uptake. This may be attributed to the primarily agricultural applications of organochlorine pesticides, which constitute the majority of POPs. However, recent studies revealed that in instances where POPs were mainly used for indoor applications, the resultant elevated indoor levels, combined with the high proportion of time spent indoors (typically 90% or more in Europe and North America) renders pathways of exposure other than diet of importance. For instance, the past widespread application of PCBs in window sealants has led to elevated PCB concentrations in indoor air resulting in significant human exposure *via* inhalation. More recently, further attention has been drawn to the potential of incidental indoor dust ingestion as a pathway of human exposure to brominated flame retardants (*e.g.* PBDEs and HBCD) and perfluorinated chemicals (*e.g.* PFOS) that migrate to dust particles as a result of their widespread deployment in domestic electronic appliances and fabrics. This pathway is of special concern for toddlers due to their lower body weights and likely high dust ingestion rates compared to adults. It should also be noted that occupational and accidental exposure incidents in developing countries are likely to be higher than in developed nations, given the greater limitations on human and financial resources.[1]

Currently, very little is known about the adverse effects arising from contamination by a "cocktail" of pollutants. Bearing in mind the multitude of exposure pathways, the variety of physico-chemical properties and the global distribution of POPs, more research is still required to fully understand the combined adverse effects of POPs on the environment and humans.

6 Long-range Transport of POPs

POPs are currently ubiquitous in their distribution across the globe. Comprehensive monitoring studies report that POPs levels are generally highest in the regions where they were previously released, or are still being released. Environmental concentrations generally decrease with increasing distance from source areas. Therefore, POPs concentrations are most dependent on past and present release rates in the immediate vicinity under investigation, but are also influenced by regional releases.[49] However, the detection of

POPs at surprisingly high concentrations in remote areas, especially the arctic has led to several studies to understand the natural transport processes responsible for such wide distribution throughout the global environment. These processes are:

6.1 Atmospheric Transport

Atmospheric transport and accumulation of POPs (*e.g.* PCBs, DDT, HCHs, and chlordanes) in the polar regions has been extensively documented. Accumulation in polar regions is partly due to global distillation followed by cold condensation of compounds within the volatility range of PCBs and pesticides. These contaminants are continually deposited and re-evaporated and fractionate according to their volatilities. The result is relatively rapid transport and deposition of POPs with intermediate volatility (*e.g.* HCB) and slower migration of less volatile substances such as DDT.[12] Rapid transport of pollutants from south to north, bringing pollutants from midlatitudes to the Arctic in a few days, can also occur. These transport events occur during winter and are caused by transient weather events that result in strong winds directed toward the north in a narrow current lying typically between a strong high-pressure region to the east and a low-pressure system to the west. Although these episodes are infrequent, the amount of POPs deposited can be substantial. To illustrate, Welch *et al.* (1991) documented a long-range transport event that deposited thousands of tonnes of Asian dust onto a region of the central Canadian Arctic over a period of 3 days. Analysis of the resulting brown snow showed high levels of various POPs including PCBs (6.9 ng g^{-1} particles), DDT (4.2 ng g^{-1}), toxaphene (3.0 ng g^{-1}), HCB (0.7 ng g^{-1}) and chlordane (0.6 ng g^{-1}). The authors estimated that this single episode may have contributed up to 10% of the annual loading of DDT to lakes in this region, and up to 1–3% for the other measured POPs (depending on loading scenario assumption).[50]

POPs are mainly classified as semi-volatile compounds. They can either exist in the vapour phase or attached to particles. The relative amounts in the gaseous and particle-associated forms depend mainly on the air temperature and the chemical structure of the compound. Generally, warmer temperatures are found close to the Earth's surface with decreasing latitude (tropics), whereas colder temperatures are found with increasing latitude (polar) or altitude. Characteristic temperatures at which half of a particular compound $(T_{1/2})$ would be present as a gas were estimated to indicate the tendency of a POP to remain as a gas or to attach to particles. The more volatile POPs, such as HCB, reach this temperature at $-36\,^{\circ}$C, DDT $+13\,^{\circ}$C and PCBs with 6 Cl atoms at $+10\,^{\circ}$C. As the number of Cl atoms increases in PCBs and PCDD/Fs, $T_{1/2}$ increases and the molecules tend to partition onto particles. Compounds such as mirex and toxaphene are expected to be found mainly attached to particles (high $T_{1/2}$). Observations show that compounds with low $T_{1/2}$ tend to have the highest potential for long-range atmospheric transport and to remain in the gas phase even at high latitudes, whereas compounds with high values of $T_{1/2}$ tend to remain concentrated close to their sources.[51] It should be noted that deposition on to the surface does not, however, mean that a POP has been permanently removed from the atmosphere. Several processes (*e.g.* photochemical surface reactions) can cause particle-bound POPs to be released and evaporate into the atmosphere. Only through processes such as reaction with OH radicals in the atmosphere, or biologically mediated reactions in soil, is a POP finally gone from the environment. These processes are temperature dependent and proceed faster at higher temperatures, leading to higher POPs degradation in tropical than in polar regions. However, a degradation product of the original POP may also be a POP, or even more toxic in the environment (*e.g.* dieldrin is formed during the degradation of aldrin).[9]

6.2 The "Grasshopper" Effect

As a result of their persistence and semi-volatile nature, POPs have the ability to volatilize in warm regions, be deposited in colder ones, and repeat this process until a location is reached where it is too cold for the POP to again revolatilise. This process of global distillation is known as the "grasshopper" effect, which was hypothesized to result in a net transport of POPs from lower latitudes to high latitudes (polar regions) in a series of jumps. This was attributed to the normal decrease of temperature with increasing latitude, causing compounds to condense on surfaces as they are transported northward by winds associated with passing weather systems.[52]

6.3 Oceanic Currents

Long-range transport of POPs can occur through hydrologic pathways, with POPs entrained on sediment, in microscopic species, or in solution (for the more water-soluble compounds). POPs released or deposited onto terrestrial areas are transported down rivers to oceans, and then potentially to remote locations through oceanic currents. POPs deposited and accumulated on ice in the Arctic can also be transported into the North Atlantic by ice floes. The contribution of hydrologic transport to global POPs pollution has not been quantified, although it is generally considered to be substantially less than atmospheric transport of these semi-volatile, hydrophobic substances. Very little is known about the latitudinal distribution of POPs in seawater. Concentrations of HCH in seawater were found to increase with latitude by roughly a factor of 20 on a transect from the Java Sea to the Beaufort Sea, although a considerable part of this increase may be attributed to atmospheric deposition.[52] Interestingly, the PCB body burdens of deep-sea (>800 m) marine biota collected from the North and South Atlantic regions were significantly higher than those found in surface-living species of the same region. This indicates that the ocean may also act as an ultimate sink for POPs, either through the deposition of dead biological organisms or *via* deep-current circulation.[53]

6.4 Migratory Animals

POPs transport *via* migratory species is considered a potential source of contaminant distribution under the Stockholm Convention. However, the bulk mass of POPs undergoing long-range transport by migratory birds, fish, and/or marine mammals is highly uncertain, and estimates are not available. However, it was estimated that the POPs loading from spawning and dying salmon swimming to localized lakes in Alaska is greater than air deposition levels, indicating the importance of this transport pathway for the studied area. POPs may also be transported by migratory birds to remote rookeries, and from there be transferred to resident species. More importantly for high trophic predators and humans, the oceanic movement of some fish and marine mammals can transfer the POPs loads obtained throughout this migratory journey directly to the end predator.[9]

7 Temporal Trends – Has Legislation Really Worked?

It was not long after Rachel Carson's *"Silent Spring"* sounded the alarms in 1962 about the toxic and persistent nature of some widely used chemicals that a growing number of scientists, international organizations, and non-governmental organizations (NGOs) have devoted a considerable amount of time, efforts, and energy to addressing the hazards and risks posed by persistent organic compounds. For the past few decades, health problems associated with toxic chemical releases into the environment have been a growing major concern in societies across the globe. This has triggered a series of legislation to restrict and/or ban the production and usage of harmful chemicals on various national and regional levels before the Stockholm Convention treaty came into force in 2004. For example, PCBs were banned in the USA by 1978, while penta-BDEs were banned by the European Union in 2003 before the Stockholm Convention legislation for these compounds came into force in 2004 and 2009, respectively. To investigate the effect of legislation on reducing the environmental levels of target compounds, several studies were performed to measure the levels of POPs in various biotic and abiotic samples collected over prolonged periods of time. Results of these studies show the time trends in concentrations of the banned chemicals over a few decades, which provide an indication of the impact of legislation on the environmental levels of POP chemicals.

7.1 Legacy POPs

The weight of available scientific evidence suggests a general decrease in the levels of pesticide POPs in the environment and biota from different parts of the world. Data on concentrations of DDT, PCB, HCHs and HCB in various biota samples collected from several locations in Northern Europe from 1967–1995 showed a significant decrease of all target compounds over the studied period of time. The levels of DDT showed the highest percent decrease with an average of 10% a year, while PCBs decreased by approximately 3% a year. While time series showed that DDT and PCB concentrations in both freshwater and the marine biota decreased concordantly, the slower rate of PCBs decrease was attributed to the continuous release of these contaminants to the environment from various existing sources. For instance, while the use of DDT in the Baltic region was banned in the early 1970s, 70–190 tonnes of PCBs were used in Swedish buildings between 1950–1972, with ongoing leakage of PCBs to the environment from these buildings until 1997. Furthermore, the greater persistence and stability of highly chlorinated PCB congeners may also play a part in their slowly decreasing environmental levels compared to other POPs.[54] A detailed study from the UK reports estimated annual PCBs emission of 799 kg year^{-1} from various sources in 2010 (see Figure 5a), which represents a 29% reduction on the 2004 estimates. PCBs have not been manufactured and used in the UK for many decades but old PCB-containing equipment continues to exist and still accounts for the majority of all air emissions at 59%. Moreover, modelling data over the past 20 years revealed that levels of PCB releases to both air and land in the UK were significantly reduced (see Figure 5b) following an awareness campaign in 2000 that resulted in the identification and removal of large PCB containing equipment from the environment.[55]

This was further confirmed by results of a long time trend study (1991–2008) of various atmospheric POPs sampled in several locations over the Great Lakes region. While the levels of DDT, endosulfan, α-HCH, lindane, chlordane and PCBs showed a significant decrease over the studied period, the overall estimated halving time for PCBs (17 ± 2 years) was the longest among all the studied POPs (see Figure 6), which was also attributed to the ongoing release of PCBs from existing sources as well as their high environmental persistence.[56]

An extended time trend (1930–2004) of environmental levels of selected POPs in Norway was presented *via* analysis of sediment core samples from Lake Ellasjoen, in conjunction with sediment layer dating. Results revealed

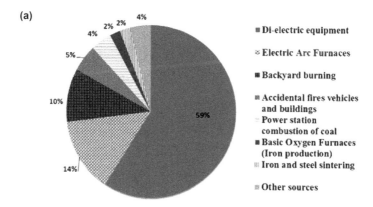

Figure 5a Estimated release of PCBs to air (kg year^{-1}) from the UK in 2010.

(b)

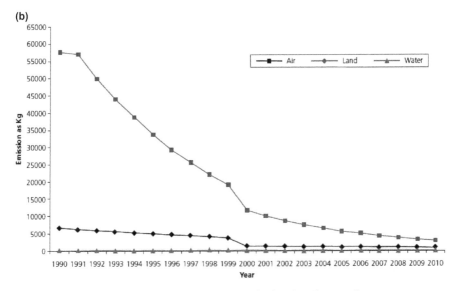

Figure 5b Time trends of PCBs emissions to air, land and water from 1990 to 2010 in the UK.

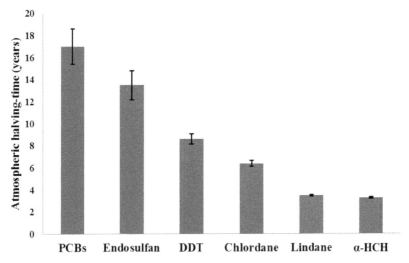

Figure 6 Estimated halving-times (in years) of various POPs in the Great Lakes atmosphere.

PCBs and DDT levels to increase from 1930 to 1970 followed by a significant continuous decrease afterwards, which was mainly attributed to the impact of restrictive legislation on both POPs.[57] The decreasing environmental levels of several legacy POPs in the past few decades was reflected by the overall decreasing trend of PCBs, HCB and DDT concentrations in serum of newborn American infants obtained *via* dried blood spot (DBS) analysis of archived samples (1997–2011).[58]

Temporal trend studies of PCDD/Fs in dated sediment cores have shown a slightly different profile, where the concentrations seem to peak in the 1980s up to the early 1990s followed by a continuous decrease afterwards. During a study period from 1978–2009, the concentrations of PCDD/Fs in Baltic herring samples from the Finnish coasts decreased by ~82%. The percentage decrease was of similar magnitude among young (<5 years) and old (>5 years) herring, although the absolute concentrations were more than twice as high in the old fish. Most of the decrease occurred before 1990s, and during 1993–2009 there was no substantial change in PCDD/Fs concentrations. This was mainly attributed to the substantial decrease in direct emissions of dioxins into water systems achieved by changes in many industrial processes during the 1990s.[59] These findings were supported by results of a UK study showing a significant decrease of PCDD/Fs emissions to the atmosphere and land during the 1990s (see Figure 7) due to stricter controls on industrial sources. The emissions started to level off since 2000, suggesting that it is likely to become increasingly difficult to achieve further significant reductions.[55] Similar results were reported from long-term monitoring programmes in various European countries, while the decrease of open-burning combustion activities has resulted in a marked decline of PCDD/F concentrations in the atmosphere and soils from Eastern China in the past decade.[60]

A recent study from Canada reported a significant decrease of PCDD/Fs in human milk samples collected from various locations between the years 1992 (1.5 ng kg^{-1} milk lipid) and 2005 (0.8 ng kg^{-1} milk lipid) showing a

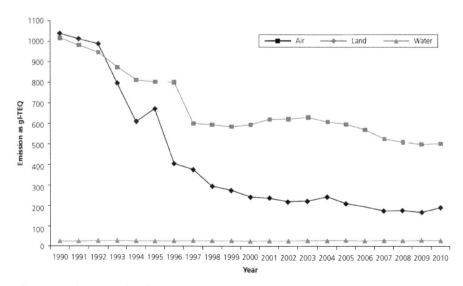

Figure 7   Time trends of PCDD/Fs emissions to air, land and water from 1990 to 2010 in the UK.

steady decline in concentrations by a factor greater than two, indicating the effect of restrictive legislation on open-burning and industrial combustion processes on reducing human body burdens of PCDD/Fs.[61]

7.2   Newer POPs

Studies on penta- and octa-BDEs show a general trend of increasing concentrations up to the early 2000s followed by levelling-off or gradual decrease. In sewage sludge samples from Sweden, levels of penta- and octa-BDEs decreased by ~20% from 2004 to 2010, while deca-BDE concentrations increased by 16% over the same period. On the other side of the Atlantic, archived sludge samples (1975–2008) showed increasing concentrations of penta-BDE that levelled-off around 2000, while deca-BDE concentrations rose from 1995–2008, doubling approximately every 5 years. A similar trend for penta- and octa-BDE congeners was observed in air samples collected from 4 different sites in the UK from 1999–2011 (see Figure 8), where the general decrease in atmospheric PBDE levels after 2003 was linked to the EU restrictions on the production and usage of penta-BDE commercial formulations in 2004.[62]

Few authors have studied the temporal trends of HBCD in the past few years. Concentrations of HBCD in blubber of harbour porpoises (*Phocoena phocoena*) collected between 1994–2003 from UK coasts showed a sharp increase from 2001 onwards. However, levels of HBCD in pooled herring gull eggs from the North and Baltic Sea coasts showed increasing concentrations from 1998 to 2000 before beginning to decline until 2008. Studies from Asian

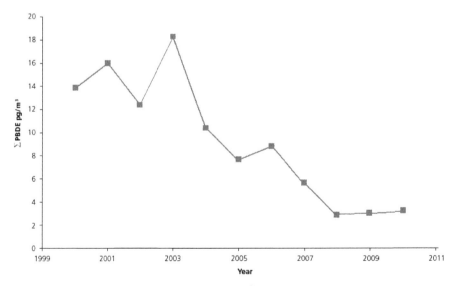

Figure 8   Time trend of PBDEs levels (pg m^{-3}) in the UK atmosphere from 2000 to 2010.

waters showed a continuous increasing trend for HBCD in the past 30 years, reflecting the high usage profile of HBCD in Asia.[63]

The effect of legislation on environmental levels of lindane was highlighted by a 12 year study (1993–2005) of α-HCH and lindane (γ-HCH) in air samples collected weekly from the two Arctic monitoring stations of Alert, Nunavut, Canada, and Zeppelin Mountain, Svalbard, Norway. Results revealed a general decline in the atmospheric concentrations of both α- and γ-HCH over the time series at both Arctic sites (see Figure 9). However, the steeper decline observed at Zeppelin (Norway) reflects the declining usage of lindane, particularly in Europe since 1998. Meanwhile, the Canadian usage actually increased during the 1990s, reaching a maximum of 558 tonnes in 1994, followed by a gradual decrease until the complete ban was introduced in 2004 in Canada.[64]

Long-term studies on PFOS have revealed contrasting temporal trends. For example, Niagara River suspended solids (1981–2006) demonstrated a peak of PFOS concentration in 2001 (1.1 ng g^{-1}) followed by a decrease from 2001 to 2006 (half-life = 9 years), While three sediment cores from Western, Central, and Eastern Lake Ontario showed increasing temporal trends of PFOS between 1984–2004. This variation was attributed to ongoing emissions from existing sources and different rates of environmental transformation/degradation of PFOS precursors.[65] Dated sediment core samples (1950–2004) showed an evident decreasing trend of PFOS concentrations from the early 1990s. This was linked to the shift from PFOS-based products to telomere-based products after the phase-out of PFOS products in 2001.[66] In Europe, PFOS concentrations in archived guillemot eggs from the Baltic

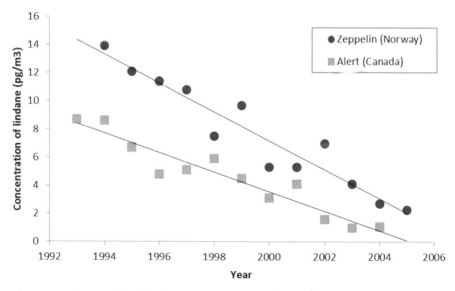

Figure 9 Time trends of lindane concentrations (pg m^{-3}) in the Arctic atmosphere from Alert (Canada) and Zeppelin (Norway) between 1993–2005.

Sea (1968–2003) showed a sharp peak in 1997 followed by slow decrease until 2002. However, this could not be linked to the PFOS phase-out, which occurred at the end of this period.[67]

8 Climate Change and POPs – Future Scenario

Climate change induced by anthropogenic activities is one of the major global problems, with significant social, economic, ecological, and health related impacts. There is a growing body of evidence that climate change will have broad negative impacts on the behaviour and fate of environmental pollutants. Climate variability forecasted by climate change scenarios shows that the atmospheric temperature is expected to increase by 1.8–4.0 °C by the end of the century. In addition, climate change will affect the atmospheric and oceanic circulation patterns, and the precipitation rate. This is likely to affect the environmental behaviour and distribution of POPs and increase exposure to POPs in some regions (*e.g.* the Arctic).[68] An increase in temperature of 1 °C increases the volatility of a typical POP by 10–15%. At the local level, atmospheric temperatures can, however, increase much more: for example, an increase from 10 °C to 15 °C can double the vapour pressure of PCB-153.[69] Global warming will therefore lead to enhanced volatilization of POPs from contaminated environments, stockpiles and open applications.

While the increase of secondary POPs emissions from contaminated environments as a result of higher temperature was documented by several studies, an elevated temperature could enhance the rate of atmospheric degradation of POPs, offsetting such increase in POPs atmospheric concentrations. These processes provide opposing impacts on POPs atmospheric levels as a result of anticipated changes in temperature and the net result is difficult to predict. It should also be noted that degradation of POPs often includes the formation of products that are structurally similar to the parent compound and may also exhibit toxic and persistence properties (*e.g.* DDE and DDD, the major products of DDT degradation).

Another important factor affecting global transport of POPs is the wind. As a result of climate change, modified wind patterns and higher wind speeds may cause faster and more efficient atmospheric long-range transport of POPs. Desertification induced by climate change might also enhance the global distribution of POPs through Sahara dust transport associated with altered wind fields.[68]

Climate change will also lead to changes in precipitation patterns. A decrease in the precipitation rate may result in enhanced volatilization of POPs to the atmosphere, while a higher intensity and frequency of rain events was found to enhance wet deposition of airborne POPs. The Intergovernmental Panel on Climate Change (IPCC) reported that extreme precipitation events are expected to become more frequent, widespread, and intense. Such extreme events are expected to have a dramatic impact on the remobilization and redistribution of POPs. As storms and rainfall events become more intense and frequent, increasing amounts of POPs bound to soil particles could be transported by erosion and transferred to rivers, lakes and oceans, making them available to the aquatic environments. Flooding events may also contribute to the dissemination and redistribution of particle-bound POPs from soil and sediment.[70]

Another major impact on global POPs levels is likely to be exerted by snow melting. Abundant snow deposition may lead to a large contaminant release during snowmelt, with the potential to affect drinking and agricultural water supplies. Of particular interest is the melting of glaciers, which cover most of the polar regions and Greenland. As a result of global distillation (see Section 6.2) and slower degradation at low temperatures, the Arctic acts as a sink for POPs. Therefore, melting of polar icecaps is likely to release more pollutants making them available for transfer to the atmosphere or to aquatic and terrestrial ecosystems. The remobilization of POPs from the Arctic snow, ice, ocean, and presumably soil reservoirs back to the atmosphere as a result of climate change has already been reported.[71] Release of POPs from mountain glaciers to Alpine lakes has also been observed.[72] The released POPs could be carried and redistributed to other geographical regions *via* the atmosphere or oceanic currents. Furthermore, polar ice melting and increased evaporation rates affect ocean salinity, which in turn influences the solubility of organic chemicals (POPs solubility decreases when water salinity is higher) and, consequently, their air–water partitioning behaviour. The changes in salinity and wind patterns are likely to affect the oceanic current which can have a major influence in the transport and distribution of more water-soluble POPs (*e.g.* HCH, lindane and PFOS).[68]

According to the IPCC, 20–30% of plant and animal species assessed so far are likely at risk if increases in global average temperature are greater than 1.5 °C–2.5 °C. Therefore, climate change is likely to affect the transfer of POPs through the food chains and alter their bioaccumulation and biomagnification in top predators.[70] This factor is also compounded by expected changes in the species migration patterns, which in turn can affect the biotic transport pattern of POPs *via* migratory animals.

9 Managing the Problem

The main approach to reduce the current levels of POPs is to prevent/control their release to the environment. With the exception of a few, very limited specific applications (*e.g.* the use of DDT for disease vector control), the production and use of intentionally produced POPs is banned under the Stockholm Convention. The production and marketing of new chemicals with POP characteristics could be prevented by more restrictive legislation. For instance, the production or import of new POPs in Europe could be prevented *via* the REACH regulations. Under this framework, companies that manufacture or import more than one tonne of a chemical per year are required to register it in a central database. Substances of very high concern, including persistent, bio-accumulating and toxic substances (PBT) and very persistent and very bio-accumulating substances (vPvB) require author-ization for particular uses. The European Chemicals Agency (ECHA) has the right to request further testing if a substance is suspected to exhibit POP characteristics.

As a priority action, the Stockholm Convention requires the identification and safe management/disposal of stockpiles containing or consisting of POPs. The parties to the Convention are requested to dispose of waste containing, consisting of, or contaminated with POPs in such a way that POPs are destroyed or irreversibly transformed. The Convention also en-courages Parties to develop strategies for identifying sites contaminated with POPs; if remediation is necessary then it must be done in an environ-mentally sound manner.

Unintentionally produced POPs are continuously formed and released in small quantities. Therefore, the Stockholm Convention requires Parties to take sufficient measures to reduce the total emissions of these POPs to the environment. This can be mainly achieved *via* applying Best Available Techniques (BAT) and Best Environmental Practices (BEP).[3]

Guidelines on BAT/BEP have been developed under the Convention and are currently being introduced to the Parties. These may include end-of-pipe solutions or the development of substitute or modified materials, products and processes that avoid the formation and release of POPs. A range of ac-tivities focused on assisting developing and transition economy countries to meet their obligations under the Stockholm Convention are currently being led by international organizations. For instance, the United Nations Indus-trial Development Organization (UNIDO) is supporting the development of demonstration projects aimed at promoting the uptake of BAT and BEP in the metal industry. However, a lot of effort is still needed to introduce and apply processes and technologies preventing POPs from being formed and transferred to air, water, soil and waste streams, especially in developing countries.

10   Concluding Remarks

Since the Stockholm Convention came into force in 2004, there has been a substantial progress in our understanding of the environmental sources, fate, behaviour, and impacts of POPs. Meanwhile, new challenges have emerged including the listing of more chemicals as POPs, alongside the development and implementation of new procedures and strategies to tackle this global problem.

While the inclusion of harmful chemicals under the POPs list is likely to have an immediate impact on the production and usage of these chemicals, the levels of POPs in the environment and humans are not expected to show dramatic decreases within a short time period. Further reductions in con-tamination are likely to be limited for as long as goods and materials pro-duced before the ban remain in use or are otherwise not safely disposed of. One example is the existence of substantial stockpiles of unused organochlorine pesticides in developing countries. Another source of POPs emission to the environment is the uncontrolled e-waste recycling activities in Asia. More pertinent to the developed world is the burden of POPs associated with consumer goods, appliances and materials contained within the built environment, during their usage life and following their disposal. Where the "turnover" or replacement time of such items is long, they will represent an ongoing emission source of POPs for some time after implementation of the ban on manufacture.

It should also be noted that while the vast majority of the current knowledge about POPs concerns the developed world, there is evidence that substantial problems exist in developing countries. The nature and magnitude of POPs-associated problems faced by developing countries is often complicated by the limited human and financial resources required to tackle these problems.

Given their potential for long-range transport and wide global distribution, no single country can solve its national POPs problems alone. Therefore, the international community needs to respond in a coherent and cost-effective fashion, with measures acceptable from public health and socio-economic perspectives. Furthermore, there is an urgent need for creating, or strengthening international coordinating mechanisms to prevent further release of POPs to the environment, provide alternatives for the banned chemicals and implement strategies for safe disposal of the existing stockpiles.

References

1. S. Harrad, *Persistent Organic Pollutants*, Wiley, 2009.
2. Y. M. Kuo, M. S. Sepulveda, T. M. Sutton, H. G. Ochoa-Acuna, A. M. Muir, B. Miller and I. Hua, *Ecotoxicology*, 2010, **19**, 751–760.
3. United Nations Environment Programme (UNEP), http://chm.pops.int/TheConvention/POPsReviewCommittee/OverviewandMandatc/tabid/2806/Default.aspx, 2014.
4. Annex D: Information requirements and criteria for the proposal and screening of proposed perssistent organic pollutants, http://www.pops.int/documents/meetings/ceg2/en/ceg2-2e.html.
5. F. Wania and D. Mackay, *Environ. Pollut.*, 1999, **100**, 223–240.
6. W. A. Schimpf, in *Chemistry of Crop Protection*, Wiley-VCH Verlag GmbH & Co. KGaA, 2004, DOI: 10.1002/3527602038, ch. 4, pp. 40–53.
7. V. Jain, *Environ. Sci. Technol.*, 1992, **26**, 226–228.
8. G. Duca and I. Barbarasa, in *The Fate of Persistent Organic Pollutants in the Environment*, ed. E. Mehmetli and B. Koumanova, Springer Netherlands, 2008, DOI: 10.1007/978-1-4020-6642-9_2, ch. 2, pp. 13–20.
9. US EPA, http://nepis.epa.gov/Exe/ZyPDF.cgi/20008M5X.PDF?Dockey=20008M5X.PDF, 2002.
10. M. S. Majewski and P. D. Capel, *Pesticides in the Atmosphere: Distribution, Trends, and Governing Factors*, Taylor & Francis, 1996.

11. J. P. Meador, *Environmental Contaminants in Wildlife: Interpreting Tissue Concentrations*, Taylor & Francis, 1996.
12. L. Ritter, K. R. Solomon and J. Forget, *A review of selected persistent organic pollutants*, www.who.int/ipcs/assessment/en/pcs_95_39_2004_05_13.pdf, 1995.
13. UNEP, *Document UNEP/POPS/INC.5/7*. http://www.chem.unep.ch/pops/POPs_Inc/INC_5/finalreport/en/inc5efinrep.PDF, 2000.
14. H. van den Berg, *Environ. Health Perspect.*, 2009, **117**, 1656–1663.
15. S. Harrad and M. A. Abdallah, in *Pollution: Causes, Effects and Control*, ed. R. M. Harrison, The Royal Society of Chemistry, 2014.
16. C. Xhrouet, C. Pirard and E. De Pauw, *Environ. Sci. Technol.*, 2001, **35**, 1616–1623.
17. R. Weber and H. Hagenmaier, *Chemosphere*, 1999, **38**, 529–549.
18. P. A. Bertazzi, I. Bernucci, G. Brambilla, D. Consonni and A. C. Pesatori, *Environ. Health Perspect.*, 1998, **106**, 625–633.
19. Y. Leon Guo, M.-L. Yu and C.-C. Hsu, in *Dioxins and Health*, John Wiley & Sons, Inc., 2005, DOI: 10.1002/0471722014.ch22, pp. 893–919.
20. R. E. Alcock and K. C. Jones, *Chemosphere*, 1997, **35**, 2317–2330.
21. V. Ivanov and E. Sandell, *Environ. Sci. Technol.*, 1992, **26**, 2012–2017.
22. G. M. Currado and S. Harrad, *Environ. Sci. Technol.*, 1998, **32**, 3043–3047.
23. S. Harrad, S. Hazrati and C. Ibarra, *Environ. Sci. Technol.*, 2006, **40**, 4633–4638.
24. A. Jamshidi, S. Hunter, S. Hazrati and S. Harrad, *Environ. Sci. Technol.*, 2007, **41**, 2153–2158.
25. US EPA, An exposure assessment of polybrominated diphenyl ethers. EPA/600/R-08/086F, 2010.
26. EU Risk Assessment Report, *CAS number: 1163-19-5, EINECS Number: 214-604-9*, 2004.
27. EU Risk Assessment Report, *European Commission, Joint Research Centre, European Chemicals Bureau, EUR20402EN, 2002.*, 2002, Vol. 17.
28. G. Suzuki, A. Kida, S. Sakai and H. Takigami, *Environ. Sci. Technol.*, 2009, **43**, 1437–1442.
29. T. F. Webster, S. Harrad, J. R. Millette, R. D. Holbrook, J. M. Davis, H. M. Stapleton, J. G. Allen, M. D. McClean, C. Ibarra, M. A. Abdallah and A. Covaci, *Environ. Sci. Technol.*, 2009, **43**, 3067–3072.
30. EU Risk Assessment Report, *CAS Number: 32534-81-9, EINECS Number: 251-084-2*, 2000.
31. EU Risk Assessment Report, *CAS No: 32536-52-0, EINECS No: 251-087-9*, 2002.
32. KEMI (National Chemicals Inspectorate), *R044_0710_env_hh.doc; Sundbyberg, Sweden* 2008.
33. C. H. Marvin, G. T. Tomy, J. M. Armitage, J. A. Arnot, L. McCarty, A. Covaci and V. Palace, *Environ. Sci. Technol.*, 2011, **45**, 8613–8623.
34. A. G. Paul, K. C. Jones and A. J. Sweetman, *Environ. Sci. Technol.*, 2009, **43**, 386–392.

35. H. Peters, D. Cripps, A. Gocmen, G. Bryan, E. Erturk and C. Morris, *Ann. N. Y. Acad. Sci.*, 1987, **514**, 183–190.
36. R. E. Bailey, *Chemosphere*, 2001, **43**, 167–182.
37. S. H. Safe, *Eur. J. Lipid Sci. Technol.*, 2000, **102**, 52–53.
38. A. Abelsohn, B. L. Gibson, M. D. Sanborn and E. Weir, *Can. Med. Assoc. J.*, 2002, **166**, 1549–1554.
39. J. R. Allen, W. A. Hargraves, M. T. S. Hsia and F. S. D. Lin, *Pharmacol. Ther.*, 1979, 7, 513–547.
40. A. Parkinson and S. Safe, in *Polychlorinated Biphenyls (PCBs): Mammalian and Environmental Toxicology*, ed. S. Safe, Springer Berlin Heidelberg, 1987, vol. 1 , ch. 3, pp. 49–75.
41. D. C. G. Muir, R. Wagemann, B. T. Hargrave, D. J. Thomas, D. B. Peakall and R. J. Norstrom, *Sci. Total Environ.*, 1992, **122**, 75–134.
42. C. D. Wren, *J. Toxicol. Environ. Health*, 1991, **33**, 549–586.
43. K. C. Jones and P. de Voogt, *Environ. Pollut.*, 1999, **100**, 209–221.
44. P. O. Darnerud, *Int. J. Androl.*, 2008, **31**, 152–160.
45. L. Ritter, The International Programme on Chemical Safety and Inter-Organization Programme for the Sound Management of Chemicals, *A Review of Selected Persistent Organic Pollutants: DDT, Aldrin, Dieldrin, Endrin, Chlordane, Heptachlor, Hexachlorobenzene, Mirex, Toxaphene, Polycholorinated Biphenyls, Dioxins and Furans*, 1995.
46. B. G. Svensson, T. Hallberg, A. Nilsson, A. Schütz and L. Hagmar, *Int. Arch. Occup. Environ. Health*, 1994, **65**, 351–358.
47. E. Dewailly, A. Nantel, J. P. Weber and F. Meyer, *Bull. Environ. Contam. Toxicol.*, 1989, **43**, 641–646.
48. K. G. Harley, A. R. Marks, J. Chevrier, A. Bradman, A. Sjodin and B. Eskenazi, *Environ. Health Perspect.*, 2010, **118**, 699–704.
49. O. I. Kalantzi, R. E. Alcock, P. A. Johnston, D. Santillo, R. L. Stringer, G. O. Thomas and K. C. Jones, *Environ. Sci. Technol*, 2001, **35**, 1013–1018.
50. H. E. Welch, D. C. G. Muir, B. N. Billeck, W. L. Lockhart, G. J. Brunskill, H. J. Kling, M. P. Olson and R. M. Lemoine, *Environ. Sci. Technol.*, 1991, **25**, 280–286.
51. F. Wania, J. Axelman and D. Broman, *Environ. Pollut.*, 1998, **102**, 3–23.
52. F. Wania and D. Mackay, *Environ. Sci. Technol.*, 1996, **30**, A390–A396.
53. O. Froescheis, R. Looser, G. M. Cailliet, W. M. Jarman and K. Ballschmiter, *Chemosphere*, 2000, **40**, 651–660.
54. A. Bignert, M. Olsson, W. Persson, S. Jensen, S. Zakrisson, K. Litzen, U. Eriksson, L. Haggberg and T. Alsberg, *Environ. Pollut.*, 1998, **99**, 177–198.
55. DEFRA, *Department for Environment, Food and Rural Affairs*, http://www.defra.gov.uk/environment/quality/chemicals, 2012.
56. M. Venier and R. A. Hites, *Environ. Sci. Technol.*, 2010, **44**, 8050–8055.
57. A. Evenset, G. N. Christensen, J. Carroll, A. Zaborska, U. Berger, D. Herzke and D. Gregor, *Environ. Pollut.*, 2007, **146**, 196–205.
58. W. L. Ma, C. Gao, E. M. Bell, C. M. Druschel, M. Caggana, K. M. Aldous, G. M. B. Louis and K. Kannan, *Environ. Res.*, 2014, **133**, 204–210.

59. R. Airaksinen, A. Hallikainen, P. Rantakokko, P. Ruokojarvi, P. J. Vuorinen, R. Parmanne, M. Verta, J. Mannio and H. Kiviranta, *Chemosphere*, 2014, **114**, 165–171.
60. P. Wang, H. Zhang, J. Fu, Y. Li, W. Thanh, Y. Wang, D. Ren, P. Ssebugere, Q. Zhang and G. Jiang, *Environ. Sci.: Processes Impacts*, 2013, **15**, 1897–1903.
61. J. J. Ryan and D. F. K. Rawn, *Chemosphere*, 2014, **102**, 76–86.
62. R. J. Law, A. Covaci, S. Harrad, D. Herzke, M. A. E. Abdallah, K. Femie, L.-M. L. Toms and H. Takigami, *Environ. Int.*, 2014, **65**, 147–158.
63. R. J. Law, D. Herzke, S. Harrad, S. Morris, P. Bersuder and C. R. Allchin, *Chemosphere*, 2008, **73**, 223–241.
64. S. Becker, C. J. Halsall, W. Tych, R. Kallenborn, Y. Su and H. Hung, *Atmos. Environ.*, 2008, **42**, 8225–8233.
65. A. L. Myers, P. W. Crozier, P. A. Helm, C. Brimacombe, V. I. Furdui, E. J. Reiner, D. Burniston and C. H. Marvin, *Environ. Int.*, 2012, **44**, 92–99.
66. Y. Zushi, M. Tamada, Y. Kanai and S. Masunaga, *Environ. Pollut.*, 2010, **158**, 756–763.
67. K. E. Holmstrom, U. Jarnberg and A. Bignert, *Environm. Sci. Technol.*, 2005, **39**, 80–84.
68. R. Kallenborn, K. Borgå, J.H. Christensen, M. Dowdall, A. Evenset, J. Ø. Odland, A. Ruus, K. Aspmo Pfaffhuber, J. Pawlak, L.-O. Reiersen, AMAP, *Combined Effects of Selected Pollutants and Climate Change in the Arctic Environment*, http://www.amap.no/documents/download/978, 2011.
69. L. Lamon, H. von Waldow, M. MacLeod, M. Scheringer, A. Marcomini and K. Hungerbuehler, *Environ. Sci. Technol.*, 2009, **43**, 5818–5824.
70. IPCC, *Intergovernmental Panel on Climate Change Fourth Assessment Report: Climate Change 2007 (AR4)*, http://www.ipcc.ch/publications_and_data/publications_and_data_reports.shtml, 2007.
71. J. Ma, H. Hung, C. Tian and R. Kallenborn, *Nat. Clim. Change*, 2011, **1**, 255–260.
72. C. Bogdal, P. Schmid, M. Zennegg, F. S. Anselmetti, M. Scheringer and K. Hungerbuehler, *Environ. Sci. Technol.*, 2009, **43**, 8173–8177.

Emerging Chemical Contaminants: How Chemical Development Outpaces Impact Assessment

SHANE A. SNYDER* AND TARUN ANUMOL

ABSTRACT

The term "emerging contaminants" is used both to describe substances that have been recently identified in the environment, as well as substances previously identified and quantified but for which new information related to health effects, persistence and/or prevalence has been discovered. Examples of such substances include pharmaceuticals and personal care products, perfluorinated compounds and endocrine disrupting chemicals. Using these classes of compounds, as well as other contaminants of emerging concern as examples, the changing regulatory environment and advances in analytical methodologies in recent decades are reviewed and discussed in the context of the implications for water sustainability.

1 Chemical Ubiquity

Chemicals are a necessary and fundamental aspect of human existence. Indeed, our bodies are composed of various types of simple and complex organic structures that have evolved over millions of years. However, synthetic chemistry has created structures that are not known to exist in nature, yet can have profound impacts on environmental and public health by

---

*Corresponding author.

Issues in Environmental Science and Technology No. 40
Still Only One Earth: Progress in the 40 Years Since the First UN Conference on the Environment
Edited by R.E. Hester and R.M. Harrison
© The Royal Society of Chemistry 2015
Published by the Royal Society of Chemistry, www.rsc.org

design or by chance. Large-scale chemical synthesis did not largely come of age until the late 19th century with the development of coal-tar feedstocks initially used for dye production. Since the 20th century, chemical synthesis has grown at a staggering rate, with more than 65 million chemical compositions available commercially.[1] Today, synthetic chemicals can be found in nearly any location on earth and in spacecraft beyond the earth's atmosphere. With continued advances in analytical technologies, increasing numbers of synthetic and natural chemicals are being identified in the environment at extremely sensitive detection limits. As progress in analytical techniques is made and additional chemistries are discovered, a conundrum exists. The very products that have been developed and utilized to improve human existence will be detected in locations unrelated to their intended use, for instance, trace levels of pharmaceuticals detected in rivers and streams. Additionally, some of the properties that make a particular chemical or chemical mixture effective for the intended use may dictate a level of persistence that has a side-effect of chemical resilience in environmental compartments. A difficult question going forward is how will we balance detectability of chemicals in the environment with increasing demands for effective chemicals to improve human lives and safety?

The term "emerging contaminants" can be vastly confusing and all encompassing. The simplest definition of an emerging contaminant is a substance (or organism) that has been recently identified in the environment. However, the term is often extended to substances that have been previously identified and quantified, but for which new information related to health effects, persistence, and/or prevalence has been discovered. The US Environmental Protection Agency (US EPA), and others, use the term "contaminants of emerging concern" or CECs.[2,3] According to the US EPA, CECs are environmental contaminants for which "the risk to human health and the environment associated with their presence, frequency of occurrence, or source may not be known".[4] As examples of CECs, the US EPA specifically names pharmaceuticals and personal care products (PPCPs) and perfluorinated compounds.[4] Endocrine disrupting chemicals (EDCs) are also commonly considered emerging contaminants; however, EDCs are substances exerting a type of toxicity rather than a finite list of chemicals. A similar term would be carcinogens, which defines the types of biological effects elicited from exposure rather than a finite term like "PPCPs" which specifies the classes of chemicals based on use. Regardless of definition employed, an emerging contaminant suggests a particular view at a specific point in time. Thus, when a substance is initially discovered in the environment it may be considered an emerging contaminant, but decades later the same substance may still be considered emerging if new information regarding toxicity or occurrence is discovered.

Some absolute certainties in moving forward will include:

1. More synthetic chemicals, and transformation products, will be discovered in the environment;

2. Analytical methods will continue to evolve, providing higher sensitivity and selectivity;

3. Chemical mixture toxicity will need to be addressed in risk assessment; and

4. Traditional animal testing paradigm will not be feasible for addressing all chemicals and chemical mixtures detected in the environment.

This chapter will provide a historical perspective, with key examples, of how synthetic chemicals came to be released and detected in the environment; a look into the regulatory framework in the US; and a view towards future opportunities to more comprehensively evaluate the relevance of detections for human and environmental health.

2   Birth of the Chemical Industry

While various early reports exist on small-scale chemical synthesis, the nascency of mass produced synthetic chemicals was in the 1800s utilizing coal-tar, which was initially regarded as a useless by-product of coking coal.[5] Synthesis of the organic dyes from coal-tar began with the discovery of aniline purple, also known as mauve, by W.H. Perkin in 1856 which was quickly patented and produced on an industrial scale.[6] Numerous other synthetic dyes rapidly emerged from various derivatives of coal-tar and scientists worked to elucidate, isolate, and purify the various components of coal-tar. By the start of the 20th century, Germany was producing 90% of the worldwide output of synthetic dyes.[7] The synthesis of organic dyes led to several discoveries of medicinal properties and the subsequent establishment of the first synthetic pharmaceuticals, which in turn gave birth to some of the first pharmaceutical companies.[8]

The continued development of coal tar industrial usage, especially for organic dyes, yielded an increased interest in the toxicological relationships to occupational exposures. In addition, several studies were underway to investigate the potential of numerous isolates from coal-tar as medicines and pesticides. One natural fit to the dye industry was the search for chemicals that would deter moths from damaging clothing. This search led to the initial discovery that malachite green dye would largely protect fabrics from moths. In 1939, 4,4'-dichlorodiphenyltrichloroethane (DDT) was discovered by Paul Müller of J.R. Geigy AG (Geigy) to be highly effective against moths and other insects after having investigated hundreds of ineffective compounds.[7] With a low cost of production and a high degree of effectiveness, DDT rapidly became consider a "miracle" in combatting malaria and typhus lice. During World War II, DDT was used extensively in applications directly to personnel for lice and for aerial spraying for mosquitos. From an initial field test in June of 1943 through use through January of 1944, DDT was effective for completely ending the typhus epidemic in Italy. DDT was also widely distributed in Asian campaigns from the allied forces, where DDT would be sprayed in advance of troops to control malaria-carrying mosquitos. Because of the great success of DDT, Paul Müller was awarded the Nobel Prize in Medicine in 1948. In his acceptance speech, Müller explained that the ideal pesticide should have seven specific qualities:[7]

1. Efficient insect toxicity;
2. Rapid onset of toxic action;
3. Little or no mammalian or plant toxicity;
4. No irritant effect and no, or only a faint, odor (in any case not an unpleasant one);
5. The range of action should be as wide as possible, and cover as many arthropods as possible;
6. Long, persistent action, *i.e.* good chemical stability; and
7. Low price.

Interestingly, the "long persistent action" and highly stable structure desired by Müller also became one of the greatest legacies of what is now broadly considered the most famous synthetic environmental contaminant in human history. By the end of the 1940s, concerns were already being expressed regarding the potential for DDT bioaccumulation and impacts to ecological health.[7] As early as the 1950s, structural derivatives and isomers of DDT were found to have significant estrogenic properties.[9] Later studies realized that DDT technical mixture contained approximately 80% p,p-DDT and 20% o,p-DDT, with the o,p-DDT exhibiting significantly greater estrogenicity.[10] While DDT and its transformation products have now been detected globally through atmospheric and food chain transfer, there are continued arguments in favor of DDT use in areas where malaria continues to cause illness and death.[11] Indeed, we face challenging questions when a product has demonstrated effectiveness to prevent/control spread of disease, yet has known environmental consequences.

3   Birth of the Pharmaceutical Industry

In ancient times, humans became aware of the medicinal and aesthetic potential of certain chemicals isolated from natural products, which were used for perfumes,[12] medicines,[12] and paints/dyes.[13] A Neanderthal skeleton from Iraq, which is believed to be approximately 60 000 years old, was found in a cave where flowers of medicinal value also were located.[14] Ginseng and other herbal remedies are believed to have been used in China for nearly 5000 years.[15] The Ebers Papyrus, which has been dated to 1550 BC, provides nearly 1000 prescriptions and remedies used in Ancient Egypt to treat a variety of health issues ranging from asthma to gonorrhoea.[16] Thus, a wide range of naturally-occurring chemicals were realized to have the capacity to alter human health and were thereby exploited for ancient remedies.

Chemicals isolated and purified from natural sources were not only used as pharmaceuticals, but also as poisons for weapons and intentionally induced toxicity. In Homer's classic poem *The Iliad*, written in approximately

1250 BC, poisoned arrows from Heracles were used in the Trojan War. As another example, batrachotoxin isolated from the skin secretions of a brightly coloured frog (*Phyllobates aurotaenia*) have been used to poison the tips of darts and arrows for centuries.[17] In 399 BC, the classical Greek philosopher Socrates died from a self-administered solution of hemlock.[18] Throughout human history, a vast wealth of potions, medicines, and poisons have been realized, yet it was not until the 19th century that synthetic organic chemistry would produce structures in large quantities that are not available, or discovered, in nature.

Although chloral hydrate is often considered the first synthetic pharmaceutical, salicylic acid (aspirin) is without question the most famous and successful of the early synthesized medicinal structures. Willow leaves have been known for their analgesic effects for thousands of years.[19] By the early 1800s willow bark had become an important commodity and was becoming increasingly scarce due to importation challenges from Napoleon's continental blockade to restrict British trade.[20] Through the work of various scientists, the medicinal agent in the willow bark was identified as the glycoside salicin, which was subsequently split to form salicylic acid. By 1874, the structure had been elucidated and commercial production was already underway.[20]

A report published in 1876 specifically linked the use of salicylic acid to relief of pain from rheumatism;[21] however, others had noted that salicylic acid caused stomach irritation and had an unpleasant taste. This inspired Felix Hoffman to develop acetylsalicylic acid, which retains the analgesic effects without the side effects noted previously.[20] The structure was registered as "aspirin" by Friedrich Bayer & Co, Elberfeld, Germany (Bayer) in 1899 and has since been produced by numerous companies around the world. Since the initial uses as an analgesic, aspirin has been determined to provide benefits in the treatment, and/or prevention, of ailments such as certain cardiac diseases and cancers.[19,22] Thus, aspirin has become an essential global medicine with an estimated consumption of more than 120 billion tablets per year.[23] Likewise, traces of aspirin have now been detected in wastewater discharges;[24] therefore, aspirin could be considered an emerging contaminant despite more than 100 years of commercial production.

4 Discovery of Estrogens in Water

Endocrine disruption has been one of the most concerning issues related to environmental and public health. Seminal publications from the late 1990s demonstrated that estrogenic substances occurred in wastewater effluents at concentrations capable of impacting fish endocrinology.[25–28] Trace levels of natural and synthetic estrogens were subsequently linked to effects observed in fish.[29,30] As conclusive evidence, an experimental lake in Canada was intentionally dosed with ng L^{-1} levels of the oral contraceptive ethynylestradiol and clear reproductive effects were documented in fish.[31] While these findings, and others, greatly surprised scientists, regulators, and the public, in retrospect, it is more surprising that we did not predict this much further in advance.

Ernest Starling first coined the term "hormone" to describe the chemical messengers which coordinate and stimulate growth and development within the body.[32] Throughout the early 1990s, scientist began to unravel the structure and functions of steroid hormones. In 1930, Walker and Janney published a manuscript which associated green pigmentation in plants with estrogenicity in animals.[33] By 1934, scientists began to experiment with various synthetic substances and had determined that chemical composition, particularly a polycyclic ring structure containing a phenolic oxygen, could induce estrogenic effects in animals.[34] Ethynylestradiol (EE2) was first synthesized in 1938, yet its application as a pharmaceutical for birth control was not realized for nearly 20 years.[35] A publication in *Science* from 1946 demonstrated that molecular geometry played a major role in the estrogenic potency of synthetic molecules, with maximum estrogenicity arising from relatively rigid molecules with hydrogens spaced by 8.55 angstroms.[36] Throughout the 1950s, various research groups experimented with synthetic progestins in attempts to develop an oral birth control pharmaceutical, yet it was not initially realized that the most effective progestins evaluated were actually contaminated with the estrogen mestranol (a precursor to EE2).[37] Research continued toward the development of an oral birth control contraceptive through the late 1950s.[38] The pharmaceutical Enovid (containing both progestin and mestranol) was approved by the US Food and Drug Administration for gynecological disorders in 1957 and for birth control in 1960.[39]

By 1965, researchers at Harvard University demonstrated that both endogenous steroid hormones and synthetic hormones used in pharmaceuticals were not completely eliminated during wastewater treatment.[40] This finding was subsequently confirmed and expanded by the US Department of the Interior, who concluded that synthetic hormones were generally less biodegradable than natural hormones.[41] In the early 1970s, the US Environmental Protection Agency (US EPA) had detected "steroids, drugs, and drug metabolites" in both influents and effluents of municipal wastewater treatment plants in the US.[42] Researchers also reported the occurrence of certain pharmaceuticals in the effluents of a large US wastewater plant in 1977.[43] The US EPA investigated the occurrence and removal efficacy of 18 natural and synthetic hormones from 14 wastewater treatment plants located within a 80 km radius of Cincinnati Ohio, US.[44] In this report, the US EPA demonstrated that both natural and synthetic steroid hormones were released into the environment through wastewater outfalls at concentrations ranging from <0.01 to 9 µg L^{-1}. Specifically, EE2 and E2 were detected at concentrations ranging from 0.25 to 1.78 µg L^{-1} and non-detectable to 0.02 µg L^{-1}, respectively. This report indicated that natural hormones exhibited greater attenuation through the treatment processes than did the synthetic hormones and that hormone concentrations were greater in the winter as compared to summer, due to lesser microbial activity in the cooler temperatures.

In 1985, researchers from the UK investigated the occurrence of the synthetic hormones norethisterone and EE2, as well as the endogenous hormone progesterone and the anti-cancer pharmaceutical methotrexate.[45] These authors also specifically considered methotrexate as an "indicator compound" because of its "widespread use" and relatively high dosages used. From the eight rivers and six potable water supplies evaluated, norethisterone and progesterone were detected in two and one sample, respectively. Ethynylestradiol and methotrexate were not detected in any of the river or potable water samples tested despite 0.005 and 0.00625 μg L^{-1} reporting limits. However, methotrexate was detected in one hospital effluent at 1 μg L^{-1}. These authors compared the concentrations of the hormones detected to prescribed dosages and concluded that "there appears to be no risk from the reuse of river water from highly active therapeutic agents", but suggested that "further analyses for other drugs are required to confirm these findings." Concerns over the loading of pharmaceuticals in the environment began to appear in the UK in the 1980s,[46] yet would not gain great international recognition until natural and synthetic hormones became implicated as potential causative agents for endocrine disruption observed in fish captured downstream of municipal wastewater treatment facilities.[27,30]

5   Ubiquity of Chemicals in the Environment

Today, there are millions of commercially available chemical formulations and the list of chemicals registered in Chemical Abstract Services (CAS) grows by approximately 15 000 chemicals day^{-1}.[1] The growth in registrations of small molecules in CAS has essentially tripled between 2005 and 2010 (see Figure 1). Within the world of chemicals discovered, there are an estimated 4000 pharmaceutical active ingredients used globally within tens of thousands of pharmaceutical formulations.[47] With such a high rate of discovery and commercialization, it is virtually impossible for all potential environmental and public health impacts to be evaluated using conventional animal testing techniques. Thus, the conventional paradigm for conducting risk assessment on a chemical by chemical basis is largely futile for highly complex mixtures of chemicals such as those found in wastewater effluents. Moreover, many chemicals will form transformation products in the environment and/or during water treatment that may be bioactive.[48]

While the number of chemicals utilized globally is staggering, an even more mind boggling figure is the innumerable combinations of these substances that are part of the "exposome" (the exposome can be defined as the measure of all the exposures of an individual in a lifetime and how those exposures relate to health).[49]

In addition to transformation products, a fundamental issue exists in the consideration of mixture toxicity. Indeed, there are no environmental exposure scenarios where a discrete chemical exposure exists in the absence of other chemicals and ambient matrix background. Thus, all mixtures are

Substances in CAS registry

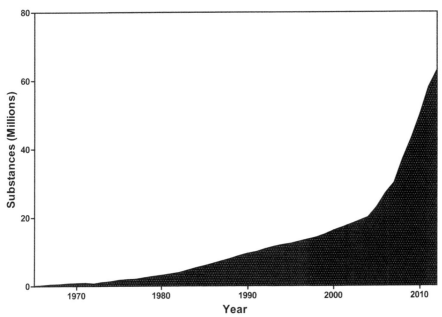

Figure 1   Number of chemicals registered in Chemical Abstract Services (CAS) from
1965 through 2013. (http://www.cas.org).

different and cannot reasonably be evaluated for safety by analytical measurements only. Conversely, analytical screening alone also cannot guarantee
safety. Rapid biological assays provide a step in the right direction in terms
of more comprehensive screening capabilities to determine the bioactivity of
chemicals and chemical mixtures occurring in the environment.[50] Future
conversations regarding chemical safety must be inclusive of transformation
products and potential mixture toxicity.

6   Formation of New and Previously Unknown Products

With the millions of chemical formulations currently in circulation and
thousands of new chemicals created daily, the probability of these chemicals
entering the environment is inevitable. Yet, it is also possible that they may
be transformed into several other similar chemical structures by nature or
anthropogenic influences in the environment. For instance, the UV transformation product of the potent synthetic androgen trenbolone can be
converted back to the parent structure under environmentally relevant
conditions.[51] As another example, the fungicide tolylfluanid was founded to
decompose to form a metabolite which could be readily transformed into the
potent carcinogen nitrosodimethylamine (NDMA) when exposed to ozone
during water treatment.[52] Carbamazepine, an anti-epileptic drug can be

metabolized in the body and excreted as 32 different metabolites.[53] Some of these metabolites are subsequently found at higher concentrations than the parent compound in nature.[54] Sometimes, transformation products can be formed inadvertently when trying to attenuate the parent compound, for example, in water treatment processes. Benzotriazole is used as a metal-corrosion inhibitor and is often found in wastewaters around the world. Application of ozone is known to remove this compound from the water, however lower ozone doses have been shown to form 1*H*-1,2,3-triazole-4,5-dicarbaldehyde, a by-product of benzotriazole.[55] Similarly, ozonation of carbamazepine was reported to form three new oxidation by-products not formed during human excretion.[56] Thus, for comprehensive consideration of human exposure to potential toxicants in the environment, one must consider not only the parent compound but also the multitude of transformation products that are formed through natural and engineered processes.

Often, these transformation products are hard to identify without sophisticated analytical equipment and may even be even more toxic than their parent compounds. Diuron, for example is an herbicide used to inhibit photosynthesis and is only slightly toxic to mammals. However, upon biotic degradation in soil its chief transformation product is 3,4-dichloroaniline which is highly toxic and genotoxic (a genotoxin is a substance which causes damage to DNA and is thereby capable of causing mutations or cancer).[57] Perhaps, the most well-known example of formation of toxic by-products is the class of chemicals termed "disinfection by-products". The final treatment stage at a water treatment plant is often a disinfection step using an oxidant like chlorine, chloramine, chlorine dioxide, ultraviolet light or ozone. This is meant to inactivate common microbes that may be present in the water and prevent common water-borne diseases. Prior to 1908, no municipal water disinfection scheme was present in the United States and incidence of typhoid fever and cholera were high. The US Centers for Disease Control and Prevention (CDC) estimates that since 1908, the number of typhoid fever cases has reduced over a thousand fold primarily due to water disinfection.[58] However, these disinfectants can also react with naturally occurring substances in the water to form other by-products commonly referred to as "disinfection by-products" (DBPs).

A large number of these compounds have been shown to cause toxic effects in wildlife and, potentially, humans.[59] Due to this, The US EPA has regulated a section of these compounds including trihalomethanes (THMs) and haloacetic acids (HAAs). However, only a few compounds in these two classes of DBPs have been regulated, while entire classes of DBPs, like halonitromethanes (HNMs) and nitrosamines, remain unregulated due in part to the labour-intensive and time-consuming methods needed to perform toxicological studies relevant to regulatory action. As a result, a number of "newer" classes of DBPs, thought to be more toxic than the regulated DBPs, are still unregulated and continue to be formed in drinking water.[60]

In the 1970s, THMs were identified as the primary DBPs formed during disinfection of water.[61] However, other classes, like HAAs, haloaldehydes

Table 1   Common categories of DBPs.

Class	Regulated by US EPA	Class	Regulated by US EPA
Trihalomethanes	✔ (4)	Bromate	✔
Haloacetic acids	✔ (5)	Chlorite	✔
Haloacetonitriles	✕	Nitrosamines	✕
Haloketones	✕	Halonitromethanes	✕
Haloaldehydes	✕	Carbonyls	✕

( ) Number of compounds regulated in the class.

and haloacetinitriles, were later identified as common DBPs present in water too. After several decades of research, less than half the total organic halogens in water during disinfection can be accounted for by known DBPs.[62] This figure is considered to be significantly lower for alternative disinfectants like ozone and ultraviolet (UV) light; however, any process that transforms organic constituents will certainly create transformation products unless carbon structures are mineralized.[1] Overall, only about 500 DBPs have been identified as yet while <2% have been regulated in the United States. Further, emerging DBP classes like nitrogenous and iodinated DBPs have been shown to be significantly more toxic that the ones regulated thus far.[59,63] Table 1 shows common categories of DBPs identified on disinfection of water.

The formation of potentially toxic by-products from naturally-occurring materials and chemically-synthesized compounds both through anthropogenic intervention and in nature is of concern and should be considered. However, municipal disinfection has resulted in prevention of many diseases and thus creates the conundrum in balancing the risks of the exposure of DBPs with the need to provide reliable disinfection. Indeed, the increase in sensitivity and further development of analytical instrumentation will lead to the discovery of previously unknown and unidentified transformation products. With identification of such compounds, authorities may have to consider transformation products and potential mixture toxicity when developing the regulatory framework.

7   Regulatory History and Current Framework in the US

The United States has a population of 319 million spread across 50 states, one federal district, and six territories (Guam, Puerto Rico, US Virgin Islands, American Samoa, Northern Mariana Islands), dispersed onto 9.39 million km^2 (3.79 million miles^2).[64] According to the US EPA, there exist almost 153 000 drinking water systems around the country with over 90% of community water systems being supplied by groundwater. However, 71% of the people using community water systems are supplied by surface water.[65] According to the US EPA, almost 95% of the public water works in the US supply smaller communities (<3300 people), while only slightly more than 5% serve larger populations.[65] These small systems account for about 7000 drinking water quality standard violations per year which amounts to nearly 85% of all drinking water violations in the country.[65]

The Water Pollution Control Act of 1948 was the first piece of legislation in the US aimed at controlling water pollution. However, it gave limited authority to the federal government and was mainly driven by local issues with the states having most of the responsibility and hence was not very effective. In 1965, the Water Quality Act was enacted which required states to develop water quality standards for waters shared between states. It also resulted in the formation of the Federal Water Pollution Control Administration, whose duty was to oversee and set standards for non-consenting states.

The 1960s led to several instances where water pollution was brought into the public's eye. In 1962, Rachel Carsons's *Silent Spring* was the first work to raise serious concerns regarding impacts to human and ecological health from environmental contamination. In June 1969, the Cuyahoga River in Ohio caught fire most likely due to a discharge of highly volatile petroleum derivatives. The resulting fire caused over $10 000 in damage and was one of several occasions in the early twentieth century when this had occurred. Such incidents forced the Federal government to take more responsibility in environmental regulation and led to the formation of the US Environmental Protection Agency in October 1970. The Clean Water Act (CWA) was established in 1972, regulating discharges of pollutants into the water bodies in the United States. The act also empowered the US EPA with setting and regulating water quality benchmarks for surface waters. The CWA prevents municipal wastewater plants, industries, and other facilities from discharging into a water body without obtaining a permit through the US EPAs National Pollution Discharge Elimination System (NPDES).

In 1974, Congress passed the Safe Drinking Water Act (SDWA) aimed at ensuring safe drinking water and protect public health of Americans. Under this act, the US EPA is required to set drinking water standards for chemical and biological contaminants in drinking water while overseeing the implementation of these regulations by states and water authorities. The US EPA regulates contaminants in drinking water through the National Primary Drinking Water Standards (NPDWS), which are legally enforceable standards employed to protect public health and apply to all public water systems. Currently more than 60 contaminants are part of the NPDWS, with each having a maximum contaminant level (MCL). The MCL is determined individually for each contaminant based on health risk and occurrence in drinking water. The US EPA decides to regulate certain contaminants based on a rigorous process involving the Contaminant Candidate List (CCL). The CCL is a list of contaminants that is not currently regulated but known or suspected to occur in drinking water systems.

In 1998, the US EPA released its first CCL, with 10 microbial contaminants and 50 chemical contaminants. Subsequently, in 2005 the US EPA released its CCL2 list with 51 contaminants, followed by CCL3 in 2009 consisting of 104 chemical contaminants or groups and 12 microbes.[66] Regulatory determination on these chemicals is very slow though. After a five year period, in 2003 the US EPA passed a decree on only nine of the contaminants in CCL1 deciding not to regulate them. Clearly, this indicates that the pace of chemical ingredients and formulations entering the market, and subsequently the environment, exponentially outweighs the ability to bring any regulatory authority on them. This poses a dilemma for regulatory authorities, who must consider the present framework of regulation to deal with the plethora of new chemical products released into the market. The advancement of new chemistries and development of better analytical techniques will lead to greater numbers and detection of compounds in the environment leading to potential regulatory concerns.

8   Evolution in Analytical Methodologies

Due to advances in analytical technology, a plethora of organic compounds have been identified and quantified in water at minute concentrations. From pharmaceuticals[67] to endogenous human hormones,[68] essentially all chemicals on earth are detectable in water if extremely sensitive analytical techniques are applied. In 1960, scientists could barely imagine methods with parts-per-billion sensitivity.[69] Today, method reporting limits at parts-per-trillion, or lower, have now been developed for numerous chemicals occurring in water.[1] A part-per-trillion is 0.000000001 grams dissolved in a liter of water. In terms of time, a part per trillion is equivalent to one second in 31 700 years. In relationship to volumes, a part per trillion is approximately one ounce in 7.5 billion gallons of water. Analytical methods can now measure extremely small levels of nearly all chemicals in water, and it is clear that detection alone is not sufficient to imply toxicological relevance.

Analytical techniques for the identification and quantification of environmental contaminants in water have historically been challenging due to the sensitivity of available instrumentation, the vast physio-chemical properties of target analytes, and the relatively low levels that occur in water. In fact, until modern analytical techniques were developed, it was extremely difficult to detect substances in water at sub-μg L^{-1} levels. Most commonly, organic contaminants are extracted from water using variations of liquid–liquid or solid–phase extraction. However, both extraction techniques isolate not only target compounds, but a broad range of organic substances that co-occur in a particular water matrix. For this reason, gas or liquid chromatography (GC or LC) are generally applied to separate the complex mixture isolated in an organic extract from water. Numerous detectors have been applied in tandem with GC and LC separations; however, mass spectrometric detection has emerged as the most commonly employed due to its selectivity and sensitivity.

As discussed previously, some of the earliest work on environmental contaminants involved the relatively non-polar organic compounds such as DDT which could be readily separated using GC. Since many pharmaceuticals and steroids have low volatility and appreciable water solubility, chemical reactions are traditionally applied in order to derivatize the compounds into a molecular structure amenable to gas chromatography.

Unfortunately, derivatization methods are inherently more prone to error and the derivatives formed may be unstable.

Some of the earliest methods demonstrating sub-µg L^{-1} sensitivity utilized GC separations with electron capture detection (GC-ECD), which is generally used for the detection of halogenated compounds.[70] The selectivity of GC-ECD methods depend on chromatographic retention time and the presence of a compound capable of capturing an electron. Thus, derivatizations of polar organic contaminants often utilized reagents that would add halogenated species to the molecule to increase sensitivity using ECD. Despite the ability of GC-ECD to quantify halogenated organics at low part-per-billion (ppb or µg L^{-1}) levels, the challenges of potentially interfering compounds caused a move to the more selective mass spectrometric (MS) detection later in the 1980s.

The ECD suffers from a general lack of selectivity since the detector utilizes a radioactive source to produce beta particles that eject electrons from electron-rich makeup gas, typically nitrogen.[71] The production of gaseous phase electrons results in an electrical current that becomes the baseline for the subsequent chromatogram. Substances exiting the GC column that are capable of capturing an electron result in a decrease of electrical current, thus actual signal from an ECD is negative and is simply inverted for consistency with other detector outputs. The degree to which a particular substance can capture electrons is related to atomic composition and orientation. Compounds containing halogens, nitrogen, sulfur, and other atoms with high electronegativity are particularly sensitive to detection using an ECD. Thus, the most prominent application of GC-ECD has been for halogenated substances, most notably chlorinated organics such as polychlorinated biphenyls, organochlorine pesticides, and chlorofluorocarbons.[72]

Despite essentially unprecedented sensitivity, the ECD suffered from lack of specificity. For instance, in a peer-reviewed publication authored by the US Department of Agriculture, an organochlorine herbicide and organochlorine flame-retardant were shown to have the same retention time, even though two different chromatography columns were used in parallel.[73] In other words, the flame retardant and organochlorine herbicide could not be differentiated using GC-ECD regardless of analytical column applied. Thus, despite the fact that the ECD is highly-sensitive for halogens and other electronegative substances, the rapid proliferation of more sensitive, selective, cost effective MS systems has largely displaced ECD applications.

The first published report demonstrating the use of GC-MS for the detection of environmental contaminants became prominent in the 1980s with the development of relatively reliable quadrupole mass spectrometers. Most recently, liquid chromatography with tandem mass spectrometric detection (LC-MS/MS) has become the method of choice for the analysis of polar organic contaminants, such as most pharmaceuticals, in water. The main advantage of this technique is the direct detection of polar organics without derivatization due to the ability to produce gas phase ions from nonvolatile species. A second important advantage is increased sensitivity due to the higher signal-to-noise ratios resulting from tandem mass spectrometry.

Of the various types of mass spectrometers that can perform tandem mass spectrometry, the triple quadrupole mass spectrometer is generally considered the most sensitive. This mass spectrometer design was first developed in the late 1970s[74] and was initially interfaced with gas chromatography. Early attempts at interfacing liquid chromatography directly to mass spectrometers were not highly successful due the difficulty of evaporating the liquid phase prior to ionization in the vacuum system of the mass spectrometer. It was not until the advent of the electrospray ionization interface[75,76] and the ion spray interface[77] in the mid- to late 1980s that LC-MS/MS became a robust analytical technique. In fact, the development of electrospray ionization yielded John Fenn a Nobel Prize in Chemistry awarded in 2002.[78] Further technological refinement of ion sources and ion optics continued through the 1990s, leading to improved sensitivity in commercially available instruments. The application of LC-MS/MS for environmental analyses was not widely established until the late 1990s and early 2000s.

Today, a multitude of analytical approaches are available which can identify and quantify nearly any imaginable contaminant in water.[79,80] For many organic contaminants, sub-ng L^{-1} identification and quantification in water is possible. As analytical technologies continue to advance, there is no doubt that the detection of more anthropogenic substances will become increasingly prevalent and at diminishingly minute concentrations.

9 Implications for Water Sustainability

Essentially all water on earth is used and reused, or will be used and reused, in time. Whether discharged to a river, ocean, or infiltrated into the ground, the water cycle continues and water will be eventually reused. Considering the dire situation of water resources in many parts of the world, water reuse offers an important means to extend limited resources provided by natural deposition.[81] Today, many regions in diverse locations across the world are considering water reuse, including potable water reuse. However, the very thought of utilizing wastewater as a source for drinking water is often appalling to members of the public. Thoughts of diseases spread through human waste and the occurrence of complex mixtures of chemicals induce fear and consternation that can often lead to rejection of potable reuse schemes. Interestingly, many cities already reuse wastewater unknowingly through wastewater contributions to their source water through "upstream" discharges.[82] Indeed, a number of reports show that wastewater indicator compounds, such as pharmaceuticals, occur widely in source waters[83] and in finished drinking water.[67]

While a systematic monitoring program for substances with known human and/or ecological health impacts is critical, targeted monitoring does not consider the more broad classes of substances that may occur in environmental matrixes. When non-targeted mass spectrometry is applied, it is apparent that thousands of chemicals are present[84] and certainly thousands more occur but are not detectable. Considering the sensitivity of modern analytical instruments, nearly all chemicals with significant commercial usage will be detectable in wastewater.[85] However, the critical question continues to be asked, are we monitoring for the important constituents?

What does the future hold for evaluating the occurrence and relevance of chemicals in the environment? There is no question at all that more chemical structures will be discovered and synthesized. These substances are critical for the advancement of the human condition, but must be counter-balanced with the potential for unintended consequence.

In the previous discussion of DDT, a discovery heralded as one of the greatest advancements in human history at the time transformed into the most famous case of environmental damages from human activities. We should never forget that today's technological triumphs will likely become tomorrow's emerging contaminants. Moreover, we must concede that there is no reasonable possibility that conventional animal testing will provide unequivocal safety data necessary to drive regulations. The fact is, our continued advancement in chemical discovery is greatly outpaced by our abilities to evaluate comprehensively the potential for unintended human and environmental impact. The way forward involves implementation of rapid biological screening of discrete chemicals and complex mixtures using bioassays. If unintended bioactivity is detected at levels of concern, characterization using high-resolution mass spectrometry and other advanced analytical tools should be used to identify and quantify the punitive substance(s). In addition, the concept of using surrogate and indicator measurements in order ensure engineered and/or natural systems are providing effective barriers necessary to be protective of health will be increasingly critical, especially in the case of potable water reuse.[84]

The vast majority of chemicals synthesized and/or purified from natural sources have significantly improved and expanded human life. However, there are numerous examples where unintended consequences have led to ailments and environmental deterioration. Growing human population and increased urbanization will dictate that we more carefully manage our fragile environmental resources. The precarious balance between human triumph and natural collapse is never easy, but rapid screening tools will better help us keep up with pace of chemical development. The final domain will be the rapid characterization of the resulting complex mixtures and transformation products using a suite of bioassays that are representative of relevant human and environmental health.

References

1. S. A. Snyder, Emerging Chemical Contaminants: Looking for Better Harmony, *J. – Am. Water Works Assoc.*, 2014, **106**(8), 38–52.
2. P. J. Novak, W. A. Arnold *et al.*, On the need for a national (US) research program to elucidate the potential risks to human health and the

environment posed by contaminants of emerging concern, *Environ. Sci. Technol.*, 2011, **45**(9), 3829–3830.

3. J. E. Drewes, P. Anderson *et al.*, Designing monitoring programs for chemicals of emerging concern in potable reuse – what to include and what not to include?, *Water Sci. Technol.*, 2013, **67**(2), 433–439.

4. EPA, Contaminants of Emerging Concern, 2015, Retrieved 19th January, 2015, from http://water.epa.gov/scitech/cec/.

5. A. S. Travis, Hofmann, August, Wilhelm (1818-1892), *Endeavour*, 1992, **16**(2), 59–65.

6. R. D. Welham, The early history of the synthetic dye industry 1. The chemical history, *J. Soc. Dyers Colour.*, 1963, **79**(3), 98–105.

7. W. M. Jarman and K. Ballschmiter, From coal to DDT: the history of the development of the pesticide DDT from synthetic dyes till Silent Spring, *Endeavour*, 2012, **36**(4), 131–142.

8. A. W. Jones, Early drug discovery and the rise of pharmaceutical chemistry, *Drug Test. Anal.*, 2011, **3**(6), 337–344.

9. A. L. Fisher, H. H. Keasling *et al.*, Estrogenic action of some DDT analogs, *Proc. Soc. Exp. Biol. Med.*, 1952, **81**, 439–441.

10. J. Bitman, H. C. Cecil *et al.*, Estrogenic activity of o,p′-DDT in the mammalian uterus and avian oviduct, *Science*, 1968, **162**(3851), 371–372.

11. A. Attaran and R. Maharaj, DDT for malaria control should not be banned., *Br. Med. J.*, 2000, **321**(7273), 1403–1404.

12. G. P. Prasad, G. P. Pratap *et al.*, Historical perspective on the usage of perfumes and scented articles in ancient Indian literatures, *Anc. sci. Life*, 2008, **28**(2), 33–39.

13. M. JoseYacaman, L. Rendon *et al.*, Maya blue paint: An ancient nanostructured material, *Science*, 1996, **273**(5272), 223–225.

14. R. S. Solecki, Shanidar-4, a neanderthal flower burial in Northern Iraq, *Science*, 1975, **190**(4217), 880–881.

15. R. Nair and S. Sriprasad, Emperor Shen-Nung's root: Ginseng in the management of erectile dysfunction in ancient China (3500-2600 BCE), *Eur. Urol., Suppl.*, 2011, **10**(2), 57–58.

16. S. G. Cohen, Asthma in antiquity – the ebers papyrus, *Allergy Proc.*, 1992, **13**(3), 147–154.

17. E. X. Albuquerque, J. W. Daly *et al.*, Batrachotoxin – chemistry and pharmacology, *Science*, 1971, **172**(3987), 995.

18. T. Reynolds, Hemlock alkaloids from Socrates to poison aloes, *Phytochemistry*, 2005, **66**(12), 1399–1406.

19. J. Miner and A. Hoffhines, The discovery of aspirin's antithrombotic effects, *Texas Heart Inst. J.*, 2007, **34**(2), 179–186.

20. D. B. Jack, One hundred years of aspirin, *Lancet*, 1997, **350**(9075), 437–439.

21. T. Maclagan, The treatment of rheumatism by salicin and salicylic acid, *Br. Med. J.*, 1876, **1**(803), 627.

22. M. J. Thun, M. M. Namboodiri *et al.*, Aspirin use and reduced risk of fatal colon cancer, *N. Engl. J. Med.*, 1991, **325**(23), 1593–1596.

23. T. D. Warner and J. A. Mitchell, Cyclooxygenase-3 (COX-3): Filling in the gaps toward a COX continuum?, *Proc. Natl. Acad. Sci. U. S. A.*, 2002, **99**(21), 13371–13373.

24. M. Rabiet, A. Togola *et al.*, Consequences of treated water recycling as regards pharmaceuticals and drugs in surface and ground waters of a medium-sized Mediterranean catchment, *Environ. Sci. Technol.*, 2006, **40**(17), 5282–5288.

25. C. E. Purdom, P. A. Hardiman *et al.*, Estrogenic effects of effluents from sewage treament works, *Chem. Ecol.*, **199**(48), 275–285.

26. H. E. Bevans, S. L. Goodbred *et al.*, Synthetic organic compounds and carp endocrinology and histology in Las Vegas Wash and Las Vegas and Callville Bays of Lake Mead, Nevada, 1992 and 1995, *Water-Resources Investigations Report*, 1996, 96-4266.

27. C. Desbrow, E. J. Routledge *et al.*, Identification of estrogenic chemicals in STW eflleunt. 1. Chemical fractionation and in vitro biological screening, *Environ. Sci. Technol.*, 1998, **32**(11), 1549–1558.

28. S. Jobling, M. Noylan *et al.*, Widespread sexual disruption in wild fish, *Environ. Sci. Technol.*, 1998, **32**, 2498–2506.

29. E. J. Routledge, D. Sheahan *et al.*, Identification of estrogenic chemicals in STW effluent. 2. *In vivo* responses in trout and roach, *Environ. Toxicol. Chem.*, 1998, **32**(11), 1559–1565.

30. S. A. Snyder, D. L. Villeneuve *et al.*, Identification and quantification of estrogen receptor agonists in wastewater effluents, *Environ. Sci. Technol.*, 2001, **35**(18), 3620–3625.

31. K. Kidd, P. J. Blanchfield *et al.*, Collapse of a fish population after exposure to a synthetic estrogen, *Proc. Natl. Acad. Sci. U. S. A.*, 2007, **104**(21), 8897–8901.

32. E. H. Starling, The Croonian Lectures on the chemical correlation of the functions of the body – Delivered before the Royal College of Physicians of London on June 20th 22nd, 27th and 29th 1905, *Lancet*, 1905, **2**, 339–341.

33. B. S. Walker and J. C. Janney, Estrogenic substances. II. Analysis of plant sources, *Endocrinology*, 1930, **14**, 389–392.

34. J. W. Cook, E. C. Dodds *et al.*, Estrogenic activity of some condensed ring compounds in relation to their other biological activities, *Proc. R. Soc. London*, 1934, **B114**, 272–286.

35. C. Djerassi, Chemical birth of the pill, *Am. J. Obstet. Gynecol.*, 2006, **194**(1), 290–298.

36. F. W. Schueler, Sex-hormonal action and chemical constitution, *Science*, 1946, **103**, 221–223.

37. J. W. Goldzieher, Estrogens in oral contraceptives: historical perspectives, *Johns Hopkins Med. J.*, 1982, **150**(5), 165–169.

38. G. Pincus, An oral contraceptive, *Lancet*, 1958, **1**(Jun7), 1230.

39. S. W. Junod and L. Marks, Women's trials: The approval of the first oral contraceptive pill in the United States and Great Britain, *J. Hist. Med. Allied Sci.*, 2002, **57**(2), 117–160.

40. E. Stumm-Zollinger and G. M. Fair, Biodegradation of steroid hormones, *J. - Water Pollut. Control Fed.*, 1965, **37**, 1506–1510.
41. H. H. Tabak and R. L. Bunch, Steroid hormones as water pollutants. I. Metabolism of natural and synthetic ovulation-inhibiting hormones by microorganisms of activated sludge and primary settled sewagem, *Dev. Ind. Microbiol.*, 1970, **11**, 367–376.
42. A. W. Garrison and J. D. Pope *et al.*, *GC/MS Analysis of Organic Compounds in Domestic Wastewaters. Identification and Analyses of Organic Pollutants in Water*, ed. L. H. Keith, Ann Arbor Publishers Inc, Ann Arbor MI, 1976, pp. 517–556.
43. C. Hignite and D. L. Azarnoff, Drugs and drug metabolites as environmental contaminants: Chlorophenoxyisobutyrate and salicylic acid in sewage water effluent, *Life Sci.*, 1977, **20**(2), 337–341.
44. H. H. Tabak, R. N. Bloomhuff *et al.*, Steroid hormones as water pollutants II. Studies on the persistence and stability of natural urinary and synthetic ovulation-inhibiting hormones in untreated and treated wastewaters, *Dev. Ind. Microbiol.*, 1981, **22**, 97–519.
45. G. W. Aherne, J. English *et al.*, The role of immunoassay in the analysis of microcontaminants in water samples, *Ecotoxicol. Environ. Saf.*, 1985, **9**, 79–83.
46. M. L. Richardson and J. M. Bowron, The fate of pharmaceutical chemicals in the aquatic environment, *J. Pharm. Pharmacol.*, 1985, **37**(1), 1–12.
47. C. G. Daughton, Eco-directed sustainable prescribing: feasibility for reducing water contamination by drugs, *Sci. Total Environ.*, 2014, **493**, 392–404.
48. D. M. Cwiertny, S. A. Snyder, D. Schlenk and E. P. Kolodziej, Environmental designer drugs: When transformation may not eliminate risk, *Environ. Sci. Technol.*, 2014, **48**(20), 11737–11745.
49. S. M. Rappaport, Implications of the exposome for exposure science, *J. Exposure Sci. Environ. Epidemiol.*, 2011, **21**(1), 5–9.
50. B. I. Escher, M. Allinson *et al.*, Benchmarking Organic Micropollutants in Wastewater, Recycled Water and Drinking Water with In Vitro Bioassays, *Environ. Sci. Technol.*, 2014, **48**(3), 1940–1956.
51. S. Qu, E. P. Kolodziej *et al.*, Product-to-parent reversion of trenbolone: unrecognized risks for endocrine disruption, *Science*, 2013, **342**(6156), 347–351.
52. C. K. Schmidt and H. J. Brauch, N,N-dimethosulfamide as precursor for N-nitrosodimethylamine (NDMA) formation upon ozonation and its fate during drinking water treatment, *Environ. Sci. Technol.*, 2008, **42**(17), 6340–6346.
53. K. Lertratanangkoon and M. G. Horning, Metabolism of carbamazepine, *Drug Metab. Dispos.*, 1982, **10**(1), 1–10.
54. M. la Farre, S. Perez *et al.*, Fate and toxicity of emerging pollutants, their metabolites and transformation products in the aquatic environment, *TrAC, Trends Anal. Chem.*, 2008, **27**(11), 991–1007.

55. D. B. Mawhinney, B. J. Vanderford *et al.*, Transformation of 1H-benzo-triazole by ozone in aqueous solution, *Environ. Sci. Technol.*, 2012, **46**(13), 7102–7111.

56. D. C. McDowell, M. M. Huber *et al.*, Ozonation of carbamazepine in drinking water: Identification and kinetic study of major oxidation products, *Environ. Sci. Technol.*, 2005, **39**(20), 8014–8022.

57. S. Giacomazzi and N. Cochet, Environmental impact of diuron transformation: a review, *Chemosphere*, 2004, **56**(11), 1021–1032.

58. USCDC, Drinking Water Week, 2014, *U. C. f. D. C. a. Prevention*.

59. S. W. Krasner, The formation and control of emerging disinfection by-products of health concern, *Philos. Trans. R. Soc., A*, 2009, **367**(1904), 4077–4095.

60. S. D. Richardson, F. Fasano *et al.*, Occurrence and Mammalian Cell Toxicity of Iodinated Disinfection Byproducts in Drinking Water, *Environ. Sci. Technol.*, 2008, **42**(22), 8330–8338.

61. T. A. Bellar, J. J. Lichtenberg *et al.*, The occurrence of organohalides in chlorinated drinking water, *J. – Am. Water Works Assoc.*, 1974, **66**, 703–706.

62. X. R. Zhang, S. Echigo *et al.*, Characterization and comparison of disinfection by-products from using four major disinfectants, *Abstr. Pap. Am. Chem. Soc.*, 1999, **217**, U736.

63. S. E. Duirk, C. Lindell *et al.*, Formation of toxic iodinated disinfection by-products from compounds used in medical imaging, *Environ. Sci. Technol.*, 2011, **45**(16), 6845–6854.

64. Census, Population Estimates: National Totals, 2014.

65. USEPA, Fiscal Year 2011, Drinking Water and Ground Water Statistics, 2011, Retrieved 03/20/2014, from http://water.epa.gov/scitech/datait/databases/drink/sdwisfed/upload/epa816r13003.pdf.

66. USEPA, Drinking Water Contaminant Candidate List and Regulatory Determination, 2014, USEP Agency.

67. M. Benotti, R. A. Trenholm *et al.*, Pharmaceuticals and endocrine disrupting compounds in U.S. drinking waters, *Environ. Sci. Technol.*, 2009, **43**(3), 597–603.

68. S. Snyder, T. L. Keith *et al.*, Analytical methods for detection of selected estrogenic compounds in aqueous mixtures, *Environ. Sci. Technol.*, 1999, **33**(16), 2814–2820.

69. L. T. Hallet, Sensitivity of analytical methods, *Anal. Chem.*, 1960, **32**(9), 1057.

70. J. Belisle and D. F. Hagen, Method for the determination of per-fluorooctanoic acid in blood and other biological samples, *Anal. Biochem.*, 1980, **101**(2), 369–376.

71. J. E. Lovelock, The electron capture detector: Theory and practice, *J. Chromatogr. A*, 1974, **99**, 3–12.

72. J. E. Lovelock, R. J. Maggs *et al.*, Halogenated Hydrocarbons in and over the Atlantic, *Nature*, 1973, **241**(5386), 194–196.

73. W. C. Koskinen, J. M. Otto *et al.*, Potential interferences in the analysis of atrazine and deethylatrazine in soil and water, *J. Environ. Sci. Health, Part B*, 1992, **27**(3), 255–268.

74. R. A. Yost and C. G. Enke, Selected ion fragmentation with a tandem quadrupole mass-spectrometer, *J. Am. Chem. Soc.*, 1978, **100**(7), 2274–2275.

75. C. M. Whitehouse, R. N. Dreyer *et al.*, Electrospray interface for liquid chromatographs and mass spectrometers, *Anal. Chem.*, 1985, **57**(3), 675–679.

76. J. B. Fenn, M. Mann *et al.*, Electrospray ionization for mass-spectrometry of large biomolecules, *Science*, 1989, **246**(4926), 64–71.

77. A. P. Bruins, T. R. Covey *et al.*, Ion spray interface for combined liquid chromatography/atmospheric pressure ionization mass-spectrometry, *Anal. Chem.*, 1987, **59**(22), 2642–2646.

78. M. A. Grayson, John Bennett Fenn: A curious road to the prize, *J. Am. Soc. Mass Spectrom.*, 2011, **22**(8), 1301–1308.

79. S. D. Richardson, Environmental mass spectrometry: Emerging contaminants and current issues, *Anal. Chem.*, 2012, **84**(2), 747–778.

80. S. D. Richardson and T. A. Ternes, Water analysis: Emerging contaminants and current issues, *Anal. Chem.*, 2014, **86**(6), 2813–2848.

81. S. B. Grant, J.-D. Saphores *et al.*, Taking the "Waste" out of "Wastewater" for human water security and ecosystem sustainability, *Science*, 2012, **337**(6095), 681–686.

82. J. Rice, A. Wutich *et al.*, Assessment of de facto wastewater reuse across the US: Trends between 1980 and 2008, *Environ. Sci. Technol.*, 2013, **47**(19), 11099–11105.

83. M. J. Focazio, D. W. Kolpin *et al.*, A national reconnaissance for pharmaceuticals and other organic wastewater contaminants in the united states— II) untreated drinking water sources, *Sci. Total Environ.*, 2008, **402**, 201–216.

84. S. Merel, T. Anumol, M. Park and S. A. Snyder, Application of surrogates, indicators, and high-resolution mass spectrometry to evaluate the efficancy of UV processes for attenuation of emerging contaminants in water, *J. Hazard. Mater.*, 2014, **282**, 75–85.

85. R. A. Trenholm, B. J. Vanderford *et al.*, Determination of household chemicals using gas chromatography and liquid chromatography with tandem mass spectrometry, *J. Chromatogr. A*, 2008, **1190**, 253–262.

A Change of Emphasis: Waste to Resource Management

I. D. WILLIAMS

ABSTRACT

It is 40 years since the waste hierarchy was introduced in the European Union. During this period the World has changed significantly, with massive changes in population, socio-economic and demographic circumstances, politics, consumption of goods and services, range and type of products available, energy and water consumption, *etc.* We have seen an increase in public awareness and education relating to environmental issues, public consultation on matters such as the siting of waste facilities, and the development of thinking and actions relating to sustainable development. Nevertheless, we still generate enormous quantities of waste that require treatment, and in some cases, disposal. This chapter will look at how waste management has evolved towards resource management during the last 40 years using the structure of the waste hierarchy as a template. It will discuss how our strategic approach to waste (resource) management has developed in the context of changing international and national policies, technology and consumption, and how our understanding of human behaviour relating to waste and resources has evolved. It will also consider how certain waste streams have altered or come on-line over this period and how the circular economy has burst onto the scene.

Issues in Environmental Science and Technology No. 40
Still Only One Earth: Progress in the 40 Years Since the First UN Conference on the Environment
Edited by R.E. Hester and R.M. Harrison
© The Royal Society of Chemistry 2015
Published by the Royal Society of Chemistry, www.rsc.org

1   Introduction

Humans have always generated wastes. Initially, the quantities generated were insignificant due to low population densities, shorter life-expectancies and limited levels of exploitation of natural resources. The wastes generated were predominantly human and animal biodegradable wastes and ashes from fires set for cooking and heating purposes. Wastes of this nature in small quantities are readily released into the ground locally, with little or no environmental impacts. This type of process, often represented as the "Circle of Life," relates to a cycle in which animals either become food for another or decompose into fertilizers for future development of plant life. The cycle in which everything has a use and nothing is wasted is reflected in Albert Einstein's famous quote that "Energy cannot be created or destroyed, it can only be changed from one form to another."

As humans became more advanced, they made tools from stone, wood and metal that were passed down through generations or reused before finally wearing out, and they gradually lived more sophisticated lives that used more resources and led to the production of a wider range of wastes. However, a step-change in the amount of waste generated occurred as a result of the industrial revolution of the mid-late 18th and early 19th centuries and the availability of modern medicine. These two major turning points in human history transformed daily life, enabling populations to grow at a rapid rate and average incomes to rise significantly. This facilitated increased purchase of goods and services, with a consequent growth in demand for raw materials, water and energy. The wastes generated became a particular problem in more densely populated urban areas, facilitating the spread of diseases and causing blight to quality of life.

The build-up of wastes in the streets of London led to the world's first recorded waste management strategy. Corbyn Morris recognized that street cleaning was going to become essential to preserve the health of the people, and proposed a Waste Strategy for London in 1751. Morris recommended an integrated London-wide approach to waste management, conveyance of wastes to sites outside the city and the use of selected wastes as a land improver. This idea rapidly spread to other cities, including Southampton, where in 1753 Messers Warwick and Minshaw "undertook to collect the waste and dung for the Council, paying a yearly rate of ten guineas plus a couple of capons".[1] By 1800, London had an organized system for collecting and disposing residual waste and an informal system for collecting and selling recyclables. The residual waste, mainly coal ash from domestic fires, was collected by private contractors who established so-called "dust yards" where materials were separated for sale as feedstock for brick-making and soil conditioner for growing crops.[1] These yards provide an early example of the circular economy in action.

However, despite the development of these systems for waste management, ongoing poor sanitation led to the spread of diseases such as cholera, smallpox and typhus, which caused a number of severe public health epidemics. The social reformer Sir Edwin Chadwick produced reports on the health of working people that led to the formation of local boards of health (the forerunners of local authorities) and the passing of the Public Health Act 1848. This focus on improving public health, combined with Victorian legislation to curb air pollution and set up local government departments, led to the establishment of municipal authorities charged with collecting and disposing of waste and sewerage. The foundations of municipal waste management had been built.

In other parts of the world, systems for waste management developed at different rates. In 1800s America, poor sanitation also led to disease epidemics and consequently regional institutions prioritized effort and funding for water treatment and human waste removal. Systems for solid waste management developed later; in the early 1900s, solid wastes were usually placed in open pits where they were frequently burned. Landfill and sanitary landfill – a modern technique that uses engineered systems to prevent pollution and the spread of disease – are still the most popular choice for waste disposal in America.

In Asia, a similar picture emerges. Prior to World War II, when cholera killed almost 100 000 people, Japan focused on providing public sanitation and then started to develop proper solid waste management and disposal. Incineration of solid waste was implemented as early as 1930 when the Waste Cleansing Law enforced incineration of solid waste as compulsory as a result of limited areas for landfills. Rapid changes occurred after World War II when almost all infrastructure and facilities were destroyed and Japan now has one of the most modern and technologically advanced systems for waste management in the world. Other Asian countries (India, China and Indonesia) have responded to the challenges of waste management as their populations have grown, become richer and more urbanized. However, large parts of the world still have little or no effective systems for safely collecting, treating and disposing of waste from anthropogenic activities, with open dumping and uncontrolled burning and composting widespread. This is a significant global challenge as the world's population crosses the 7 billion mark and rapidly heads towards 8 billion.

The management of wastes is vital in terms of preserving a clean and healthy environment for citizens. Historically, the importance of the waste management sector to society's quality of life has not really been widely recognized and acknowledged. In fact, it has suffered from a poor image and has been characterized as: "dirty work in dirty and dangerous places carried out by people with little or no qualifications for little or no wages, sometimes just for subsistence on discarded materials".[2] Waste is managed in many different ways across the globe, and indeed, many different definitions of waste exist (see Table 1).

In general, wastes can be regarded as all the items that people no longer have any use for, which they either intend to get rid of or they have already discarded. Wastes include all the items which people are required to discard, for example because of their hazardous properties. All modern daily

Table 1   Examples of definitions of waste from around the world.

Organisation	Definition
European Union	Waste shall mean any substance or object in the categories set out in Annex I which the holder discards or is required to discard.[3]
OECD	Materials that are not prime products (that is, products produced for the market) for which the generator has no further use in terms of his/her own purposes of production, transformation or consumption, and of which he/she wants to dispose.[4]
United States of America	Waste is defined as any garbage or refuse, sludge from a wastewater treatment plant, water material, including solid, liquid, semi-solid, or contained gaseous supply treatment plant, or air pollution control facility and other discarded material resulting from industrial, commercial, mining, and agricultural operations, and from community activities.[5]
China	Waste in solid or semi-solid state generated in production, construction, daily life and other activities, which might pollute the environment. (Law on the Prevention and Control of Solid Waste Pollution to the Environment, 1996).
Republic of Korea	Useless materials generated from human and business activities, such as refuse, burnable waste, sludge, waste oil, waste acid, waste alkaline and dead animals. (Waste Management Law, 1991 amended).
Malaysia	Waste includes any scrap material or other unwanted surplus substance or rejected products arising from the application of any process; any substance required to be disposed of as being broken, worn out, contaminated or otherwise spoiled; or any other material that … is required by the authority to be disposed of.

activities can give rise to a large variety of different waste flows from different sources, including from households, healthcare and education establishments, agriculture, commercial and industrial activities, construction and demolition projects. A small fraction of the waste generated is hazardous, *i.e.* it poses substantial or potential threats to human health or to the environment.

It is 40 years since the waste hierarchy was introduced in the European Union (EU). During this period the World has changed significantly, with massive changes in population, socio-economic and demographic circumstances (including a huge growth in the so-called "consumer class"), politics, consumption of goods and services, range and type of products available, energy and water consumption, *etc.* We have seen an increase in public awareness and education relating to environmental issues, public consultation on matters such as the siting of waste facilities, and the development of thinking and actions relating to sustainable development. Nevertheless, the waste we generate is enormous: in 2010, the total generation of waste in the EU-27 amounted to 2.5 billion tonnes whilst in the USA the total amount of municipal solid waste alone exceeded 250 million tonnes. These numbers are likely to escalate as the world's human population increases by 1.1% annually (approximately 75 million people per year); it is expected to reach 10 billion by the end of the century. The simultaneous surge in human numbers and consumption has led to calls for increased production and threatens to offset any savings in resource use from improved efficiency, as well as any gains in reducing per capita consumption. In reality, our global economy has evolved with little consideration towards the residuals of production and consumption and their impacts on the environment, which include resource depletion; air, water and land contamination; and the deterioration of related ecosystem goods and services.

Since the 1970s, waste management has changed significantly in some parts of the world and stayed broadly the same in others. In developed countries, the modern waste industry has become progressively more sophisticated and technological. Wastes are increasingly regarded as valuable resources to be utilised and exploited commercially rather than dumped and forgotten. This relatively recent change of emphasis reflects society's desire to secure and manage resources in a more sustainable fashion and to protect the environment, locally as well as globally.

This chapter will look at how waste management has evolved towards resource management during the last 40 years using the structure of the waste hierarchy as a template. It will also consider how certain waste streams have altered or come on-line over this period and how the circular economy has burst onto the scene.

2 Application of the Waste Hierarchy

The waste hierarchy is a process that indicates an order of preference for action to reduce and manage waste. The aims of the waste hierarchy are to protect the environment, conserve resources and to minimise the amount of waste generated. It is typically presented diagrammatically in the form of a pyramid (see Figure 1), although numerous slightly different versions are utilised. The waste hierarchy guides the progression of a material or product through consecutive stages of waste management treatments, and represents the latter part of the life-cycle for a material/product. The hierarchy

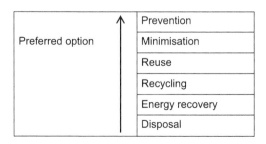

Preferred option	Prevention
	Minimisation
	Reuse
	Recycling
	Energy recovery
	Disposal

Figure 1 Waste hierarchy.

gives top priority to preventing waste; when waste is created, it gives priority to direct re-use, recycling, recovery methods, such as energy recovery, and last of all disposal.

The European Union's Waste Framework Directive of 1975 (Council Directive 75/442/EEC) introduced the waste hierarchy into European waste policy for the first time. It was formalized as an ordered system of preferred management options in the European Commission's Community Strategy for Waste Management in 1989 and this approach was endorsed in the Commission's review in 1996. Application of the hierarchy was intended to have multiple benefits, including conservation of resources, reduction of pollution (including prevention of greenhouse gas emissions), saving of energy, creation of sustainable employment and stimulation of the development of green technologies. Implementation of the waste hierarchy was optional to Member States; but there was an expectation that it would be included within national waste management legislation.

In practice most countries have viewed the hierarchy as a "ladder" and have sought to climb it step-by-step from the bottom (landfill) to the top (waste prevention). In fact, there is theoretically nothing to stop countries heading straight for the top, skipping multiple steps in one bound. However, making such a step change requires all parts of society – government, industry, commerce, the charity and community sectors, and individuals – to work together in an integrated and coherent fashion. Policies and strategies have to be agreed; when should we use an incentive (a carrot) or a sanction (a stick) to generate an action that moves a waste stream up the hierarchy? Once a strategy has been developed and agreed, a policy has to be formulated to deliver the strategy – this requires legislation to be created, passed and enforced; infrastructure to be built; services and training to be provided; markets to be created and developed; products to be redesigned; entrenched values and behaviours to be changed. On top of this, technological change has been so fast that society has struggled to keep up and waste management is just one of multiple issues that authorities need to address, including healthcare, education, housing, policing, transportation, social welfare, and so on. An integrated approach for waste reduction using the hierarchy as a guiding principle requires all these factors to come together. With so many other competing issues to address, it is probably no surprise that many countries have taken a slow, steady and stepwise approach to introducing the principles of the waste hierarchy into their systems for waste management.

EU-27 municipal waste recycling and composting rates increased to 40% in 2008 compared to 16% in 1995, whilst waste landfilling rates decreased from 62% to 40% over the same period (source: Eurostat). Table 2 summarises and categorises the contemporary dominant waste management practices in the EU. It can be seen that northern European countries have made most progress in terms of moving away from landfill, whilst countries in the east and south have made little or no progress. The reasons for this are complicated but include the availability of finance, political and social will,

Table 2 Categories of European Union countries by waste destination which dominated their waste management practices in 2011.

Category	Criteria	Countries (% landfilled in 2011)
Recovery society	> 50% incinerated or composted in 2011	Austria (3.4) Belgium (1.3) Denmark (3.5) France (27.7) Germany (0.5) Luxembourg (15.4) Netherlands (0.8) Norway (2.3) Sweden (0.9) Switzerland (0)
Non-destination specific	All destinations <50% in 2011	Finland (40.2) Italy (49.2) United Kingdom (49.2)
Landfill society	> 50% landfilled in 2011	Bulgaria (94.3) Czech Republic (64.8) Cyprus (80.2) Estonia (69.5) Greece (82.3) Hungary (67.2) Ireland (54.5) Latvia (88.3) Lithuania (78.9) Malta (88.3) Poland (70.7) Portugal (58.8) Romania (98.6) Slovenia (58.1) Slovakia (78.8) Spain (63.1)

technical skills, suitable planning and legal frameworks, and a wide range of other social, demographic, cultural and administrative factors. Many countries in eastern Europe have only recently joined the EU and so have not been required to use the waste hierarchy as a guiding principle. In addition, the principle of subsidiarity, which is fundamental to European decision-making, determines that decisions should be taken as closely as possible to the citizen, meaning that national strategies for waste management vary enormously between EU Member States.

The complexities of managing waste can be illustrated by looking at how waste strategies have changed in a single Member State. The UK has had multiple different strategies for waste management in the last 40 years. In the early 1970s, separate waste disposal and collection authorities were created in England, and new county-level authorities were required to produce 5–10 year Waste Disposal Plans. However, authorities in Wales, Scotland and Northern Ireland retained collection and disposal responsibilities.

Waste was mainly disposed *via* unlined and poorly engineered landfill sites using the "dilute and disperse" approach, which assumed that the effects of pollution could be mitigated by mere dilution. In addition, about 40 incinerators operated across the UK, although only 5 utilised significant energy recovery. Waste management was a hidden subject (out of sight, out of mind) until public uproar about the fly-tipping of hazardous wastes in children's playgrounds led to an immediate response from Parliament, with legislation passed within 3 months. Further public concerns about the impact of emissions from waste incinerators on human health and the environment meant that no new incinerators were built from the mid-1970s until a plant was completed in south east London in 1994. The Department of the Environment established the Landfill Practices Review Group, and this led to more than 30 technical so-called "Waste Management Papers", which provided a guidance for improved landfill practices for the next 20 or so years.

Changes in the political landscape in the 1980s meant that local authorities had to compete with private sector providers for the provision of waste collection and cleansing services. The intention was to facilitate the delivery of more cost-effective services. The ability of major urban areas to manage waste in a strategic fashion was further impacted by the abolition of large metropolitan authorities in the mid-1980s. Maintaining an overall national direction of travel for waste management thus became more difficult as individual authorities tended to choose the cheapest approach for their social and demographic circumstances rather than working together in the national interest.

In the 1990s, burgeoning public interest in sustainable development and concern about environmental degradation, combined with new treaties and powers for the European Commission, led to momentous and extensive changes to environmental and waste legislation. More stringent regulatory regimes for waste management were introduced, including registration procedures for waste carriers and Duty of Care responsibilities for waste producers and handlers. EU Directives led to adoption of recovery targets and compliance schemes for packaging, the introduction of stringent emission standards for municipal incinerators, and regulations for the transportation of hazardous wastes. All of this had little overall impact on municipal solid waste (MSW) management in the UK, where landfill continued as the dominant disposal method, although the overarching philosophy had changed to containment of leachate and minimisation of emissions to air.

The most significant changes occurred in the 2000s, when the impacts of the EU's Landfill Directive (99/31) started to become apparent. The Directive set steadily increasing targets for reducing the amount of biodegradable MSW disposed *via* landfill combined with similar incremental targets for increased composting and recycling. The Landfill Tax, introduced in 1996 as a "polluter pays" tax, increased incrementally every year and started to reach values that meant landfill was becoming uneconomic as a viable approach to managing wastes. Individual local authorities introduced separate collections for recyclables and garden waste combined with roll-outs of home composting schemes. Local authorities and organisations such as the Waste and Resources Action Programme (WRAP) and Waste Watch encouraged people to take control of recycling in their own homes by providing public education and awareness raising programmes and incentives, such as prize draws, alongside a greater emphasis on separating waste into different recyclable materials.

In this decade, the UK made substantial changes to the way in which it approached municipal waste management. The EU's Landfill Directive was a key driver for the Waste Strategy 2000 for England and Wales, which in turn was a catalyst for the development of municipal waste management strategies by local authorities. There was a general movement away from disposal *via* landfill and an increase in recycling and composting; >40% of household waste was recycled in England in 2010/11, compared to just 11% in 2000/01. Nevertheless, most performance indicators (*e.g.* disposal to landfill, recycling rates) show that the UK does not perform well when compared to similarly developed countries in Europe. The EC's official statistics show that Switzerland achieved zero waste to landfill in 2007, with Germany and Sweden close behind. In contrast, in 2010 Bulgaria recycled nothing and Romania just 1%.

However, a combination of factors has meant that the rapid progress of the 2000s has not been maintained. The household waste recycling rate has only increased slightly since 2010 and the prevailing rate of increase is probably insufficient to meet the 50% EU target by 2020. There are marked differences in the proportions of MSW destined for landfill, recycling and incineration at national, regional and sub-regional scales in the UK. There has been an increase in the number of incinerators and anaerobic digestion plants planned but little infrastructure has been built. To complicate matters, the government announced (in 2013) a cut-back of resources to departments that support waste policies, effectively leaving waste policy and strategy to the whims of the market.

This timeline shows that change is a key feature of UK waste policy. It is apparent that UK waste management policy, practice and infrastructure have not been consistently aligned with the aims and principles of the waste hierarchy. The wide range of strategies and actions in place demonstrates a lack of a long-term overarching policy and strategy; the picture is similar in many other European countries.

The task of implementing the waste hierarchy in waste management practices within a country is difficult. Many challenges must be met, including:

- The development and implementation of a suitable waste management strategy;
- The establishment of separate collection and sorting systems for different waste streams;

- Funding and construction of appropriate treatment and disposal facilities;
- The development of partnerships involving government, local authorities, the private and third sectors;
- The establishment of systems for data collection and monitoring to ensure successful delivery of waste management systems;
- The establishment of effective systems for enforcement and control of legal frameworks; and
- The development of human resources at all levels to manage the administrative, financial, information and technical systems put in place.

However, these challenges have clearly been met in several, mainly northern European, countries that have established long-term over-arching waste policies which provide clarity, stability and direction for the waste sector. Positive and straightforward political guidance in countries such as Germany, Denmark, Norway, Switzerland, Austria and Ireland, has enabled private companies, the investment community and municipal authorities to swiftly build new infrastructure that provides a better fit to the European waste strategy. Ternary plots that demonstrate how changes in waste management practices have occurred in Denmark, Germany, Norway and the UK are shown in Figure 2(a–d).

In 2008, the European Parliament introduced a new version of the waste hierarchy to its waste legislation, Directive 2008/98/EC, which Member States must introduce into national waste management laws. The Waste Framework Directive, which repealed other Directives, provides a general framework of waste management requirements and sets the basic waste management definitions for the European Union (EU). Article 4 of the Directive lays down a five-step hierarchy of waste management options which must be applied by Member States in this priority order:

1. Prevention
2. Preparing for re-use
3. Recycling
4. Other recovery (*e.g.* energy recovery)
5. Disposal

In line with the waste hierarchy, the 7th Environment Action Programme (EAP) was set out to guide European environment policy until 2020 and to set out a vision until 2050. The EAP identifies three key objectives:

- To protect, conserve and enhance the Union's natural capital;
- To turn the Union into a resource-efficient, green and competitive low-carbon economy; and
- To safeguard the Union's citizens from environment-related pressures and risks to health and wellbeing.

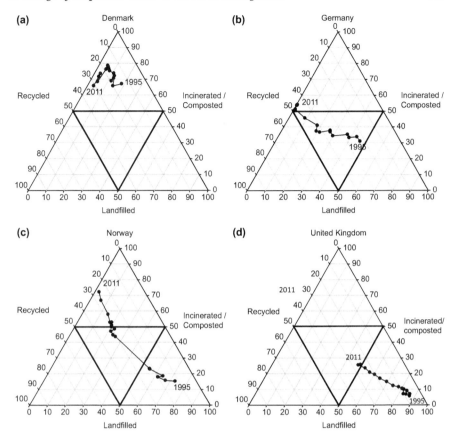

Figure 2 (a–d) Changes to selected countries municipal waste management practices 1995 to 2011, (% by weight).
(Data from European Commission).[6]

The changes we have seen over the last 40 years have been driven by a combination of factors, including European strategy and legislation, increased environmental awareness, the need to decouple waste production from economic growth, and a general drive to promote a more sustainable way of living. Even so, there will clearly have to be a considerable ramping up of activities in many Member States if the EAP's objectives are to be universally achieved.

3   Regulatory and Legal Aspects

In most developed countries, regulatory and legal mechanisms control the transport, storage, treatment and disposal of all types of wastes, including municipal solid, hazardous and nuclear wastes. Waste laws are usually intended to minimize or eradicate the uncontrolled dispersal of waste materials into the environment. Waste laws are often designed to reduce the generation of waste and promote or support recycling. Regulatory activities include identifying and categorizing waste types, and authorising transport, storage, treatment and disposal practices. Outside Europe, countries set their own national regulatory and legal mechanisms for waste management.

In the last 40 years, two EU strategies have been instrumental in driving waste management practices and regulation:

- The Waste Management Strategy and
- The EU Sustainable Development Strategy.

Even though the EC was first asked to draw up an environmental action programme in 1972, it was not until 1989 that its first waste management policy was set out. This formalized the waste hierarchy into management options in the European Commission's Community Strategy for Waste Management, which was further endorsed in the Commission's review in 1996.

Once strategies have been approved and adopted, EU Member States must implement them. Typical methods used to implement waste strategies include:

- Legislation;
- Financial incentives such as taxes or tax breaks (carrots);
- Coercive measures such as bans and restrictions (sticks);
- Education and awareness raising; and
- Voluntary measures.

Legislation is the act or process of making or enacting laws. The key sources of law that affect waste management activities in the EU are regulations, directives and decisions. Regulations are the most direct form of EU law. As soon as a regulation is passed, it has binding legal force throughout every Member State, on a par with national laws. National governments do not have to take action themselves to implement EU regulations. Regulations are passed either jointly by the EU Council and European Parliament, or by the Commission alone. Examples of EU waste management regulations include:

- 880/92 Ecolabelling;
- 259/93 Waste Shipment Regulations;
- 93/98 Control of Transboundary Movements of Hazardous Waste and their Disposal (Basle Convention);
- 94/3 European Waste Catalogue;
- 94/904 Hazardous Waste List;
- 97/138 Packaging and Packaging Waste;
- 2037/2000 Ozone Depleting Substances;
- 2150/2002 Waste Statistics;
- 2635/2003 End-of-Life Vehicles Regulations; and
- 2003/33 Waste Acceptance Criteria.

Directives are addressed to national authorities, who must then take action to make them part of national law. Examples of EU waste management Directives include:

- 75/442 Waste Framework Directive;
- 91/156 Waste Directive;
- 99/31 Landfill Directive;
- 2000/76 Waste Incineration Directive;
- 2000/53 End of Life Vehicles Directive;
- 2002/95 Restrictions of Hazardous Substances Directive;
- 2002/96 Waste Electrical and Electronic Equipment Directive;
- 2008/98 Waste Framework Directive; and
- 2013/56 Batteries and Accumulators Directive.

Decisions are used when the EC concludes that a particular activity is contrary to a particular EU policy. Decisions only apply in certain cases, involving specific authorities or individuals, and can come from the EU Council (sometimes jointly with the European Parliament) or the Commission. They can require authorities and individuals in Member States either to do something or to stop doing something, and can also confer rights on them. EU decisions are addressed to specific parties (unlike regulations) and fully binding. As an example, Decision 2000/532/EC established a classification system for wastes, including a distinction between hazardous and non-hazardous wastes.

The EU has recently adopted ambitious targets to move to virtually zero waste to landfill by 2020 as part of its "Roadmap to a Resource Efficient Europe." A series of medium and long-term objectives are included in the Roadmap and the Commission plans to prepare policy and legislation to implement these aims, as well as focusing its grant-supported initiatives (*e.g.* the LIFE+ and Horizon 2020 programmes) to drive these changes.

4   Collection of Wastes

Household waste collection has evolved considerably since the mid-1970s. In Europe, wastes were typically collected weekly in a single (metal) bin and transported for disposal. There was very little organised separate collection of recyclables or organic (garden or food) wastes. Indeed, many European countries had no system for collecting MSW and wastes were often disposed by household burning.

Now in many European countries, it is common for residual waste to be collected separately from recyclable and organic materials. However, the collection systems used vary considerably in terms of materials collected, types of collection vessels and vehicles used, logistics and frequency of collection, and ultimate destinations for collected materials. Different forms of collection and disposal are used since each regional or municipal authority has to adapt to its own circumstances; each has their own infrastructure, service provision and socio-economic conditions.

Collection is defined under EC Directive 2008/98/EC as "the gathering of waste, including the preliminary sorting and preliminary storage of waste for the purposes of transport to a waste treatment facility." A further definition is provided for separate collection: "a collection where a waste stream is kept separately by type and nature so as to facilitate a specific treatment."

Residual wastes are normally collected in a single receptacle that is usually a wheeled or non-wheeled bin or a plastic bag. Large quantities of residual wastes can be collected in skips. However, the range of containers and approaches used to collect recyclables and compostables has grown considerably in the last 20 years. Containers used include wheeled bins of various sizes (from 180 l upwards) and colours (for identification of materials) for both separate and co-mingled collections, plastic (and sometimes cardboard) boxes of various sizes, colours and shapes for recyclables, clear and opaque plastic sacks, and small "caddies" for food waste. Boxes can be lidded or unlidded. The size and type of containers used depends upon location and type of property. Recyclables and compostables that are collected co-mingled and separately from households can include: glass, cans and tins, aerosols, foil, paper, card, plastic bottles only, mixed plastics, batteries, textiles, shoes, small items of waste electrical and electronic equipment (WEEE) *e.g.* mobile phones, food and garden waste (including Christmas trees). Items that are collected less frequently as "bulky wastes" include: household and garden furniture, large WEEE, bric-a-brac (including toys), bicycles, paints, solvents, mattresses, wood, scrap metal, kitchen units and bathroom suites, carpets, fluorescent tubes, "Do-it-yourself" (DIY) materials (construction and building wastes).

Household bulky waste includes large and heavy items such as furniture and electrical appliances. Bulky waste may be defined as any article of waste which exceeds 25 kg in weight or any article of waste which cannot be fitted into a receptacle of 750 mm diameter and 1 m length. In the UK, collected bulky items represent less than 5% of total household waste, and often contain several different material types that are difficult to separate.[7] Curran *et al.*[8] estimated that UK local authorities recycle approximately 30% of the bulky waste they receive (collected from households and at recycling centres), but only re-use 2–3%. Cameron-Beaumont and Lee-Smith[9] estimated potential re-use rates for household bulky waste of up to 51% for collected items. Bulky wastes are typically collected by local authorities and by authorised groups from the charity and community sector in countries such as the UK. However, across continental Europe, the development of sustainable solutions for formalising non-authorised (informal) waste collections of bulky goods has proved problematic, although the TransWaste project (www.transwaste.eu) has provided some useful insights into the scale, structure, operation, activities and socio-economic consequences of these collections.

Waste collection has become an expensive public service and in developed countries, typically accounts for 75–80% of an authority's municipal solid waste management budget. Authorities that collect waste have limited budgets to operate their services and those that operate separate collections tend to aim for high capture of recyclates at a reasonable cost to house-holders. Achieving a cost-effective and highly efficient dual (or multi-material) collection service requires a high percentage of residents actively participating and presenting the full range of materials accepted without any incorrect materials.

A number of studies have looked at the various systems used for collecting household waste and some key messages have emerged. Residents are typically not allowed to over-fill their wheeled bins or boxes, limiting the amount of non-recyclable waste which they can dispose and forcing them to reduce and recycle. To gain and maintain householder's cooperation, a scheme needs to be both user- and operator-friendly. Simple to understand schemes with frequent collections using containers appropriate to the property's available storage space have a better chance of achieving high participation levels. However, there is no standard design model that can be successfully duplicated nationally or internationally. Each recycling scheme operates in its local context and takes into account situational, demographic, political and cultural factors. This makes planning and designing a waste collection scheme very challenging, especially with local variables such as available disposal points, amendments to existing schemes and contractual issues.

One of the most controversial approaches to waste management has been the introduction of alternate weekly collection (AWC) systems for residual waste and recyclables. In England in particular, there has been vociferous political opposition and a hostile campaign against their implementation. Supporters of AWC argue that providing refuse is stored appropriately and that the scheme is carefully explained to the public, providing only one collection of residual waste each fortnight encourages people to manage their waste more effectively. In theory, an effective AWC system could lead to a range of benefits, including:

- Easier collection of waste;
- Increased waste minimisation, reuse and recycling;
- Reduced collection, disposal and transportation costs; and
- Reduced transport emissions.

It could also lead to increased public awareness about the downstream impacts of purchasing decisions and reduced wear-and-tear on the bodies of operatives in an industry with a poor reputation for the health, safety and welfare of its workforce.

However, opponents of AWC are adamant that leaving refuse uncollected for two weeks is unpopular amongst householders and a "basic right." They claim that fortnightly collection is complicated; unhygienic; attracts vermin, flies and maggots; leads to increased fly-tipping; can be cheaper than weekly collections; and is favoured by "industry bin barons who smell the whiff of easy money in fortnightly collections and municipal bin bullies" (quote from a British Government Minister).

A detailed study by Williams and Cole[10] sought to investigate how implementing an AWC system could impact directly on those factors that affect recycling and waste minimisation behaviour and ultimately the amount of waste generated. Household waste collection trials examined changes to frequency of collection, type of container issued and the amount of sorting required of residents for recyclables. The results of the study are summarised in Table 3. The introduction of the AWC scheme did not have any adverse effects on public participation, household waste arisings or the local environment and no public health problems were reported; this supports the findings of previous studies in the UK and elsewhere. Regardless of collection method, the introduction of an AWC scheme clearly resulted in an increase in the amount of recyclates collected from households. There was a reduction in residual waste collected in both trial areas over the 10 collection dates in 2009 when the AWC was implemented, compared to the same weeks in 2008, although external factors such as the downturn in the global economy need to be considered. In addition, the AWC system was cheaper than the previous waste management service and also resulted in a reduction of transport and employee costs. This study, and others, clearly showed that the arguments put forward by supporters of AWC can be practically demonstrated and realized in independently conducted and

Table 3 Estimated best performance by collection frequency and method (the best performing system is flagged by ★).

	Health, safety and welfare of crew	Productivity (hours worked)	Productivity (weight of recyclates collected h^{-1})	Maximum yield of recyclates	Cost-effectiveness	Simplicity
Weekly collection; kerbside box recycling						
AWC;[a] single stream fully comingled	★	★	★		★	★
AWC;[a] dual stream			★			

[a]AWC = alternative weekly collections of residual waste and recyclables.

evaluated field trials, with no adverse consequences for the public or operatives. It has re-opened the debate about whether we should root our waste management policies on an evidence base or on the political views of democratically elected representatives.

5 Trends in Waste Arisings

5.1 Sources and Composition

All daily activities can give rise to a large variety of different waste flows from different sources. These sources include:

- Household activities (*e.g.* packaging and food preparation wastes);
- Commercial activities (*e.g.* cardboard packaging waste from shops, food waste from restaurants, paper from offices);
- Industrial activities (*e.g.* fly ashes from thermal processes of energy generation, textile waste and tanning liquor from clothes manufacturers);
- Agricultural activities (*e.g.* slurry); and
- Construction and demolition projects (*e.g.* wood, concrete, steel, glass, plastics and wire).

A small part of the waste which is generated is hazardous, that is, it poses substantial or potential threats to human health or to the environment.

In 2010, the total generation of waste from economic activities and households in the EU-27 amounted to 2500 million tonnes. The most important sectors in terms of waste generation were construction (860 million tonnes or 34.4% of the total), mining/quarrying (672 million tonnes or 26.8% of the total) and manufacturing industry (275 million tonnes or 11% of the total) (data from www.ec.europa.eu/eurostat). By comparison, households contributed just 219 million tonnes or 8.7%. The proportional composition of materials in general household waste for different studies in the UK is shown in Table 4.

There is considerable variation in the amount of municipal waste (waste collected by or on behalf of municipal authorities and disposed of through the waste management system) generated in different countries. The highest share of the EU-27 total (2010) was accounted for by Germany (14.5%). However, a different way of looking at these data is to relate municipal waste arisings to population. Using this metric, Latvia generated the lowest level of municipal waste per inhabitant in 2010 (304 kg) among the EU Member States, with the amount of municipal waste generated peaking at 760 kg per inhabitant in Cyprus. The amount of waste generated in any given year depends upon economic activity and prevailing social circumstances. For example, waste generation per inhabitant within the EU was 474 kg in 1995, 527 kg in 2002 and 488 kg in 2012. This means that, on average, each person in European households produced 1.2 kg of waste each day in 2012 (data from www.ec.europa.eu/eurostat).

Table 4 Proportion of materials in general household waste for different UK studies (%).[a]

	England[11] (2006/7)	Haringey[12] (2010)	North London Region[12] (2010)	Hounslow[15] (2012)	Surrey Councils[13] (2010)	Greater Manchester[14] (N/A)
Paper & cardboard	23	20	21	10	14	17
Plastics	10	18	17	14	14	15
Glass	7	9	5	2	4	3
Metals	4	2	4	2	3	3
Food waste	18	28	26	30	29	33
Garden waste	14	6	6	1	7	N/A
Textiles	3	3	3	4	3	N/A
WEEE	2	1	1	1	1	1
Other	20	13	17	N/A	N/A	N/A

[a]Sources: DEFRA,[11] ENTEC UK Ltd,[12] Murphy et al.[13] Wells,[14] Widdowson et al.[15] Note: N/A stands for Not Available.

A significant change in the last 40 years has been that wasted materials are starting to be seen as valuable resources that we can no longer afford to discard. As demand for key resources such as metals and minerals increases, driven by relentless growth in the emerging economies of Asia and South America, competition for resources is growing. Human factors such as geopolitics, resource nationalism, strikes and accidents are most likely to disrupt supply. Policy-makers, industry and consumers are concerned about supply risk, the need to diversify supply from the Earth's resources and about the environmental implications of burgeoning consumption. In response, the EU has drawn up a list of 14 key raw materials which we need to maintain our economy and lifestyle and initiated plans for a "resource efficient Europe". These strategies highlight the need for effective closed-loop resource management systems, especially within industry.

6 Methods of Waste Treatment and Disposal

6.1 Disposal

Landfill
Landfill is a site for the deposit of waste onto or into land (i.e. underground). Landfill also encompasses "landraise" – the creation of a hill from waste where this suits the local topography. It has been the principal method of waste disposal in the EU and elsewhere for the last 100 years or more. Even though recent legislation and a transformation in philosophy from waste to resource management is under way, it is likely that there will be a role for landfill in the disposal of residual waste for some time yet, especially in developing countries and those that have been slow to adopt the waste hierarchy.

In the early 1970s, landfill sites were typically just massive holes in the ground – depressions or voids created by, for example, mineral extraction – in which rubbish was squashed and buried with little regard for the consequences: "out of sight, out of mind". Landfills were generally designed on the "dilute and disperse" principle, which assumes that pollutants are generated slowly and migrate into the surrounding environment *via* chemical, physical, biological and microbiological processes that render them less concentrated until they become harmless. In such sites, wastes were buried with no lining or cap so that leachate and gas were free to seep through the ground and escape into either groundwater or the air. In addition, organic wastes were often mixed with liberal amounts of earth and other non-reactive waste and hence breakdown of waste could be quite slow.

Over the last 40 years, a combination of regulation and research has enabled the best landfill sites to be transformed into highly engineered containment vessels in which the deposited wastes have stabilised physically, chemically and biologically to a state in which the undisturbed contents are unlikely to pose a pollution risk – so-called "completion". Modern landfills are usually divided into carefully engineered cells and lined with an impervious material (clay or plastic) to stop leachate escaping and polluting the nearby land or water. Individual cells are relatively small and, after filling, are sealed, with the disposal of waste being transferred to a fresh cell on the site. Further cells may be engineered horizontally and vertically until the completed site consists of a honeycomb arrangement of cells. Soil or compost is used as daily cover to stop rubbish blowing away. When a site is full, it is sealed with a final impermeable layer, landscaped with soil and topsoil, and returned to society in a suitable condition for development. Advanced landfills are designed to facilitate flushing by rainfall or to recirculate leachate in order to accelerate waste degradation. Whilst flushing can provide landfill gas as a source of energy in the short-term, in the longer term, it can flush out contaminants and bring forward the date of completion. Stringent regulation is applied in many countries to ensure that landfill operators cannot walk away from their sites unless they have demonstrated that they no longer pose a threat to human health or the environment. Operators typically have to complete a risk assessment alongside absolute requirements; most long-term models of risk assume that the pollutants on-site degrade or are reduced in mass over time and balance this against deteriorating engineering systems.

Whilst landfill as a disposal option has a limited future in the EU, it is widely recognised that the polluting potential for landfills is likely to persist for centuries rather than decades. This is significant because there is a legacy of historical landfills across the world. For example, it is estimated that there are approximately 6000–7000 closed (often undocumented and unlicensed) landfill sites across the UK that have received a variety of waste types in the past, including hazardous wastes. Many of these sites have the potential to cause environmental pollution, primarily from the migration of leachate to ground and surface waters and fugitive methane emissions from the ongoing degradation of wastes. In the coming years, there will be a need to develop remediation strategies and technologies for such sites.

Dumping at Sea

In the 1970s, the dumping of waste at sea started to become a global concern. This prompted the Inter-Governmental Conference on the Convention on the Dumping of Wastes at Sea in 1972, known as the "London Convention," which prohibited the dumping of certain hazardous materials into the oceans. However, in 1975, the National Academy of Sciences reported that approximately 14 billion pounds of refuse was dumped every year by sea-going vessels into the oceans. Furthermore, the Great Pacific Garbage Patch – a huge area of plastics, chemical sludge and other debris in the central North Pacific Ocean – was identified in 1988. Despite its huge size, it is not visible to the naked eye because it consists of tiny particles suspended below the ocean's surface. Similar patches have been found in the Atlantic and Indian oceans. The plastic waste in these patches has had significant detrimental impacts on wildlife, including killing marine birds and animals by entering their digestive systems *via* feeding, and facilitating the spread of invasive species that attach to floating plastic and drift long distances to colonize ecosystems in other regions.

More stringent international protocols have been agreed to tackle this massive global problem. In 1988, the International Convention for the Prevention of Pollution from Ships (MARPOL 73/78) came into force, placing restrictions on the types of waste that could be dumped into the oceans and where they could be dumped. In 1993, amendments were adopted that banned the dumping of low-level radioactive wastes, phased out the dumping of industrial wastes (by December 1995) and banned the incineration at sea of industrial wastes. In 1996, a protocol that restricts all dumping of wastes at sea (except for a permitted list, which still require permits) was adopted, although it didn't enter into force until 2006.

Ongoing research suggests that we still don't really understand the quantities and impacts of waste dumped at sea. A recent paper by researchers at the Monterey Bay Aquarium Research Institute[16] shows that refuse is also accumulating in the deep sea.

6.2 Energy Recovery

Incineration

Incineration is a process that is used to combust the organic materials found in waste. Whilst incinerators were originally often used to destruct and dispose of hazardous materials, over the last 40 years they have become a key method to dispose of municipal waste whilst simultaneously recovering value *via* energy recovery. The waste materials are converted into flue gas, ash and heat.

Historically, incineration has been used as the technology of choice for dealing with hazardous wastes that may contain carcinogens, teratogens, mutagens and pathological materials. However, incinerators cannot destroy inorganic substances, although it can concentrate them in ash for disposal *via* landfill or use as part of road aggregate. Incinerators have been controversial in many parts of the world because of public concerns that they may release potentially harmful substances such as dioxins and poly-chlorinated biphenyls to air if combustion does not occur properly. Consequently, the flue gases must be cleaned before release by carefully controlled combustion processes and the use of efficient end-of-pipe technologies such as baghouse filters and electrostatic precipitators.

Modern incineration and other high temperature waste treatment systems have become known as thermal treatment. Alongside other waste-to-energy technologies such as pyrolysis and gasification, incineration with energy recovery, sometimes preceded by in-line recovery of recyclables, has become particularly popular in developed countries with an "engineering culture" (Denmark, Sweden and Germany) or where land is a scarce resource (Japan). Nevertheless, energy recovery and incineration remain relatively uncommon within the EU-27; in 2008, 49% of total waste was disposed and 46% was recovered, with energy recovery and incineration accounting for just 3% and 2%, respectively (data from www.ec.europa.eu/eurostat).

Since the 1970s, countries with robust and strong long-term national waste plans have tended to move from landfill towards an energy-from-waste infrastructure that includes regional incinerators; this has created their status as "recovery societies" (see Table 2 and Figure 2). In Ireland, where the population has historically been against the building of incinerators, the country is moving directly from being a "landfill society" towards becoming a "recycling society." In the UK, the home nations have totally different plans for the future. For example, Wales has detailed and ambitious long-term plans to become a high recycling society by 2025 and a zero waste nation by 2050. In contrast, England has no clear long-term waste management strategy and it is not dominated by a single municipal waste management destination. Here there are over 20 planned, or in construction, incinerators (as of 2013) and therefore England is effectively committing to incineration in the absence of a stated governmental preference.

Incineration with or without energy recovery remains controversial, and because of the up-front costs, once a country commits to this infrastructure, it tends to stick with it for at least 20–30 years.

Anaerobic Digestion

Anaerobic digestion (AD) is a collection of processes in which biodegradable matter is broken down by microorganisms in the absence of oxygen. Historically, it has mainly been used to treat sewage sludge. However, interest in its use to deal with other organic wastes, particularly food and agricultural wastes, has soared in the last 10–15 years. Anaerobic digesters can also be fed by purposefully grown crops such as maize.

AD is thought to be a more sustainable method of dealing with organic waste than composting because it returns nutrients to land and extracts energy that may be used to generate electricity. AD offers significant potential for diversion of organic waste from landfill (it is estimated that in the UK alone almost 17 million tonnes of food is wasted per annum) and energy generation from food waste, *via* biogas, has an energy-generating capability of 12–15 times that of conventional animal wastes. This is significant because AD is deemed a renewable energy and many countries have created incentives for the use of renewable energy in order to simultaneously address the challenges of the likely adverse impacts of climate change and concerns about energy security.

AD has emerged as a new standard for biomass valorisation and in the EU it is attracting increasing investment. Life-cycle assessments of different waste disposal strategies for utilisation of the organic fraction of the MSW have shown AD as inducing significant resource savings and being the most environmentally favourable solid waste management option in terms of both greenhouse gas saving and environmental toxicity to the terrestrial and aquatic environments when compared to aerobic composting, incineration or landfilling.[17,18] Revenues for AD can come from energy (gas, heat and electricity), tipping or service fees (landfill disposal offset), secondary products (digestate, liquid fertiliser, and feedstock for downstream processes), carbon offset credits, and government incentives (renewable energy tax credits and price supports).

The scale of development varies considerably; there are over 4000 anaerobic digesters for non-sewage-derived organic wastes in Germany (many of which are farm-scale), but only about 50 in the UK, and hardly any in countries such as Bulgaria. However, there are a lot of planning applications in the pipeline for AD plants in the UK, with the trend likely to be repeated elsewhere. Public anxieties about the development of AD plants tend to centre on concerns about excessive traffic movements, and malodours. Consequently, there are now strict guidelines and protocols established for the AD sector; in the UK, the British Standards Institution's "Publicly Available Specification" (PAS 110) sets out strict criteria for collection, heat treatment and disposal of the waste material. For these reasons, it is likely that AD will continue to grow as a waste treatment process for the foreseeable future.

6.3   Recycling and Composting

Recycling

Recycling is defined under EC Directive 2008/98/EC as "any recovery operation by which waste materials are reprocessed into products, materials or substances whether for the original or other purposes. It includes the reprocessing of organic material but does not include energy recovery and the reprocessing into materials that are to be used as fuels or for backfilling operations." In the EU-27, the share of municipal waste recycled has grown steadily from 10% in 1995 to 22% in 2008 (source: Eurostat).

Household recycling is justified by four main points:[19]

- It reduces demand for virgin raw materials;
- There are fewer environmental impacts from material extraction, processing and transportation;

- Products made from recyclates rather than virgin materials generally consume less energy in manufacturing; and
- Lower down the hierarchy, less waste is disposed of by the more environmentally damaging methods.

Factors that Influence Recycling

Over the last 40 years, a considerable body of authoritative information has been published on factors that influence recycling (positively or negatively). Most of this information relates to household waste recycling, although there are some reports about specific waste streams from industrial or business sources. These factors are summarised below.

Factors that impact recycling may be categorised as situational variables – infrastructure (I) and service provision (S) – as well as variables that influence behaviour (B) (*e.g.* psychological factors such as values and attitudes), and the complex relationship between the two. The ISB model[20] recognises the range of, and interaction between, individual elements of a recycling system. In the ISB model, infrastructure, service and behaviour has been defined as follows:

(a) Infrastructure – is the built environment, products and objects (*e.g.* buildings, bins, collection vehicles, treatment facilities, waste (composition), packaging, material recovery facilities, incinerators, landfill, recycling re-processing facilities and technologies).
(b) Service – is the systems, providers and enablers that allow people to participate in a particular environmental practice (*e.g.* collections (frequency, method), role of crews, communication materials, perception of customer service and service provider, economic incentives, penalties, markets for reusable and recyclable materials).
(c) Behaviour – relates to people: who we are and our disposition towards the environmental practice (*e.g.* values, attitudes, knowledge, awareness, personalities, lifestyles, communities, social status and norms).

Examples of variables that influence recycling, categorised by I, S and B, are summarised in Table 5.

Infrastructure – Key Recycling Factors

Physical infrastructure for all types of waste can often take a significant period of time (multiple years, sometimes over a decade) to plan, design and build, especially where the process for approval of waste management facilities is complex and involves many stakeholders. Infrastructure such as the most appropriate containers for waste materials and vehicles for transportation should ideally be designed to fit seamlessly alongside and in operational harmony with treatment facilities but in practice there is often a time-lag. There is often significant public opposition to the creation of new waste management infrastructure (especially buildings and outdoor treatment facilities) – the "not-in-my-backyard (NIMBY) syndrome".

Table 5 Examples from the literature of key variables that influence recycling.

Infrastructure
- Property type and space to store waste (MORI,[21] Williams and Timlett,[22] Garcia and Baird[23]).
- Access to a recycling container (Barr and Gilg[24]).
- Type and size of containers provided (Wilson and Williams[25]).
- Type of waste produced (Giorgi *et al.*[26] Timlett and Williams[27]).

Service
- Convenience and making refuse more inconvenient through use of limiting bin size and AWC (Barr and Gilg,[24] MORI[21]).
- Collection scheme design, type and frequency (Martin *et al.*[28] Noehammer and Byer,[29] Reid *et al.*[30] Shaw and Maynard[31]).
- Communication, doorstepping, feedback (Timlett and Williams,[32] Read,[33] Mee *et al.*[34])
- Publicity and promotion.
- Economic incentives, rewards for recycling (Maunder,[35] Woodard *et al.*[36]).
- Economic disincentives *e.g.* pay-per-throw and fines (Price,[37] Martin *et al.*[28] McAnea *et al.*[38])

Behaviour
- Environmental values or attitude *e.g.* pro-environment or pro-recycling (Perrin and Barton,[39] Barr and Gilg[24]).
- Perceived convenience *e.g.* time, effort, safety and cleanliness (MORI[21]).
- Perception of recycling as a social norm *e.g.* high visibility of behaviour, "everyone's doing it", social learning, peer pressure, nearest neighbour effects (Barr and Gilg,[24] Tucker,[40] WWF,[41] Shaw[42]).
- Past behaviour and habit (Knussen *et al.*[43]).
- Personal circumstance *i.e.* income, ethnicity (Perry and Williams).[44]
- Feeling of duty.
- Parental influence (Gunton and Williams[45]).
- Education, awareness raising and information (Perrin and Barton[39]).

There is typically no single, agreed "ideal" design for the process of recycling specific materials. Regulatory requirements and the characteristics and needs of the business or community typically dictate a scheme's design.[29,46] In addition, there are national, regional and local variations in the quantity and composition of household or business waste which have a bearing upon the infrastructure required and the subsequent performance of a recycling scheme.[37]

The development of an effective and efficient infrastructure for solid waste management requires that an integrated view of all relevant waste diversion activities must be taken. The local and regional context is important and any successful change in the design of one scheme may not necessarily be replicable elsewhere.[47] I and S variables generally have to be planned together in order to facilitate the development of an operationally and financially viable recycling scheme. There is considerable evidence that when recycling any material (or co-mingled materials), the more convenient and higher quality the infrastructure, the higher the recycling rates.

Areas with dense housing (includes mixed residential and commercial areas) present infrastructural difficulties for recyclates storage and collections, and reduced service and participation often results. In high-rise estates, spatial ownership needs to be clearly demarcated and maintained. Solutions must be tailored to existing exigencies of the built environment (such as poor vehicular access) and need to include broader infrastructural factors such as functioning lifts and convenient, safe storage facilities.[48] In most medium- and high-density households, storage space for waste, both inside and outside the home is limited or non-existent. Hence, implementing recycling schemes that rely on source segregation of wastes by the householder is problematic and participation is typically low as established "best practice" collection systems (such as requiring households to use and store two or more wheeled bins) used for a majority of (lower density) housing in England cannot be easily implemented.[49]

Recycling is also facilitated by the provision of public recycling centres for household waste, known by various names, including "civic amenity sites" and "household waste recycling centres" (HWRCs). Civic amenity (CA) sites have historically been known rather disparagingly as "the tip" or "the dump" by the public. They were first introduced into the UK in the late 1960s under the Civic Amenity Act 1967, mainly to help reduce fly-tipping, by giving people somewhere to take their bulky waste that would not normally be collected by the refuse collection service. In reality, early sites were nothing more than available areas set aside at landfill sites for waste to be dumped, and very little effort went into site design, services, facilities or promotion. However, it has been progressively recognised that household recycling centres have an important role to play in increasing recycling levels and a lot of effort has been put into improving facilities, including the provision of a wide range of separate containers for recyclables, improved signage and access, and on-site reuse shops. Many UK sites now regularly boast recycling rates of over 90% for deposited materials.

Other public infrastructure that has emerged over the last 40 years to aid recycling includes "bring banks". These are fixed recycling site facilities that allow members of the public to bring their dry recovered materials for recycling whenever they want to. They are usually positioned in easily accessible locations and may be run by operated by local authorities, private and third sector organisations. Bring banks can be used to collect a variety of materials/products, including: glass, metal cans, plastics, paper and card, textiles (clothes, shoes, rags *etc.*), media items (books, CDs, DVDs), and small electrical equipment (*e.g.* lamps, telephones, toasters, kettles, smoke alarms, irons, electric toothbrushes, hair dryers, DVD players, printers, games consoles).

Service Provision – Key Recycling Factors
The service provision for any recycling scheme needs to match adequately with the available infrastructure and needs to be carefully planned with the users in mind. Important design variables for a recycling scheme

include: whether participation is mandatory or voluntary, the type and range of materials collected, the hazard presented by the materials, the degree of sorting required, whether the collection container is provided free of charge, the collection frequency, collection day, whether financial incentives are available and the type of publicity and promotion employed in advertising the scheme.[29]

Mandatory schemes generally achieve higher participation rates than voluntary ones[29] and are most effective when there is active enforcement, increased education, financial incentives and socio-economic factors conducive to law-abiding behaviours.[50] The provision of free containers and degree and costs of enforcement are important factors and well-designed voluntary schemes can still achieve comparable results to mandatory ones.[29,50] However, Jenkins et al.[51] found no difference in the amount of recyclables collected from mandatory and voluntary schemes in the USA.

The amount of effort demanded for recycling will increase with the degree of sorting and preparation of materials prescribed by the scheme, with binary sorting i.e. separation of recyclables from non-recyclables, being more popular than multi-sorting, i.e. separating different recyclables.[29,52,53]

Reducing or simplifying the range of recyclables collected can not only lead to greatly reduced costs but also a slight increase in recovery; conversely, an increase in the range can produce lower diversion due to the increased complexity of the scheme.[54] Reducing the collection frequency can reduce costs without necessarily having a huge impact upon recovery.[47,55] As in the study by Noehammer and Byer,[29] the findings of Everett and Peirce[50] point to a slight gain in recovery when the collection frequency is increased, perhaps because the householder has an increased opportunity to recycle material that would otherwise have been discarded due to a lack of storage space; the same would apply to most businesses. More recent research suggests that alternative collections of household residual wastes and recyclates appear to increase both recycling and set-out rates.[10,25]

Whether or not the collection day for recyclables and general waste is the same appears, according to Noehammer and Byer,[29] to make no difference to participation rates, though the study of Everett and Peirce[50] detected a small increase in recovery when collection days were dissimilar.

A high housing density combined with a highly transient population can impact adversely upon the establishment of sustainable recycling behaviour.[49] Waste Watch conducted case studies and identified different ways to deliver services, including door-to-door and kerbside collections (commonly for low-rise, smaller blocks) and near-entrance and central communal bins. The best collection methods were those that were convenient to use, felt safe and secure. Successful schemes were well designed (in consultation with local residents) and tailored to the location. Good communication was important (e.g. leaflets, branded bins, outreach workers and involvement with residents groups). Even when flats are provided with recycling services, participation is typically low, mainly because of the inconvenience of many schemes. Waste Watch[19] concluded that better recycling performance in flats is more likely to occur when occupants:

(a) Live in a property that is well maintained providing a clean, safe environment;
(b) Are likely to be older or retired (generally have more spare time and socially aware lifestyles);
(c) Have a positive attitude and feel informed about recycling;
(d) Have a sense of ownership (permanent home);
(e) Have a good relationship with housing management; and
(f) Feel part of a community with a central, organising figure.

Behaviour – Key Recycling Factors
In the main, recycling relies on the separation of wastes into different material streams at the point of generation (*i.e.* by households or businesses). Typically, local authorities that recycle higher proportions of household waste require residents to segregate, store and present two or more waste streams for collection. These types of scheme rely heavily on voluntary participation – actual behaviour rather than behavioural intention – by the householder.

A number of studies over the last 40 years have identified and reviewed the key determinants of household recycling behaviour. These determinants include:

(a) Attitude – an individual's organised, stable and usually strongly held view about an issue;[56,57]
(b) Affect – an individual's feelings about performing an action;[58,59]
(c) Agency – an individual's sense that they can carry out an action successfully and that their action will assist in the achievement of an expected outcome;[60–62]
(d) Behavioural intention – a measure of the strength of an individual's attention to perform a specified behaviour;[63]
(e) Cognition – an individual's positive or negative thoughts and beliefs associated with a particular object or system;[59]
(f) Habit and routine – an individual's behaviour that occurs automatically;[64,65]
(g) Personal norms – self-expectation for behaviour;[66]
(h) Self-identity – the extent to which performing a specific behaviour is an important component of an individual's self-concept;[67]
(i) Situational factors – *e.g.* location, logistics, barriers, infrastructure, service, awareness;[64,65]
(j) Social norms – sets of beliefs regarding what other people are doing or what they approve of doing;[68–70] and
(k) Values – an individual's guiding principles; broad preferences concerning appropriate courses of action or outcomes.[71]

The literature on individual determinants of household recycling behaviour is very complex, diverse, contradictory and generally inconclusive, with no agreed consensus about priority or how these determinants interact. In order to understand how such determinants interact in influencing behaviour, researchers have developed models to either illustrate behaviour or to test the suitability of a model for illustrating behaviour. The main (practical) purpose of the models is to enable the development of policy and initiate behavioural change alongside intervention programmes.

There are many different socio-psychological models that seek to explain recycling behaviour. Behavioural models can provide the means to identify the driving forces behind recycling behaviour and in a given area determine the main likely success factors. Many of the models tend to take into account economic and behavioural factors and are often built on psychological theories.

Darnton[72] provides a comprehensive review of models of behaviour and theories of change. Each behavioural change theory or model focuses on different factors in attempting to explain changes of recycling behaviour. Examples of the main models that have been applied to recycling behaviour are listed below:[72,73]

(a) Rational Choice Theory;[74,75]
(b) Subjective Expected Utility;[67,68]
(c) Elaboration–Likelihood Model;[76]
(d) Expectancy–Value Theory;[68,77]
(e) Theory of Reasoned Action;[68]
(f) Norm Activation Theory;[78,79]
(g) Value–Belief–Norm Theory;[65,80]
(h) Structuration Theory;[81]
(i) Structuration and Social Practice;[82]
(j) Theory of Planned Behaviour;[62]
(k) Interpersonal Behaviour;[64]
(l) Attitude–Behaviour–Context Theory;[65,83]
(m) Motivation–Ability–Opportunity Theory;[84]
(n) Subjective Utility and Behaviour;[85]
(o) Theory of Trying and Consumer Action Model;[86,87] and
(p) Interaction and Level of Conflict between Personal and Situational Variables.[88]

Some non-psychological models have also been applied to recycling behaviour, such as comparing the differences between recyclers and non-recyclers, simplifying such profiles into simple conceptual models to clarify understanding. Similar models have considered *e.g.* knowledge, motives, and demographic characteristics,[89] income, gender and education. The behavioural models most commonly applied to household recycling behaviour are the Theory of Reasoned Action and the Theory of Planned Behaviour;

however, none of the models listed above adequately explain recycling behaviour and hence are not really applied in practice. There is no similar body of research pertaining to the recycling behaviours of businesses.

Summary

One of the most significant changes in how we manage our waste in the last 40 years has been the development of systems for capturing recyclable materials. The information in the literature on factors that influence recycling is sometimes complex, often contradictory and difficult to interpret, giving significant problems to those responsible for developing waste strategies, policies and operational plans. Nevertheless, the following factors have emerged as vital to the implementation of successful recycling programmes:

(a) Adequate understanding of key external factors, including the prevailing political and socio-economic climate (especially key market forces, technical, administrative, legal and financial factors);
(b) Adequate understanding of how organisations work, especially how internal decisions are made;
(c) Commitment and demonstrated support for environmental actions;
(d) Sufficient funding;
(e) Adequate publicity, communication and knowledge;
(f) Well planned infrastructure and service provision; and
(g) Reliable contractors.

The availability of an effective recycling infrastructure that enables householders and businesses to recycle their waste is clearly a crucial part of any recycling programme. There is a clear need for new and existing recycling schemes to learn from previous studies and to be carefully matched to the needs of the relevant community. The lack of an overarching national strategy for waste management infrastructure in a country is a significant problem for both local authorities and the business community (producers and consumers), especially with respect to hazardous or infectious materials.

Composting

Composting is an aerobic process where organic material is transformed through decomposition into a soil-like material called compost. It is a process that occurs naturally in the environment but as a controlled process, composting can be an invaluable waste management tool, causing a volume and weight reduction in the raw materials and producing a potentially valuable end product. The product is rendered more stable and made suitable for application to gardens and productive land as a soil improver. When carried out under ideal conditions the only outputs to the atmosphere from composting are carbon dioxide and water. When the same plant matter is disposed of through landfill however, its degradation is far from ideal, with significant potential for harm to the environment, as the main source of methane emissions and a contributor to leachate.

The overall material balance for composting approximates to:

$$C_qH_qO_rN_s \cdot aH_2O + bO_2 \rightarrow C_tH_uO_vN_w \cdot dH_2O + dH_2O + eH_2O + CO_2$$

Organic matter + oxygen → compost + water + carbon dioxide

Over the last 40 years, composting has been carried out at different scales, by a variety of techniques. The main categories of composting and the sources of organic wastes available for feeding them are listed in Tables 6 and 7, respectively. Although all the organic wastes listed in Table 7 could potentially be composted, there are limitations due to health, safety and environmental concerns, the physical and chemical requirements of the process being used and impacts on the quality of the resulting compost. Currently these limitations are only enforced with concern for health and environmental issues, by legislation on waste management licensing, licence-exemptions and composting of animal by-products. In the EU, there are often statutory or voluntary standards for finished compost products. In the UK, PAS 100 sets out a minimum compost quality baseline which composters use as appropriate to the product types and markets targeted.

Materials can be unsuitable for composting for two broad reasons: concerns about health & safety and the balance of materials present.

Some organic materials may carry significant health, safety and environmental risks:

- They are likely to cause odours. Materials rich in nitrogen, such as grass, food waste and manures can lead to emissions of ammonia and volatile organic compounds that have strong, unpleasant odours.
- They attract pests such as rats, birds and flies. Food wastes and manures will attract pests if there are sufficient quantities present.
- They may lead to the growth of bacteria, parasites, pathogens and viruses that are harmful to humans, animals or plants. Animal products in any form including food wastes, manures and animal carcasses could carry harmful bacteria *etc.* that could survive the composting process and be spread with the final material. This could lead to the contamination of grown food in gardens or arable land, nearby water sources and the soil potentially putting animals and people at risk of exposure. The same is true of plant materials carrying diseases but there is an added concern of perennial weeds and weed seeds being spread if they are added to compost.
- They may cause the build-up of environmentally harmful chemicals. Organic wastes can be contaminated with harmful chemicals from fertilisers, pesticides, traffic exhausts, household cleaning products and wastes. If these materials are composted the mass and volume loss during the process can increase the concentrations of the harmful chemicals. Again, this could lead to the contamination of grown food in gardens or arable land, nearby water sources and the soil potentially putting animals and people at risk of exposure.

Table 6 The main categories of composting. (Adapted from McKinley).[90]

Composting category	Compost material sources	Key points
Centralised composting	Large-scale commercial composting fed from municipal sources such as parks and landscaping garden waste, civic amenity site garden waste and any separately collected household garden and kitchen wastes.	Legislature requires monitoring of the process and compost produced to control environmental and public health impacts. Transport intensive due to limits on proximity to the public and large amount and sources of waste. Varying levels of cost and equipment intensity depending on particular system used. Can produce marketable product.
Community composting	Medium-scale, volunteer-based composting fed from locally generated sources of garden waste.	Low transport requirements due to close proximity to waste source. Possible social benefits to community. Not as strictly monitored as centralised composting and typically lower equipment efficiency due to smaller scale.
Supermarket composting	Medium-scale composting to deal with waste fruit, vegetable and flowers produced by supermarkets.	Transportation requirements as produce must be shipped to farms from the supermarkets.
Farm composting	Medium-scale composting dealing with wastes produced on farms including agricultural wastes, animal manure, food production wastes.	Close proximity to waste source.
Home composting	Small-scale composting fed with fractions of household kitchen and garden wastes suitable for composting at that scale.	Proximity to waste source and very low costs with only the optional requirement of a home composting bin. No legislation or monitoring means poor management by individuals may lead to harmful emissions.

The magnitude of the risks above depend on how well the compost process is managed and other factors such as the type and quantities of each feed component, the type of compost bin or heap, the bin location and the temperature the bin reaches during composting. The key composting parameter than can mitigate pathogen-related problems is temperature. Different pathogens require various temperatures for different periods of time to ensure their destruction. Due to this complexity, and the difficulty of reaching and maintaining high temperatures at the home composting scale, it is generally agreed that all potentially hazardous materials should be excluded from home composting. Materials that are typically deemed unsuitable for home composting include dead animals, human and carnivorous animal

Table 7 Sources of organic waste. (Adapted from McKinley).[90]

The domestic waste stream	The local authority waste stream	The commercial waste stream

Kitchen wastes
Garden wastes
	Municipal/park and
	  landscape
	  garden wastes
	Sewage sludge
		Golf course and general
		  commercial garden wastes
		Food leftovers
		Food processing and
		  market wastes
		Manures
		Agriculture, *e.g.* straw
		Abattoir and other animal
		  by-product wastes
		Manufacturing processes,
		  e.g. furniture

excrement, meat (cooked or uncooked), any cooked food, fish, oils and fats, dairy products, diseased plants, contaminated garden wastes, pernicious weeds, non-organic materials such as glass, metals and plastics. In addition, the composting process requires a balance of green and brown materials in order to perform well in terms of odours emitted, the time taken for composting and the quality of the compost produced.

The actual waste diverted by home composting from other sources of disposal is difficult to monitor satisfactorily for a range of reasons including:

- The intensive data collection and logistical issues required to monitor individual households;
- Significant variation in composition and quantities of waste across households and seasonally due to different eating habits and seasonal changes in garden waste;
- Variation in levels of participation at individual households over time due to social factors – enthusiasm for composting, need for compost, forgetfulness; and
- The variable nature of Municipal Solid Waste makes monitoring reductions in collection volumes due to home composting extremely difficult.

The number of households participating in home composting has been estimated several times, mainly by using data on the number of compost bins distributed by local authorities. As an indication of scale, WRAP[91] reported a central estimate of 150 kg household^{-1} year^{-1} diversion from municipal collection *via* home composting in the UK.

6.4 Reuse

Reuse appears after prevention and reduction in the waste hierarchy. It effectively means any operation by which products or components that are not waste are used again for the same purpose for which they were conceived.

Since the 1970s, socio-economic enterprises such as charities, voluntary organisations and not-for-profit companies have increasingly become involved in the repair, refurbishment and reuse of various products.[92] In Europe in particular, there has been a growth of charity and reuse shops, including specialist shops for books, furniture, textiles and bric-a-brac. There is an increasing interest in activities such as "Swap Shops" where people swap items they don't want for items they would like. "Bring and buy" and car boot sales have been hugely popular in the UK for two decades and of course, since the widespread use of the Internet started in the 1990s huge numbers of people have become involved in the online redistributions of resources *via* eBay, Swapit, Freecycle, *etc.*

6.5   Prevention and Minimisation

Waste prevention focuses on actions taken before something becomes waste that reduce the:

- Quantity of waste produced, including the extension of product life through design, repair or reuse;
- Adverse impacts of waste produced on the environment and human health; or
- Content of harmful substances in materials and products.

The revised Waste Framework Directive 2008 placed a requirement on each Member State to develop a waste prevention plan by December 2013. To be really effective, waste must be designed out of the system as much as possible. This will require collaboration between governments, the research community, manufacturers, retailers and distributors. Together with reuse, waste prevention is clearly an area that will see considerable development in the next 40 years.

7   Specific Waste Streams

A number of specific waste streams have emerged as significant problems over the last 40 years. These include waste electrical and electronic equipment (WEEE), food waste, end-of-life vehicles and batteries. Systems for effectively collecting, treating and disposing of these materials are still emerging and being trialled.

Large quantities of natural resources are used in the production of Electrical and Electronic Equipment (EEE). Modern lifestyles increasingly demand EEE which can lead to negative impacts on the environment *via* a life-cycle that includes manufacture, retail, consumption and end-of-life waste management. WEEE represents a large and increasingly diverse waste that must be carefully managed for the benefit of: (a) society as a whole and (b) the environment and public health.[93] It is the fastest growing waste stream worldwide, with an estimated global arising of 20–25 Mt and an

EU-27 total forecast of 12.3 Mt by 2020. It is an extremely difficult waste to handle because it contains a heterogeneous mix of plastics, metals and composites and a number of hazardous substances that pose significant risks to human health and the environment. Furthermore, miniaturisation and design changes in emerging electronic equipment (*e.g.* tablets) will inevitably lead to additional challenges in value recovery at end-of-life. The increasing incorporation of technology into electronic products, their affordability and the consequences of in-built obsolescence, means that they are replaced regularly. WEEE is a resource-rich stream that also represents a source of valuable and increasingly scarce strategic materials, with 7 of the EU's 14 critical raw materials (CRMs) being present in WEEE. However, only a small proportion of WEEE is collected through official channels, with significant quantities being hoarded, sent to landfill or exported. This presents both challenges and opportunities for a paradigm shift towards zero residual WEEE *via* development of a circular economy.

When WEEE is disposed of alongside non-hazardous waste forms and taken to landfill, leaching of metals and plastics can cause environmental contamination. Disposing of WEEE in landfill also means the raw materials and finite metals used in the production of EEE are lost, instead of being recovered and put back into a closed-loop system. Other issues include the adverse environmental and health impacts when WEEE is sold illegally to developing countries, where health and safety regulation is lacking and labour is poorly trained.

We generate huge quantities of food waste. Monier *et al.*[94] estimated that the amount of food waste generated in the EU-27 on an annual basis is in the region of 90 million tons. In the UK, the total amount of food waste generated annually is 16 million tons, with approximately 50% of this coming from household food waste.[91] According to 2006 EUROSTAT data, the UK has the highest amount of household food waste of the EU-27, followed by Germany. In addition, food consumption and production is estimated to contribute 20–30% of the UK's greenhouse gas emissions and to an estimated 20–30% of total household consumption impacts across the EU. Various statistics are quoted on the average amount of food wasted per house, but one stand-out statistic is that UK households throw away about one third of the food they buy.

Whilst food waste can effectively be treated and value can be recovered *via* technologies such as AD, it has become a high social and political priority for attention. The prevailing key opinions may be summarised thus:

- It should be a priority to prevent food waste. Wasting food in a global society where so many are starving is a sign that our society is at best apathetic to the circumstances of others or at worst morally and ethically bankrupt. Effective international awareness, education and prevention campaigns are needed urgently.
- We have a conundrum to solve: the widespread availability of "cheap food" in some parts of the world is important in terms of preventing

food poverty but it also contributes significantly to: (a) the food waste problem (food is not properly valued) and (b) food being grown and exported from regions that have high food poverty to those that have high food wastage and a simultaneous obesity epidemic.

- "End-of-pipe" treatment of food waste is not sensible or cost-effective but we could easily end up building lots of infrastructure to treat and dispose of food waste unless we are vigilant. We should spend money on prevention before we spend money on disposal and treatment.

8 Environmental Impacts of Waste Management

The waste management sector has come under increasing pressure to improve its environmental performance, particularly in terms of measuring, reporting and managing its greenhouse gas emissions and clearly this will affect future strategies and policies. Waste is currently responsible for the emission of 1.4 billion tonnes of carbon dioxide-equivalent climate change-causing emissions, half of which comes from landfill sites; reduced landfilling will aid emissions reduction. In addition, recycling has higher environmental benefits and lower environmental impacts than any other method of waste management. One of the main environmental benefits is the reduced need for resources; a wasteful linear economy of "take-make-dispose" is extremely resource inefficient. Sixty five billion tonnes of raw material entered the economic system in 2010 (data from the Ellen MacArthur Foundation); this amount is expected to rise as demand for goods grows, but could be reduced by creating a circular economy, whereby resource loops are closed, bringing social and economic benefits.

9 The Emergence of the Circular Economy

A circular economy is an alternative to a traditional linear economy in which products are made, used, and finally disposed. In a circular economy, resources are kept in use for as long as possible in order to extract the maximum value from them whilst in use, before recovery and regeneration of products and materials at the end of a resource's life. Material flows through the economy are often visualised as two types: biological nutrients, designed to re-enter the biosphere safely, and technical nutrients, designed to circulate at high quality without entering the biosphere.

The circular economy may have started out as a theoretical construct but it has recently and suddenly become an idea that is accepted by policy makers and businesses. Its key aspirations, to (a) keep resources in economic use for as long as possible; (b) drive greater resource productivity; (c) address emerging resource security/scarcity issues; and (d) reduce the environmental impacts of production and consumption, are a significant departure from the previous focus on managing wastes safely and cheaply. Indeed, many academics argue that a more circular economy would be good for the economy as well as for the environment.

Of course, a global or even regional circular economy will not happen overnight; zero waste to landfill in Europe can only be achieved over a 10–30 year timescale, with a huge co-ordinated and concerted effort focused on waste prevention, minimisation and re-use. It is important to recognize that zero waste is a target to be strived for, not an absolute, and it is perfectly possible that landfill may ultimately be the best option for some types of waste. It is likely that multiple strategies and complex technical interventions will be required to realise the dream of shifting from the current one-way linear resource use and disposal culture to a "closed-loop" circular system modelled on Nature's successful strategies. An indicative list of potential strategies, methods, tools and approaches to improve resource efficiency is provided in Box 1.

Box 1   Potential strategies to improve resource efficiency.

BROAD APPROACHES

- Zero waste
- Industrial ecology
- Eco-design
  - Prolongation of product use (upgrade, re-use, refurbishment)
  - De-materialisation
  - Green chemistry
- Cleaner production methods
- Pollution prevention methods
- Zero emissions
- Natural Capitalism

METHODS AND TOOLS

- Eco-industrial parks
- Industrial symbiosis
- Product stewardship
  - Extended Producer Responsibility
  - Individual Producer Responsibility
- Supply chain management
  - Reverse logistics
  - Remanufacturing
- Selling service rather than product
- End of life management
- Eco-labelling

QUANTIFICATION/ASSESSMENT/MONITORING TOOLS

- Life-Cycle Assessment
- Carbon footprinting
- Environmental Impact Assessment

- Environmental Management Systems
- Industrial metabolism
  - Material Flows Analysis
  - Energy Flows Analysis
- Social networks

OTHER

- Precautionary principle
- Proximity principle
- Social enterprises

In 2014, the EC adopted the communication, *Towards a Circular Economy: A Zero Waste Programme for Europe,* and an annex to establish a common and coherent EU framework to promote the circular economy. The EC also adopted a legislative proposal to review recycling and other waste-related targets in the EU and it estimates that the new targets would create an additional 180 000 jobs. The proposal includes targets to:

- Increase recycling/re-use of municipal waste to 70% in 2030;
- Increase packaging waste recycling/re-use to 80% in 2030 with material-specific targets set to gradually increase between 2020 and 2030 (to reach 90% for paper by 2025 and 60% for plastics, 80% for wood, 90% of ferrous metal, aluminium and glass by the end of 2030);
- Phase out landfilling by 2025 for recyclable (including plastics, paper, metals, glass and bio-waste) waste in non-hazardous waste landfills – corresponding to a maximum landfilling rate of 25%; and
- Reduce food waste generation by 30% by 2025.

Some evidence has already emerged – from organisations such as the Ellen MacArthur Foundation – to indicate that movement towards a circular economy in Europe will bring the anticipated benefits. An example of how research outputs are providing evidence to support policies is shown in Box 2.

A key part of the project was to create a common vision[95] and a production model. The key concepts, guiding principles, technologies, methods and tools that have been distilled into the key strategies that underpin the ZeroWIN approach are: designing waste out of the system; industrial symbiosis and closed-loop supply chain management; use of effective waste prevention methods and new technologies; applying Individual Producer Responsibility (IPR); and accurate monitoring and assessment of results.

These key concepts formed the foundation for 10 demonstration case studies, which were completed in 2013. The crucial concept is an "Industrial Network" – a physical framework for cooperation between network members aimed at zero waste and resource conservation. The development of an industrial symbiosis is a long-term process requiring extensive data collection,

Box 2   Project ZeroWIN.

Project ZeroWIN – Towards Zero Waste in Industrial Networks – was an ambitious EU-funded project researching – and trialling by means of case studies with industrial partners – methods and strategies to eliminate the wasteful consumption of resources in key industrial sectors in Europe, primarily *via* the formation of industrial networks. The project ran from 2009–2014. It had 30 academic, research and industrial partners across Europe, and one partner in Taiwan. The companies involved ranged from small charitable organisations to multi-nationals *e.g.* Continental, Hewlett Packard.

The ZeroWIN project has shown how existing approaches and tools can be improved and combined to best effect in an industrial network and how innovative technologies and design innovations can contribute to achieving a circular economy. It focused on two key waste types in four sectors: high-tech waste from three sectors: electrical and electronic equipment, automotive, photovoltaic; and construction and demolition waste. The main aims of the project were to show that the approach could facilitate a circular economy and enable industrial networks to meet at least two of the following stringent targets:

- 30% reduction of greenhouse gas emissions;
- 70% overall re-use and recycling of waste;
- 75% reduction of fresh water use.

---

data analysis, facilitating contacts between various non-related industries and overcoming various barriers. The project's website (www.zerowin.eu) outlines the methodology that was developed and utilized.

The key results from the project are a quantitative assessment of the performance of the ZeroWIN approach by 10 case studies applying the production model. Other outputs have included recommendations to policy-making; the creation of a Resource Exchange Platform; the delivery of education, training and support services; and a guide to industry and business on how to save resources. Table 8 summarises the achievements of each project against the agreed targets, and shows that the ZeroWIN project met its overall environmental targets.

The consortium has successfully developed a vision that others can follow, delivered their demonstration case studies and met their ambitious environmental targets. The ZeroWIN case studies provide insight into the challenges and obstacles existing in the studied industry sectors. They can be seen as good lessons learned from the merger of academic theory and vision with industrial practice, and provide both objective evidence and inspiration for the future development of a resource efficient Europe. ZeroWIN was an ambitious project set with difficult goals, but the demonstration activities have shown that these challenges can be met and that society can move towards a circular resource economy whilst simultaneously addressing

Table 8 Quantification of environmental improvements achieved by the ZeroWIN project, by case study. (Data from the Institute of Waste Management, University of Natural Resources and Life Sciences, Vienna, Austria).

Target	Decrease of 30% GHG emissions	Reduction of 75% of fresh water utilisation	70% of overall reuse and recycling of waste
D4R laptop	66%	65%	87%
PV – stand alone	45%	41%	91%
PV – smart grid	>100%	>100%	91%
Reuse network – desktop computer	66%	64%	100%
Reuse network – laptop computer	69%	66%	91%
Reuse network – data logger	50%	50%	100%
New building construction – UK	58%	43%	93%
New building construction – Portugal	35%	26%	97%
Refurbishment of building – Germany 1	19%	14%	78%
Refurbishment of building – Germany 2	38%	>100%	85%
Demolition of pre 1950s building – UK	>100%	>100%	99%
Demolition of 1950–1980 buildings – UK	>100%	37%	99%
Demolition of buildings – Portugal	>100%	>100%	99%

the adverse impacts of industrial pollution and sustainably managing resources.

10 Summary

Since the introduction of the waste hierarchy in 1975, we have slowly but steadily started to take collective responsibility for our environmental assets – the resources upon which our standard of living depends, and the eco-systems fundamental to life. We are being forced to act by the realisation that the availability of the resources is declining, the climate impacts are real, and in some areas the physical space for traditional waste management practices is becoming scarce. In place of "bury or burn" waste disposal practices, waste prevention, reuse, recycling and the circular economy are perceptibly gaining traction, ensuring that the burden of waste is reduced, the economy is developed sustainably and today's products can form tomorrow's raw materials by design.

There is widespread acknowledgement that European countries have led the resource revolution, with an emphasis on increased producer responsibility; design of products for recycling, repair, refurbishment and reuse

(D4R); sustainable (or green) procurement; waste minimization; recycling; high technology incineration and landfill management.

Over this period, waste (now resource) management has become a multi-disciplinary subject, incorporating all engineering and science disciplines; politics; economics; urban and rural planning; law; social sciences; media and communications; information technology; advertising; marketing; design; transportation; logistics and operational management; business studies; management; and even the creative arts. Resource management has never had a higher public profile, but its complexity and multi-disciplinary nature is still not properly recognized by a society that continues to under-value its importance to its quality of life.

References

1. L. Herbert, *Centenary History of Waste and Waste Managers in London and Southeast England*, CIWM, Northampton, England, 2010.
2. I. D. Williams, Editorial: The future for waste (resource) management education, *Waste Manage.*, 2014, **34**(11), 1909–1910.
3. DEFRA, Guidance on the legal definition of waste and its application. 2012, retrieved September 08 2014, from https://www.gov.uk/government/uploads/system/uploads/attachment_data/file/69590/pb13813-waste-legal-def-guide.pdf.
4. OECD, *OECD Glossary of Statistical Terms – Waste Definition*, 2014, Retrieved September 08, 2014, from http://stats.oecd.org/glossary/detail.asp?ID=2896.
5. US EPA, *Non-Hazardous Waste*, 2013, Retrieved September 08, 2014, from http://www.epa.gov/epawaste/nonhaz/index.htm.
6. European Commission, *Waste Generation and Treatment Data Tables*, 2011, Available: http://epp.eurostat.ec.europa.eu/portal/page/portal/environment/data/main_tables. Last accessed 17/01/2014.
7. A. Curran and I. D. Williams, The role of furniture and appliance re-use organisations in England and Wales, *Resour., Conserv. Recycl.*, 2010, **54**, 692–703.
8. A. Curran, I. D. Williams and S. Heaven, Management of household bulky waste in England, *Resour., Conserv. Recycl.*, 2007, **51**(1), 78–92.
9. C. Cameron-Beaumont and C. Lee-Smith, *Bulky Waste Collections—Maximising Re-use & Recycling: A Step-by-step Guide*, Network Recycling, in association with the Furniture Reuse Network, Bristol, UK, December 2005.
10. I. D. Williams and C. Cole, The impact of alternate weekly collections on waste arisings, *Sci. Total Environ.*, 2013, **445–446**, 29–40.
11. DEFRA, Municipal Waste Composition: A Review of Municipal Waste Component Analyses, *Department for Environment, Food and Rural Affairs*, Research Project Final Report, 2009, London.
12. ENTEC UK Ltd, North London Waste Authority, Waste Composition Analysis Project for NLWA, 2010, Final Report, Cheshire.

13. L. Murphy, D. Pawson and D. Parrott, Greater Manchester Waste Disposal Authority, Greater Manchester Waste Composition Analysis and Survey, 2012, Final Report, Manchester.
14. P. Wells, Surrey Councils Residual Waste Composition Analysis, 2010, Final Report, Birmingham.
15. S. Widdowson, J. Fitzgerald and C. Martinez, *The Composition of Municipal Solid Waste – Season 3 Results*, Hounslow Council, Didcot, 2012.
16. K. Schlining, S. von Thun, L. Kuhnz, B. Schlining, L. Lundsten, N. Jacobsen Stout, L. Chaney and J. Connor, Debris in the deep: Using a 22-year video annotation database to survey marine litter in Monterey Canyon, Central California, USA, *Deep Sea Res., Part I*, 2013, **79**, 96–105.
17. H. H. Khoo, T. Z. Lim and R. B. H. Tam, Food waste conversion options in Singapore: Environmental impacts based on an LCA perspective, *Sci. Total Environ.*, 2010, **408**, 1367–1373.
18. S. Evangelisti, P. Lettieri, D. Borello and R. Clift, Life cycle assessment of energy from waste via anaerobic digestion: A UK case study, *Waste Manage.*, 2014, **34**(1), 226–237.
19. Waste Watch, 2003, Available at www.wastewatch.org.uk [Accessed 6/10/05].
20. R. Timlett and I. D. Williams, The ISB Model (Infrastructure, Service, Behaviour): A tool for waste practitioners, *Waste Manage.*, 2011, **31**(6), 1381–1392.
21. MORI. *Recycling and Waste Minimisation Communications Strategy: Research Study Conducted for Project Integra*, MORI, London, UK, 2004.
22. I. D. Williams and R. E. Timlett, Recycling behaviour in high-density housing: a case study, in *Proceedings of Waste 2006; Sustainable Waste and Resource Management*, ed. B. Vegh and J. Royle, Stratford-upon-Avon, 19–21 September, Session 3, The Waste Conference Ltd, Coventry, UK, 2006, pp. 133–144.
23. S. Garcia and J. Baird, Accessing options for increasing recycling rates in Scotland. Case study: City of Edinburgh Council, in *Proceedings of Waste 2008: Waste & Resource Management – A Shared Responsibility*, ed. M. Townshend, Stratford-upon-Avon, UK, 16–17 September, Session 14, The Waste Conference Ltd, Stanton-on-the-Wolds, UK, 2008, pp. 515–526.
24. S. Barr and A. Gilg, Conceptualising and analysing household attitudes and actions to a growing environmental problem: Development and application of a framework to guide local waste policy, *Appl. Geogr.*, 2005, **25**, 226–247.
25. C. Wilson and I. Williams, Kerbside Collection: a case study from the north-west of England, *Resour., Conserv. Recycl.*, 2007, **52**, 381–394.
26. S. Giorgi, C. Wilkins and J. Cox, Enhancing participation in food waste collection schemes: household behaviour and motivations, in *Proceedings of Waste 2008: Waste & Resource Management – A Shared Responsibility*, ed. M. Townshend, Stratford-upon-Avon, UK, 16–17 September, Session 5, The Waste Conference Ltd, Stanton-on-the-Wolds, UK, 2008, pp. 171–180.

27. R. Timlett and I. Williams, *Recycling in a densely populated urban environment*. Conference proceedings of *Waste: the Social Context*, 2008a, Alberta, Canada, May 11–15.

28. M. Martin, I. D. Williams and M. Clark, Social, cultural and structural influences on household waste recycling: a case study, *Resour., Conserv. Recycl.*, 2006, **48**(4), 357–395.

29. H. C. Noehammer and P. H. Byer, Effect of design variables on participation in residential curbside recycling programs, *Waste Manage. Res.*, 1997, **15**, 407–427.

30. T. Reid, T. Brindley and J. Baird, Improving performance of local authority recycling programmes in Scotland, in *Proceedings of Waste 2008: Waste & Resource Management – A Shared Responsibility*, ed. M. Townshend, Stratford-upon-Avon, UK, 16-17 September, Session 14, The Waste Conference Ltd, Stanton-on-the-Wolds, UK, 2008, pp. 527–536.

31. P. J. Shaw and S. J. Maynard, The potential of financial incentives to enhance householders' kerbside recycling behaviour, *Waste Manage.*, 2008, **28**, 1732–1741.

32. R. Timlett and I. Williams, Public participation & recycling performance in England: a comparison of tools for behaviour change, *Resour., Conserv. Recycl.*, 2008b, **52**(4), 622–634.

33. A. Read, A weekly doorstep recycling collection, "I had no idea we could!" Overcoming the barriers to participation, *Resour., Conserv. Recycl.*, 1999, **26**, 217–249.

34. N. Mee, D. Clewes, P. S. Phillips and A. D. Read, Effective implementation of a marketing communications strategy for kerbside recycling: a case study from Rushcliffe, UK, *Resour., Conserv. Recycl.*, 2004, **42**, 1–26.

35. A. Maunder, The use of incentives to change householder behaviour to improved waste management practices in England, in *Proceedings of Waste 2006; Sustainable Waste and Resource Management*, ed. B. Vegh and J. Royle, Stratford-upon-Avon, 19–21 September, Session 11, The Waste Conference Ltd, Coventry, UK, 2006, pp. 459–468.

36. R. Woodard, F. Firoozmand and M. K. Harder, The use of cash vouchers to incentivise householders to recycle, in *Proceedings of Waste 2006; Sustainable Waste and Resource Management*, ed. B. Vegh and J. Royle, Stratford-upon-Avon, 19–21 September, Session 11, The Waste Conference Ltd, Coventry, UK, 2006, pp. 477–486.

37. J. Price, The landfill directive and the challenge ahead: demands and pressures on the UK householder, *Resour., Conserv. Recycl.*, 2001, **32**(3–4), 333–348.

38. P. McAnea, K. Williams, C. Lowe and M. Clark, Public attitudes and behaviour to using free to use and pay by use civic amenity sites, in *Proceedings of Waste 2008: Waste & Resource Management – A Shared Responsibility*, ed. M. Townshend, Stratford-upon-Avon, UK, 16–17 September, Session 5, The Waste Conference Ltd, Stanton-on-the-Wolds, UK, 2008, pp. 149–158.

39. D. Perrin and J. Barton, Issues associated with transforming household attitudes and opinions into material recovery: a review of two kerbside recycling schemes, *Resour., Conserv. Recycl.*, 2001, **33**, 61–74.

40. P. Tucker, Normative influences in household waste recycling, *J. Environ. Plann. Manage.*, 1999, **42**, 63–82.

41. WWF, *Community learning and action for sustainable living (CLASL) Final report*, 2008. Available at: http://www.wwflearning.org.uk/localmatters/resources/community-learning,1346,AR.html [accessed 20/02/09].

42. P. J. Shaw, Nearest neighbour effects in kerbside household waste recycling, *Resour., Conserv. Recycl.*, 2008, **52**, 775–784.

43. C. Knussen, F. Yule, J. MacKenzie and M. Wells, An analysis of intentions to recycle household waste: The roles of past behaviour, perceived habit, and perceived lack of facilities, *J. Environ. Psychol.*, 2004, **24**, 237–246.

44. G. D. R. Perry and I. D. Williams, The participation of ethnic minorities in kerbside recycling: a case study, *Resour., Conserv. Recycl.*, 2007, **49**(3), 308–323.

45. H. B. Gunton and I. D. Williams, *Waste minimisation using behavioural change techniques: a case study for students*, in *Waste Matters: Integrating Issues*, ed. P. Lechner, Proceedings of 2nd BOKU Waste Conference, Vienna, 2007, 303-14.

46. I. D. Williams and J. Kelly, Green waste collection and the public's recycling behaviour in the Borough of Wyre, England, *Resour., Conserv. Recycl.*, 2003, **38**(2), 139–159.

47. P. Tucker, J. Grayson and D. Speirs, Integrated effects of a reduction in collection frequency for a kerbside newspaper recycling scheme, *Resour., Conserv. Recycl.*, 2000, **31**, 149–170.

48. C. Alexander, C. Smaje, R. Timlett and I. D. Williams, Improving social technologies for recycling, *ICE Proc., Waste Resour. Manage.*, 2009, **162**(WR1), 15–28.

49. R. Timlett and I. D. Williams, The impact of transient populations on recycling behaviour in a densely populated urban environment, *Resour., Conserv. Recycl.*, 2009, **53**(9), 498–506.

50. J. W. Everett and J. J. Peirce, Curbside recycling in the USA: convenience and mandatory participation, *Waste Manage. Res.*, 1993, **11**, 49–61.

51. R. R. Jenkins, S. A. Martinez, K. Palmer and M. J. Podolsky, The determinants of household recycling: a material-specific analysis of recycling program features and unit pricing, *J. Environ. Econ. Manage.*, 2003, **45**(2), 294–318.

52. S. S. Chung and C. S. Poon, Hong Kong citizens' attitude towards waste recycling and waste minimization measures, *Resour., Conserv. Recycl.*, 1994, **10**(4), 377–400.

53. A. Bruvoll, B. Halvorsen and K. Nyborg, Households' recycling efforts, *Resour., Conserv. Recycl.*, 2002, **36**, 337–354.

54. C. Thomas, Public understanding and its effect on recycling performance in Hampshire and Milton Keynes, *Resour., Conserv. Recycl.*, 2001, **32**, 259–274.

55. R. Woodard, M. K. Harder, M. Bench and M. Philip, Evaluating the performance of a fortnightly collection of household waste separated into compostables, recyclates and refuse in the south of England, *Resour., Conserv. Recycl.*, 2001, **31**(3), 265–284.

56. G. W. Allport and H. S. Odbert, Attitudes, in *Handbook of Social Psychology*, ed. E. M. Murchison, Clark University Press, Massachusetts, USA, 1935.

57. l. Ajzen, Nature and Operation of Attitudes, *Annu. Rev. Psychol.*, 2001, **58**, 27–52.

58. R. B. Zajonc, Feeling and thinking: preferences need no inferences, *Am. Psychol.*, 1980, **35**, 117–123.

59. M. Thompson, M. Zanna and D. Griffin, Lets not be indifferent about (attitudinal) ambivalence, in *Attitude Strength. Antecedents and Consequences*, ed. R. E. Petty and J. A. Krosnick, Ohio State University Press on Attitudes and Persuasion, Erlbaum, Mahwah, NJ, 1995, vol. 4 , pp. 361–386.

60. D. J. Bem, *Beliefs, Attitudes, and Human Affairs*, Brooks/Cole, Belmont, CA, 1970.

61. A. Bandura, *Self-efficacy: The Exercise of Control*, WH Freeman, New York, 1977.

62. I. Ajzen, The theory of planned behaviour, *Organ. Behav. Hum. Decis. Processes*, 1991, **50**, 1–33.

63. S. Han, *Individual Adoption of Information Systems in Organisations: A Literature Review of the Intention-based Theories, TUCS Technical Report No 539*, Turku Centre for Computer Science, Finland, 2003.

64. H. C. Triandis, Values, Attitudes and Interpersonal Behaviour, in *Nebraska Symposium on Motivation*, University of Nebraska Press, Lincoln, 1980, pp. 195259.

65. P. Stern, Toward a Coherent Theory of Environmentally Significant Behaviour, *J. Soc. Issues*, 2000, **53**(3), 407–424.

66. S. Schwartz and J. Fleishman, Personal norms and the mediation of legitimacy effects on helping, *Soc. Psychol.*, 1978, **41**, 306–315.

67. A. H. Eagly and S. Chaiken, *The Psychology of Attitudes*, Harcourt Brace Jovanich, Fort Worth, TX, 1993.

68. I. Ajzen and M. Fishbein, *Understanding Attitudes and Predicting Social Behaviour*, Prentice Hall, Englewood Cliffs, N.J., 1980.

69. R. B. Cialdini, R. R. Reno and C. C. Kallgren, A focus theory of normative conduct: Recycling the concept of norms to reduce littering in public places, *J. Pers. Soc. Psychol.*, 1990, **58**, 1015–1026.

70. P. W. Schultz, Changing Behaviour with Normative Feedback Interventions: A Field Experiment on Curbside Recycling, *Basic Appl. Soc. Psychol.*, 1998, **21**(1), 25–36.

71. S. Barr, A. W. Gilg and N. J. Ford, Differences between household waste reduction, reuse and recycling behaviour: a study of reported behaviours, intentions and explanatory variables, *Environ. Waste Manage.*, 2001, **4**(2), 69–82.

72. A. Darnton, Reference Report: An overview of behaviour change models and their uses, Centre for Sustainable Development, 2008, University of Westminster, July 2008.

73. T. Jackson, Motivating sustainable consumption—A review of models of consumer behaviour and behavioural change, *A Report to the Sustainable Development Research Network*, Policy Studies Institute, London, 2005.

74. G. Homans, *Social Behaviour: Its Elementary Forms*, Routledge and Kegan Paul, London, 1961.

75. J. Elster, *Rational Choice*, Basil Blackwell, Oxford, 1986.

76. R. E. Petty and J. T. Cacioppo, *Communication and Persuasion: Central and Peripheral Routes to Attitude Change*, Springer-Verlag, New York, 1986.

77. M. Fishbein, The prediction of behaviour from attitudinal variables, in *Advances in Communications Research*, ed. C. Mortenson and K. Serene, Harper and Row, New York, 1973, pp. 3–31.

78. S. Schwartz, Normative Influences on Altruism, in *Advances in Experimental Social Psychology*, ed. L. Berkowitz, Academic Press, New York, 1977, vol. 10.

79. S. Schwartz, Universals in the Context and Structure of Human Values: theoretical advances and empirical tests in 20 countries, in *Advances in Experimental Psychology 25*, ed. M. Zanna, Academic Press, San Diego, 1992, pp. 1–65.

80. R. Stern, T. Dietz, G. Abel and L. Kalof, A value-belief norm theory of support for social movements: the case of environmental concern, *Hum. Ecol. Rev.*, 1999, **6**, 81–97.

81. A. Giddens, *The Constitution of Society: Outline of the Theory of Structuration*, Polity Press, Cambridge, 1984.

82. G. Spaargaren and B. van Vliet, Lifestyle, consumption and the environment: the ecological modernisation of domestic consumption, *Soc. Nat. Resour.*, 2000, **9**, 50–76.

83. P. Stern, S. Oskamp, Managing Scarce Environmental Resources. in *Handbook of Environmental Psychology*, ed. D. Stokois and I. Altman, Wiley, New York, 1987, pp. 1043–1088.

84. F. Olander and J. Thogerson, Understanding Consumer Behaviour as Prerequisite for Environmental Protection, *J. Consumer Policy*, 1995, **18**, 345–385.

85. J. F. Franco and E. Huerta, Modelling household behaviour in municipal solid waste recycling programmes, *J. Waste Manage. Resour. Recovery*, 1997, **3**(4), 145–155.

86. R. Bagozzi and P. Warshaw, Trying to Consume, *J. Consumer Res.*, 1990, **17**, 127–140.

87. R. Bagozzi, Z. Gurnao-Canli and J. Priester, *The Social Psychology of Consumer Behaviour*, Open University Press, Buckingham, UK, 2002.

88. J. Corraliza and J. Berenguer, Environmental Values, Beliefs and Actions. A Situational Approach, *Environ. Behav.*, 2000, **32**(6), 832–848.

89. J. Vining and A. Ebreo, Predicting recycling behaviour from global and specific environmental attitudes and changes in recycling opportunities, *J. Appl. Soc. Psychol.*, 2001, **22**(20), 1580–1607.

90. S. P. McKinley, Physical chemical processes and environmental impacts associated with home composting, PhD Thesis, University of Southampton, UK, 2008.

91. WRAP, *New Estimates for Household Food and Drink Waste in the UK*, Waste and Resources Action Programme, Banbury, United Kingdom, 2011.

92. F. O. Ongondo, I. D. Williams, J. Dietrich and C. Carroll, ICT Reuse in socio-economic enterprises, *Waste Manage.*, 2013, **33**, 2600–2606.

93. F. Ongondo, I. D. Williams and T. J. Cherrett, How are WEEE doing? A global review of the management of electrical and electronic wastes, *Waste Manage.*, 2011, **31**(4), 714–730.

94. V. Monier, S. Mudgal, V. Escalon, C. O'Connor, G. Anderson, H. Montoux, H. Reisinger, P. Dolley, S. Olgivie and G. Morton, Preparatory study on Food Waste across EU 27, 2010, Final Report, European Commission, Brussels.

95. T. Curran and I. D. Williams, A zero waste vision for industrial networks in Europe, *J. Hazard. Mater.*, 2012, **207–208**, 3–7.

From "Dilute and Disperse" to "Recycle & Reuse" – Changes in Attitude and Practice of Effluent Management

DAVID TAYLOR

ABSTRACT

At the beginning of the 1970s, the time of the first UN Conference on the Human Environment, most of the major rivers in the developed world were in a parlous state with their natural purification capacity far exceeded by the volume of effluents that they were receiving from both industrial and domestic discharges. Since then there have been radical changes both in environmental governance and our understanding of the issues. During the last 40 years many of the problems of the 1970's caused by effluent discharges have been resolved and a dramatic improvement in the aquatic environment has occurred. However, both incremental advances and step changes in scientific knowledge will continue to create new challenges for effluent management in the decades to come.

1 Effluent Discharges Before 1972 – The Historical Background

The human race has always had a problem about what to do with its waste. When we were hunter gatherers the solution was easy, waste was simply left behind when the community moved on. However, as societies became less nomadic and more pastoralist this simple solution rapidly became

Issues in Environmental Science and Technology No. 40
Still Only One Earth: Progress in the 40 Years Since the First UN Conference on the Environment
Edited by R.E. Hester and R.M. Harrison
© The Royal Society of Chemistry 2015
Published by the Royal Society of Chemistry, www.rsc.org

intolerable and alternative methods of waste disposal, which ensured a separation of the waste from the community, became necessary. An early example can be found in the history of the Israelites after the exodus from Egypt.[1]

> *"Thou shalt have a place also without the camp, whither thou shalt go forth abroad; and thou shalt have a paddle upon thy weapon; and it shall be, when thou wilt ease thyself abroad, thou shalt dig therewith, and shalt turn back and cover that which cometh from thee."*

Rivers and streams quickly became an obvious solution to the problem since anything released into the water would be carried downstream and away from current human habitation. However it soon became apparent that the capacity of rivers was not infinite and that this practice needed to be regulated to ensure satisfactory conditions were maintained. In Great Britain, the first known regulations relating to effluent disposal into streams and rivers appeared in 1388[2] during the reign of Richard II. This piece of statute law specified:

> *"That none of what condition so ever he be, cause to be cast or thrown, any such annoyance, garbage, dung, intrails, nor any other ordure into the ditches, rivers, water and other places aforesaid."*

Despite the strict nature of the legislation this method of waste disposal remained commonplace across the globe for several more centuries. Waste material was frequently simply thrown into the street, often from first and second floor windows with a cheery cry of "gardyloo", and subsequently washed into the nearest watercourse by the rain.

The sentiments expressed in the 1388 UK legislation are, in fact, not that dissimilar to those contained six centuries later in the European Union Urban Wastewater Directive.[3] This is a classic example which demonstrates that it is the *implementation* of legislation that leads to environmental improvement, not simply the passing of the legal instrument.

Although the practice of tipping human waste directly into rivers and streams was tolerated well into the twentieth century, the problems associated with this practice had become acute in large conurbations such as cities, by the early nineteenth century. In 1854, John Snow first demonstrated that diseases such as cholera could be linked directly to contaminated water supplies.[4] In London the problems finally came to a peak in the "Great Stink" of 1858.[5] Such was the overpowering smell from the Thames, that the curtains of the House of Commons were soaked in chloride of lime in a vain attempt to protect the sensitivities of MPs. It is no surprise that a bill was rushed through parliament and became law in 18 days, to provide more money to construct a massive new sewer scheme for London and to build the Embankment along the Thames in order to improve the flow of water and of traffic.

The system devised and installed in London by the great Victorian engineer John Bazalgette was a model subsequently widely adopted in cities across the world. Bazalgette and other Victorian engineers recognised that the rivers had an inherent purification capacity which would ameliorate the impact of organic matter such as domestic sewage. However they recognised that it was essential not to exceed this natural capacity and thus the amount of organic waste that each section of a river could deal with needed to be constrained. In London, he therefore constructed over 85 miles of new sewers, and associated pumping stations, to intercept sewers that drained into the Thames and its many tributaries. The combined flows were then transported *via* two very large sewers, one on each bank of the river, to discharge into deeper faster flowing water further down the estuary at Beckton and Crossness. This scheme resolved the immediate problems and was seen as a major improvement; however problems with river quality were still found especially during hot weather and in 1887 treatment works to enable treatment of the sewage by chemical precipitation were constructed and brought into use at both Beckton and Crossness.

Problems were not restricted to the impact of human sewage. In the UK the rapid industrialization of the late 18th and 19th centuries led to increasing discharges of untreated industrial waste into rivers. Conditions became so bad in some Lancashire rivers that a Royal Commission was established in 1868 to enquire into the pollution of rivers. Their report in 1874[6] subsequently led, in 1876 to the first in a long series of Rivers Pollution Prevention Acts.[7] However like previous attempts, simply passing legislation did not, of itself, cure the problems.

The Cuyahoga River in Ohio, USA, for example was one of the most polluted rivers in the United States. The reach from Akron to Cleveland was devoid of fish and the oil contamination of the surface waters caused the river to catch fire on several occasions between 1868 and 1969 with a particularly serious fire in 1952 that caused over \$1 m of damage.[8] In Europe major rivers such as the Danube and the Rhine were severely polluted. Water quality problems in the Rhine had been identified in the 15th century, but the severe deterioration in the water quality of the river did not become apparent until the late 1960s. By that time, the pollution of the Rhine with organic substances had led to acute oxygen problems in the river and almost all aquatic life had disappeared. In the United Kingdom one of the first reports to emerge from the newly established Royal Commission on the Environment in 1972[9] was devoted to the problems of pollution in British Estuaries and coastal waters and this was followed in 1973 by a more detailed review of the unacceptable state of four typical UK estuaries, the Clyde, Humber, Mersey and Tees.[10]

So despite growing appreciation of the issues, by the beginning of the 1970s, the time of the first UN Conference on the Human Environment,[11] most of the major rivers in the developed world were in a parlous state with their natural purification capacity far exceeded by the volume of effluents that they were receiving from both industrial and domestic discharges.

2 Developments in the Regulatory Environment

The management of the environmental impact of effluents is only partially a technical matter. The design and implementation of regulation and the surrounding governance structures are equally, if not more important. Major changes have taken place in the regulatory environment since the 1970s that have facilitated the relatively rapid improvement in effluent treatment and disposal.

2.1 Pollutants Don't Recognise Political Boundaries

It had been recognised by the time of the 1972 UN Conference in Stockholm[11] that the management of environmental pollution needed to address the issue that pollutants will pass over political boundaries regardless of any legal sanctions that might exist and regardless of the water quality standards that might be required on either side of the boundary. There is, for example, little point in a community exerting strict limits on discharges to the river flowing through it if communities upstream have more lenient standards. This applies both at local, national, regional, international and global levels and thus an essential aspect of effective effluent management is wide ranging co-operation between all the stakeholders that have an interest in the water body concerned.

In the United Kingdom, this had already been recognised in the first Rivers Prevention of Pollution Acts in 1876[7] although it wasn't until 1948, under the River Boards Act, that 32 statutory authorities based on river catchments in England and Wales were granted powers to control polluting discharges.[12] These were further streamlined into 27 river authorities in 1965.[13] This enabled a single regulator, for the first time, to take a holistic view of all the effluents discharging into a watercourse.

In many ways the United State of America led the world in the development of environmental control policies, including those related to effluent discharges. The Nixon administration created the world's first environmental protection agency, the USEPA in 1970.[14] This Federal Agency was intended to deal not only with problems caused by effluent discharges but other issues such as those associated with air pollution, pesticides and oil exploration. Its federal authority enabled it to take a holistic approach to environmental problems across the whole country. Prior to the establishment of the USEPA pollution problems were often addressed at county level.

In mainland Europe, the problems in the Rhine River motivated the Netherlands, in which country the Rhine enters the North Sea, to begin discussions in 1950 with the other riparian states (Switzerland, France, Luxembourg and Germany) on a management strategy for the river. The initial discussions led to a co-ordinated monitoring strategy and eventually led to the establishment in 1963 of the International Commission on the Protection of the Rhine against Pollution.[15] A similar body was eventually established for the Danube in 1994.[16]

Over the 40 year period since the 1972 UN Conference in Stockholm,[11] the European Union has grown from its six original Member States in 1970 to 28 in 2014. The community now extends from Norway in the north to Malta in the south and Portugal in the east to Greece in the west. In this time the EU has gradually developed a comprehensive strategy to deal with effluent management culminating in 2000 with the Water Framework Directive.[17] The directive is based on catchment management planning and drives improvement in effluent management by having an overall objective that all water courses within the EU should be of good ecological and chemical quality. It is supported by a number of other directives targeted at specific emissions and the policy has recently been extended into the marine environment with the Marine Strategy Directive.[18]

Rivers and streams lie entirely within the jurisdiction of individual countries, or occasionally, like the Rhine and the Danube, small groups of countries. However, all rivers eventually drain into the ocean which is part of the global commons and thus of potential interest to everyone. In the 1960s the ocean was being seen by many people as a convenient, sustainable and unregulated waste disposal site. Wastes entering the sea *via* rivers, direct sea outfalls and direct dumping from ships would mix with the enormous oceanic volume where natural processes would degrade the waste materials and any toxic effects would be diluted. However, by 1970 increasing concerns were being expressed about the ethics of treating the ocean as a dustbin and more specifically about the consequences of the constant input to the oceans of persistent and bioaccumulative substances such as chlorinated hydrocarbons and metals such as cadmium, lead and mercury.

One direct outcome of the 1972 UN Conference in Stockholm was the establishment of the London Convention on the Prevention of Marine Pollution by Dumping of Wastes and Other Matter.[19] This was drafted in late 1972 and came into force, managed by the International Maritime Organization, in 1975. This international agreement seeks to eliminate or at least minimise the dumping of waste into the global ocean and is supported by a number of separate regional treaties such as the Oslo and Paris treaties which concern the NE Atlantic.[20]

2.2   Integration across the Water Cycle is Desirable

The second major change that occurred in effluent management following the Stockholm Conference was the gradual recognition that integrated approaches would be needed if problems were to be solved efficiently. This integration was of two types: firstly, integration across the water cycle from the initial abstraction of clean water from the environment *via* multiple human uses to the subsequent return of that used water back into the environment and, secondly, the need, in environmental pollution controls, to simultaneously consider ALL sources of the respective pollutants.

In 1975 the United Kingdom tried to bring all aspects of the water cycle within a single catchment under the control of a single body. It reorganised the 27 River Authorities created in 1965 into ten Regional Water Authorities each of which had responsibilities for water treatment and supply, sewage disposal, land drainage, flood control, fisheries and river pollution.[21] This was a laudable attempt at integrating the various aspects of the management of the water cycle. However, a structure where the organisation that regulated effluent discharges to the aquatic environment was the same body that was responsible for some of the major discharges of treated domestic effluent (*i.e.* poacher and gamekeeper combined) was difficult to justify and, consequently, this policy was reversed in 1989 when the regulatory responsibilities of the Regional Water Authorities were combined into a single national body for England and Wales called the National Rivers Authority.[22] Subsequently, six years later,[23] the National Rivers Authority was merged with Her Majesties Inspectorate of Pollution (HMIP), the body responsible for regulating major industrial discharges in order to facilitate the introduction of its Integrated Pollution Control Policy.

In 2000 the European Union introduced a new holistic framework directive on water quality[17] which also tried to bring the management of the complete water cycle within one piece of legislation. This Directive subsumed a number of previous Directives into a single piece of legislation, the Water Framework Directive. Prior to 2000, EU legislation on water quality had developed in a somewhat disjointed manner, due to political problems (see Section 3.3), with the introduction of a series of directives relating to different parts of the water cycle; Bathing Waters,[24] Drinking Water,[25] Water for the Abstraction of Drinking Water,[26] Discharge of Dangerous Substances,[27] Freshwater Fisheries,[28] Shellfisheries,[29] and Groundwater.[30] The last five were incorporated into the new Water Framework Directive.[17]

2.3   Pollutants have Many Sources and Don't Stay Where They are Put

The third advance made in the regulatory environment in the last 40 years resulted from the appreciation that the same pollutant could arise from a wide range of sources and that once introduced into the environment the substance might migrate into different compartments. For example, chlorinated pesticides used as seed dressings on Swedish farms might eventually reappear in the ice in Antarctica. Before the middle of the 20th century pollution control tended to be compartmentalised. Discharges to the atmosphere, watercourses, and the soil and groundwater were dealt with under separate and largely unconnected pieces of regulation often administered by different parts of government. This was not only inefficient but it produced overlaps and gaps in dealing with particular issues. In addition, legislation was frequently poorly enforced.

In 1991 the United Kingdom Government introduced the concept of Integrated Pollution Control, a radical new approach to industrial emissions control.[31] This in turn formed the basis for the EU Directive on Integrated

Pollution Prevention and Control[32] which has since had a major revision to produce the Industrial Emissions Directive.[33] The fundamental change involved in this legislation was that ALL emissions from a facility would be controlled by a single regulator. In addition, the operator, instead of having arbitrary limits imposed on specific emissions, would have to demonstrate that the total emissions, to all compartments of the environment, and from all sources were being reduced to the optimum level. Operators are expected to use the Best Available Techniques to minimise emissions for all new installations and to bring existing installations up to these standards in a reasonable time.

Thus at the beginning of the 1970s the governance of the aquatic environment and the regulatory control of effluent discharges into it was highly fragmented and consequently disjointed, inefficient and ineffective. However, by the beginning of the 21st century there was near universal acceptance of the need for catchment management controls and the desirability for integration of these controls both across the whole of the water cycle and to encompassing all potential sources of contaminants whether they were released initially to air or water. These changes have facilitated significant improvements in the aquatic environment.

3 Putting the Theories into Practice

As Richard II in 1388 and politicians in all subsequent centuries discovered, passing a law to prevent pollution doesn't achieve anything unless it is enforced. The major legislative changes that have occurred since the 1972 UN Conference in Stockholm[11] certainly made environmental improvements easier to achieve from the administrative point of view, but what happened on the ground?

In the 1970s most people agreed that the condition of rivers and coastal waters across the developed and developing world was unsatisfactory, that effluent discharges were the major cause of the problem and that something should be done. The question was, what should this "something" be? The complete elimination of effluent discharges was impractical but two competing viewpoints arose as to what was the most satisfactory solution. This eventually became known as the Emission Limits *vs.* Quality Standards argument, but the underlying problem was more a cultural than a scientific one.

3.1 *"Dilution is the solution to pollution"*

It had been known for centuries that the natural environment had the capability to absorb some types of waste, for example domestic sewage, without detriment provided that this inherent natural capacity was not exceeded. It was this theory that Bazalgette used in the initial improvements that were made to the Thames Estuary in 1858[5] after the "Great Stink". In the 1960s, some countries such as the United Kingdom and the United States of America were strongly of the opinion that this natural "treatment" capacity should continue to be used, always taking care to provide a large safety margin to prevent the natural assimilative capacity from being exceeded. However, many other countries disagreed, insisting that all wastes should be subjected to the best currently available treatment method before being released into the environment. This difference of opinion led to serious disagreements on water policy within the EU, which were only eventually resolved following the implementation of the Maastricht Treaty in 1992.[34]

Although there are some significant technical arguments, as shown below, the primary reason for this disagreement stemmed from the, often unappreciated, conflict between two different value systems. This difference is also manifest in the respective legislative and regulatory management systems. The United Kingdom and North America adopt a common law approach to regulation, and this Anglo-Saxon approach is supported by Common Law principles. These countries tend to be less risk averse than their neighbours on the continent of Europe who follow the Napoleonic Code and take a more precautionary approach to risk management. In essence mainland Europeans tend to consider the natural environment primarily as something that should be cherished and protected whereas the Anglo-Saxons see the environment as a resource for the human race to utilise for its own purposes.

Consequently in relation to effluent management, the United States and the United Kingdom considered that it was perfectly acceptable to discharge untreated effluents into the aquatic environment where its natural purification capacity would take care of any problems. The continental Europeans were horrified by this callous disregard for environmental protection since they required all operators to use best technical means to treat effluent discharges before the residual water was returned to the environment.

This European disdain led the United Kingdom to be called "the dirty man of Europe" by our European Union Partners,[35] an epithet which was used frequently by cartoonists. It is somewhat ironic therefore to note that the City of Brussels, the location of the principal European Union Institutions, discharged all its sewage completely untreated into the River Senne until 2000 and continued to discharge >60% of its untreated sewage into the river for a further seven years.[36]

Despite the political differences the actual environmental outcomes of using the two approaches were, in many cases, not dissimilar. In essence a waste water treatment plant for domestic sewage is relatively unsophisticated. After passing through preliminary screening to remove large objects and settlement tanks to remove solids, the liquid effluent is exposed to bacteria in an aerobic reactor configured to maximise the contact between bacteria, air and effluent. The "treated" effluent can then be returned to the environment. However, the operator incurs significant running costs, uses a substantial amount of energy and must invest substantial capital to construct the plant. In addition, a conventional sewage works will produce large quantities of sewage sludge for which an outlet must be found. Large quantities of surplus sludge used to be dumped at sea, but this is now prohibited by various "Dumping at Sea" treaties. Landfill was also an outlet but this is not now possible in Europe due to the provisions of the Landfill Directive.[37] Some of this material can find use as an agricultural fertiliser, although its use on food crops is becoming increasingly more difficult due to the trace contaminants that it may contain and the associated public concern. This leaves incineration (with or without heat recovery) as the only solution, but this is also of serious concern to many citizens.

However, if after initial screening and maceration to eliminate floating solids, the sewage effluent is discharged into the environment without further treatment, the bacterial breakdown of organic matter will take place naturally at zero cost and without the generation of large quantities of concentrated sewage sludge. This self-purification property of rivers has been understood for over a century and guidelines for using this capacity date back to the UK Royal Commission on Sewage Disposal established in 1898.[38] In their 8th Report, the Commissioners recommended that to maintain satisfactory river quality, effluents from sewage works should contain no more than $30 \, mg \, l^{-1}$ of Suspended Solids and $20 \, mg \, l^{-1}$ of 5 day Biochemical Oxygen Demand. The Commissioners were also very clear that the actual standards needed to vary according to the available dilution in the receiving waters. This guidance was used successfully in the design of sewage schemes until the 1980s.[38]

The marine and estuarine environments are also very effective at neutralising acid wastes due to the vast buffering capacity of the alkaline water. Taylor[39] showed that a very large estuarine discharge of waste at pH <1.5 did not cause any perturbation in pH of the receiving water within a few metres of the discharge point. Lewis and Riddle[40] reported very similar results in the wake of a waste tanker dumping highly acid waste from methyl methacrylate production into the North Sea.

Despite extensive research demonstrating the absence of any environmental impact this disposal route was terminated by decisions at the 1990 Ministerial Conference on the North Sea to stop the dumping of waste at sea. However, the subsequent £66m sulfuric acid recovery plant needed a supply of 40 000 tonnes of natural gas each year.[41]

These two parallel approaches to the management of effluent discharges continued in Europe until the 1990s, although they were further challenged in the European Community in the mid 1970s, when the first quality standards were imposed on waters used for bathing.[24] In addition to some basic physico-chemical parameters, this legislation also included a microbiological standard. To comply with the Directive, 95% of all samples taken from the bathing water concerned had to contain <10 000 coliforms 100 ml^{-1} and <2000 faecal coliforms 100 ml^{-1}. In addition, the Directive included a set of targets, "Guideline Standards", that were 20× more stringent. The 1976 Directive was revised in 2006, with the bacteriological standards becoming even more stringent.[42]

These new requirements created problems for both types of sewage disposal: direct discharge of screened but otherwise untreated sewage to the environment, particularly in the UK through traditional short sea outfalls led to bacterial contamination of bathing waters well in excess of even the mandatory standards. However, conventional sewage treatment plants found that they had a similar compliance problem. It is generally assumed by the public that a sewage treatment plant receives domestic sewage at the inlet and produces clean water at the outlet; however, this is far from the case. For example, typical abundance of total and faecal coliforms in raw sewage is respectively, 10^7–10^9 and 10^6–10^8 (100 ml)$^{-1}$. Classical treatments, which do not include any specific disinfection step, only reduce these faecal micro-organisms densities by 1–3 orders of magnitude.[43] As a consequence outfalls from sewage treatment plants also produced non-compliant bathing waters.

In the UK, which continued to follow the "dilute and disperse" philosophy until the beginning of the century, the response was to extend sewage discharge points into deeper and more distant water, the Long Sea Outfall (LSO) solution. Advances in mathematical modelling and computer systems had revolutionised the design and siting of these outfalls to minimise environmental and human health effects.[44] The LSO ensured that initial dilution was increased, the sewage field remained out of sight below the surface for longer and there was an increased die-off rate for pathogens due to the longer distance of travel before reaching bathing waters. Many such outfalls often several kilometres in length were built around the UK coast in the 1990s. United Utilities in NW England constructed 29 such outfalls in that period.

However, in many cases non-compliance with the Directive was traceable, not to a discharge of untreated sewage but to a short outfall from a conventional wastewater treatment plant. In some parts of the UK and in continental Europe the initial solution to this problem was to install disinfection on the effluent prior to its return to the environment, however, the most economic disinfectants were chlorine and its derivatives which then raised the issue of chlorinated residues being formed during the process with potential adverse effects on the ecology of the receiving water.[45] Consequently outfall pipes from conventional treatment plants were also lengthened to discharge into deeper water further from bathing waters.

3.2   Increasing Concern about Persistence

In the 1800s, effluent discharges were considered to consist mainly of two types of material: organic matter, most of which would rapidly degrade in the environment, and inorganic matter which, although not degradable, was largely inert. Thus the UK Royal Commission in 1912, in its eighth report[38] only recommended two emission standards: the 30 mg l^{-1} criterion for Biochemical Oxygen Demand to limit organic matter discharge and the 20 mg l^{-1} for suspended solids as a surrogate for particulate inorganic materials. The consensus was that the majority of organic materials would eventually degrade and since the large dilution available in the global oceans was thought to be effectively infinite, any organic materials that did persist would be of no consequence.

In the 1970s, a number of environmental problems emerged that showed that some substances such as the heavily chlorinated insecticides might persist long enough to become widely distributed in the global environment. Furthermore, even at very high dilution some of these substances could still exert toxic effects especially those that could bioaccumulate or biomagnify in animal tissue. The number of such persistent, bioaccumulative and toxic substances (PBTs) was initially thought to be very small, and specific legislation was introduced in the 1970s to control their release into the marine environment from waste dumping and direct pipeline discharges.[19,20] However, it gradually became apparent during the latter part of the century that there were many more of these substances than had been first appreciated and that more restrictive controls would be necessary to protect the global environment leading eventually in the EU to the Water Framework Directive[17] and the REACH Regulation[46] and internationally to the Stockholm[47] and Minamata Conventions.[48] The increasing evidence, that dilution alone was incapable of safeguarding the environment from persistent chemicals gradually led to the acceptance, albeit reluctantly, that most effluent discharges needed to be subjected to some effluent treatment prior to their release to the aquatic environment.

In addition to the technical arguments about which solution is better, waste disposal continues to be a source of public controversy. At the beginning of the 1970s raw sewage was still being discharged into rivers and coastal waters at many locations and there was general agreement that this situation needed to be improved.

Despite the technical and financial arguments, many people objected in principle to the discharge of untreated sewage into the sea with large numbers being concerned about the possible health effects to people bathing. Some very large LSO schemes eventually had to be abandoned because of public pressure. In the 1960s the city of Sydney decided to build two 4 km long sea outfalls to discharge screened sewage at a cost of US$300 million. Despite the controversy this caused the scheme went ahead but before the outfalls were commissioned in the 1980s, the government agreed to upgrade the coastal treatment plants so that sewage would be treated to at least secondary treatment standards before discharge into the ocean.[49] However, although the general public clearly shows a preference for sewage to be treated in land based wastewater treatment plants they usually object strongly to any proposal to build one in their locality.

So despite the fact that in most of the developed world there are few remaining discharges of untreated sewage into rivers and coastal waters, its disposal remains a matter of controversy amongst the general public.

3.3   A European Consensus

Progress on environmental regulation within the European Union began slowly. This was because the founding treaty, the Treaty of Rome,[50] made no provision for dealing with environmental affairs. Consequently, for the first 30 years of its existence the European Commission could only introduce environmental legislation under Article 100 of the Treaty of Rome which concerns "the approximation of laws affecting the Common Market" and requires unanimity for approval. In 1987 the "Single European Act"[51] incorporated environmental affairs into the treaties as Article 130 and subsequently the Treaty of Maastrict[34] in 1992 extended "Qualified Majority Voting" to environmental matters thus removing the veto on such legislation by individual Member States. Finally in 1997 the Treaty of Amsterdam[52] gave the European Parliament co-decision powers with the European Council on environmental matters. Thus after 1992, the UK no longer had a veto on EU environmental legislation and was thus forced to adopt a more conciliatory approach to its partners.

The immediate result was the passing of two major pieces of environmental legislation in which the UK sought to align its own legislation with that of the other Member States. The first, in 1996, concerned an integrated approach to pollution control[23] and the second in 2000 provided a holistic approach to water quality management.[17] These two pieces of legislation finally resolved the philosophical argument between the UK and its European Partners about the correct way to deal with effluent discharges.

The 1996 Integrated Pollution Prevention and Control Directive was important in two key areas: it recognised for the first time that effective pollution control needed to take into account releases from all sources and to all receptors because frequently trade-offs were needed to obtain the optimum improvement. Secondly, there was a recognition that emissions should be minimised using "Best Available Techniques" but that the receiving waters should also meet appropriate environmental quality standards. In other words the solution to the disagreement about quality standards and Best Available Treatment was "both/and" rather than "either/or".

The 2000 Water Framework Directive[17] was novel in that it was focussed on outcomes rather than on the more usual command and control of inputs. In essence the Directive simply requires all surface waters within the 28 Member States to be of "Good Ecological Quality" by a certain date. The methods by which this is achieved are not specified, the "means" being left to the Member State. This is primarily a Directive driven by water quality requirements, although significantly it also includes sections dealing with limitations on the discharge of substances considered to be particularly dangerous to the environment.

These two pieces of legislation, together with the 2006 REACH Regulation,[46] are accelerating the improvements in effluent disposal and the resulting water quality in the 28 Member States of the European Union. Furthermore they are being seen as a model for other countries and for regional and global governance bodies such as the United Nations.

4 Effluent Management in the 21st Century

Improvements in effluent management, since 1970, have not all been driven by regulation. Two other major drivers have gradually become important. The first was a financial imperative, whilst the second has been the growing recognition of the importance of the concept of sustainability for long-term business success.

4.1 Water Conservation and Reuse

In the 1970s, water, in the industrialised world, was considered to be a very low cost and limitless resource. Water shortages were acknowledged in some of the more arid parts of the world but since industrialisation of these regions was slight this was not seen as any great concern. However, in the economic downturn in the early 1980s, cost control was aggressively pursued in all areas, and what became known as "cost of waste analysis" was actively pursued for the first time.[53] This brought about a transformation in how water, amongst other things, was viewed and used within industry.

Before the use of "Cost of Waste" analysis, water appeared in the corporate accounts as a simple commodity purchase. However, this failed to account for the consequences of water use. Every litre of water entering a manufacturing site eventually leaves either as part of the products, as water vapour to atmosphere or as liquid effluent. Simple modelling showed that in most cases >90% of the incoming water was subsequently discharged as part of the effluent stream. This discharge created additional revenue costs if the effluent was sent to a Publically Owned Treatment Works and both a revenue and capital cost if the waste was treated before discharge in an on-site treatment plant. These secondary costs could be very substantial, depending on the complexity of the effluent, and capable of far exceeding the initial purchase costs for the water. In one of the first of these exercises at the Huddersfield manufacturing site of Zeneca Ltd, it was found that effluent treatment costs could account for up to 15% of the total production cost. In another part of Zeneca Ltd, the cost of installing a new site effluent treatment plant in the 1980s was reduced by a factor of $5\times$, following the introduction of a site-wide programme of water conservation, which dramatically reduced the volume of effluent to be treated.

In addition, as many businesses became more global it was gradually appreciated that water, far from being an unlimited resource, was actually in short supply in many regions of the world, not just in arid areas. For example, even in the south east of England, where rainfall would not be considered to be low, water supply is currently constrained due to excessive demand, and projections indicate that demand will begin to exceed supply in the region between 2020 and 2040.[54] The recognition that water use results in substantial costs, additional to its purchase price coupled with the appreciation that in many places water is a limited resource, has led to significant reductions in raw water use and subsequent effluent volumes. For example, the pharmaceutical company AstraZeneca reduced its water consumption by 76% in 10 years, moving from a consumption of 410 m^3 $million sales^{-1} in 2001 to 99 m^3 $million sales^{-1} in 2011.[55]

Increasing attention is being given to recycling and water reuse and there is also a gradual increase in interest in the appropriate use of "greywater". As improved methods are used to treat sewage effluents, it is also now becoming feasible to produce waste water treatment plant effluents of drinking water quality[56] although there is still considerable public resistance to making the physical connection between the output of the treatment plant and the drinking water supply.[57]

4.2 Advanced Effluent Treatment

Effluent treatment prior to the 1970s usually involved simple screening of the effluent stream to remove debris, settlement tanks to remove heavy particulates followed by a simple biological filtering process. It had been known for a long time that a large proportion of organic matter would break down naturally as a result of interaction with aerobic bacteria. The first "biological treatment" plants were therefore designed to provide an opportunity for appropriate bacteria to interact with the organic matter. When dealing with industrial effluents it rapidly became apparent that neutralisation of acidic or alkaline effluents was also an essential part of the process to avoid killing the biological slimes and, where the effluent load or composition was highly variable, it was necessary to install balancing tanks to avoid shock loads being experienced by the organisms.

Although the basic principles of biological treatment have remained constant,[58] the engineering process configuration has evolved over the years.[59] Initially trickling filter beds were used: circular beds of clinker onto which the effluent was continuously sprayed *via* a set of rotating arms, gradually a bacterial slime grew on the surface of the clinker which then degraded the organic matter in the waste stream. Numerous different configurations have subsequently been used, for example Rotating Biological Contact plants (RBC), in which the biological slime grows on the surface of a series of large rotating discs dipping into the effluent stream, became very popular in North America in the 1980s.[60] High rate trickling filters were introduced in the 1970s in which the slime supporting surface was supported within vertical towers containing either a prefabricated open plastic packing structure or alternatively a random media system constructed of small plastic components. In the latter case, the individual components can be packed into a tower and the effluent sprayed onto them or alternatively the individual components can be suspended in an aerated fluid.[61] This type of system was often used as a pre-treatment step before sending an industrial waste to a publically owned treatment plant for further treatment.

In 1914 the activated sludge process was invented[62] and this is now the dominant process. Previous configurations had all used biofilms attached to a solid substrate. However, the activated sludge process involves air or oxygen being introduced into a mixture of screened, and primary treated wastewater combined with organisms to develop a suspended biological floc which then interacts with the waste reducing its organic content. There are many variants to the conventional continuous-flow tank technology including aeration ditches,[63] where land is not a constraint, to deep shaft technology, where it is.[64] The sequencing batch reactor is a more recent variant,[65] particularly useful where the effluent loading is high and consistent. In some cases, a fungal stage has now been included in conventional activated sludge plants prior to the bacterial reactor to improve the overall degradation efficiency.[66]

Considering the opprobrium heaped on the UK as the "Dirty Man of Europe",[35] it is perhaps surprising that reed beds or constructed wetlands[67] also became very popular in the UK at the end of the 20th century as a "green" technology for effluent treatment. Although superficially a low technology solution, extensive engineering work is necessary in order to construct an industrial scale facility with appropriate effluent distribution and drainage. Once completed, effluent, either raw or partially treated, is allowed to percolate through the reeds, where the resident bacteria in plant and soil/sediment carry out the treatment process. However, heavy metals and persistent polar organic compounds tend to be accumulated in the root systems and associated soil/sediment, thus ensuring that the wetland eventually becomes a contaminated land site.

In addition to biological treatment, a range of more sophisticated techniques have gradually been included in effluent treatment plants in order to produce higher quality effluents. In a previous section, disinfection of the treated effluent stream was shown to be necessary for some sewage treatment works to enable compliance with the EU Bathing Waters Directive (see Section 3.1). This was done either by the use of chlorination or high-intensity UV irradiation of the final effluent stream.

Although biological treatment can be very effective at reducing the biological oxygen demand (BOD) of an effluent it is frequently only partly successful in removing micropollutants. Consequently, tertiary treatment may often be necessary to achieve compliance with regulatory standards. Typically, either activated charcoal adsorption or reverse osmosis units have been employed. However, these are both expensive to purchase and operate and they also have their own waste disposal problems. Novel techniques involving photo-degradation,[68] use of ultrasonics,[69] and nanomaterials[70] are all currently under development.

4.3   Green Chemistry

Increasingly sophisticated effluent treatment plants are capable of producing high quality effluent streams but "end-of-pipe treatment" comes at a high and increasing cost both financially and in terms of energy consumption. This came to prominence in the early applications of "Cost of Waste" analysis[53] (See Section 4.1) and led to greater attention being paid to "control at source". We have already discussed the dramatic improvements in effluent volumes brought about by water conservation measures, similar approaches to materials efficiency, including reuse and recycling have also led to reductions in effluent loadings in addition to lower volumes. In the AstraZeneca example quoted previously, the reduction of 75% in water consumption was accompanied by an 83% reduction in organic load.[55]

In the chemical and allied industrial sectors there is now increasing emphasis on the application of green chemistry principles. These 12 basic principles, originally proposed by Anastas and Warner in 1998,[71] provide a code of practice to guide developments in manufacturing:

1. **Prevention.** It is better to prevent waste than to treat or clean up waste after it is formed.
2. **Atom Economy.** Synthetic methods should be designed to maximize the incorporation of all materials used in the process into the final product.
3. **Less Hazardous Chemical Synthesis.** Whenever practicable, synthetic methodologies should be designed to use and generate substances that possess little or no toxicity to human health and the environment.
4. **Designing Safer Chemicals.** Chemical products should be designed to preserve efficacy of the function while reducing toxicity.
5. **Safer Solvents and Auxiliaries.** The use of auxiliary substances (solvents, separation agents, *etc.*) should be made unnecessary whenever possible and, when used, innocuous.
6. **Design for Energy Efficiency.** Energy requirements should be recognized for their environmental and economic impacts and should be minimized. Synthetic methods should be conducted at ambient temperature and pressure.
7. **Use of Renewable Feedstocks.** A raw material or feedstock should be renewable rather than depleting whenever technically and economically practical.
8. **Reduce Derivatives.** Unnecessary derivatization (blocking group, protection/deprotection, temporary modification of physical/chemical processes) should be avoided whenever possible.
9. **Catalysis.** Catalytic reagents (as selective as possible) are superior to stoichiometric reagents.
10. **Design for Degradation.** Chemical products should be designed so that at the end of their function they do not persist in the environment and instead break down into innocuous degradation products.
11. **Real-time Analysis for Pollution Prevention.** Analytical methodologies need to be further developed to allow for real-time in-process monitoring and control prior to the formation of hazardous substances.
12. **Inherently Safer Chemistry for Accident Prevention.** Substance and the form of a substance used in a chemical process should be chosen so as to minimize the potential for chemical accidents, including releases, explosions, and fires.

Half of these principles (1, 2, 6, 7, 8 and 9) have a direct impact on the amount of waste produced per unit of output. However, it is not just the amount of waste that is important, some individual components of the waste, if they are strictly regulated, can be critical to the cost of treatment even if present in very small quantities. An extreme, but current, example of this arises from the use of ethinyl estradiol in the contraceptive pill. As a result of this use, ethinyl estradiol is found in sewage and since it is not completely removed by current effluent treatment methods it is also found in sewage works effluents. Ethinyl estradiol has been implicated in the feminisation of male fish in some UK rivers[72] although present in sewage effluent discharges at concentrations of only a few ng l^{-1}. In order to eliminate the environmental impact it is necessary to reduce the environmental exposure. Control at source would clearly be controversial, however, it has been estimated that to reduce ethinyl estradiol to harmless concentrations in sewage works effluents in England and Wales could cost *ca.* £30b.[73]

Consequently, one of the prime targets for effluent managers is to reduce the presence of strictly regulated substances before they are discharged to sewer or the environment. Good housekeeping can make a contribution, but the most effective way to minimise the discharge of such a substance is not to use it at all and replace it with a suitable substitute. At first sight substitution sounds both logical and straightforward, however, in practice it is usually far from simple and great care needs to be taken to avoid the substitution producing a worse environmental outcome.[74]

For example, in the 1960s, chlorinated insecticides were found to be having a damaging effect on raptoral birds.[75] Consequently, the chlorinated insecticides were replaced with organophosphate insecticides. However in the 1970s, it became clear that these organophosphate insecticides were causing chronic health problems in farmworkers.[76] As a result, in the 1990s they were replaced with pyrethroid insecticides which are harmless to man and birds but which it subsequently transpired are extremely toxic to fish and invertebrates resulting in a number of major environmental problems caused by the inappropriate disposal of sheep dip liquor.[77]

4.4   21st Century Regulatory Processes

All regulation of residual effluent discharges follows the same basic principles, the discharger is granted a licence to discharge limited amounts of certain substances to the environment by a regulator, compliance with the licence is monitored by the discharger and/or the regulator and the regulator has legal sanctions that can be applied in the case of a compliance failure. In the last 40 years a number of different methods have been developed and deployed to determine how the licence conditions should be expressed: Brutal, Social, Fiscal, Technological and Environmental Controls.

"Brutal" Controls are based on the elimination of the discharge of specific substances which is in turn predicated on the elimination of the use of the substance. The Stockholm Convention of Persistent Organic Pollutants[47] and the recently agreed Minamata Convention[48] are examples of this type. It is precautionary and has the advantage of simplicity and certainty. However, the complete elimination of substance use also eliminates any benefits the substance might have had so the use of this control mechanism has to be considered in the light of societal cost–benefit.

The Authorisation procedure within the EU REACH regulation[46] seeks to standardise this process. Substances for possible authorisation are identified on the basis of their potentially harmful properties. If the substance cannot be replaced in all its current uses by a suitable alternative, an authorisation for continued use for specified purposes may be granted provided that the substance is used under strictly controlled conditions or where the socio-economic benefit is deemed to be sufficient.

Social Controls are based on the principle of public shame and embarrassment. The regulator merely requests that dischargers should put their discharge monitoring data into the public domain. Current examples are the USEPA Toxic Release Inventory (TRI)[78] and the EU Pollutant Release and Transfer Register (E-PRTR).[79] No Corporate CEO appreciates being labelled as a despoiler of the environment and shareholders don't like it. Consequently, such registers promote the voluntary reduction of the discharge of substances, especially those that are considered hazardous. This technique also has the advantage of simplicity with the majority of the implementation costs falling on the discharger rather than the regulator. One drawback that was noticed after the introduction of the USEPA TRI was the propensity for emissions reductions to be concentrated on the largest mass emissions rather than those with greatest environmental impact. The public perception being more favourable if the "numbers" were reduced regardless of the impact of the particular reduction.

Fiscal Controls are based on market forces, *i.e.* creating a charge per unit of pollution will induce industry to act in such a way as to minimise the charges. There are many examples such as product taxes, fuel duty, carbon taxes, effluent charges such as those imposed in some EU member states[80] and tradeable permits such as those that very successfully reduced sulfur dioxide emissions in the USA in the 1970s.[81] Such controls can be simple, *e.g.* product taxes, but can become extremely complex especially if they involve emissions trading.

Fiscal controls can, by regular and publicised escalation, significantly drive behaviour as has happened with the UK Landfill Taxation system. When introduced in 1996, the tax rate was only £7 tonne^{-1} but with the stated intention to increase the rate every year. Today, in 2014, the rate has risen to £72 tonne^{-1}, rising by £8 tonne^{-1} on an annual basis and this provides a powerful incentive to minimise waste going to landfill. One major advantage that fiscal controls have compared to other methods is their ability to drive continuous improvement since every reduction in emissions leads to a reduction in pollution taxes. They can be economically efficient if the financial penalty is designed to equal the environmental externality since, at least in theory, the discharger will operate at the optimum economic threshold. However, such taxes can be inflationary unless steps are taken to preserve fiscal neutrality.

Technological Controls are based on the largest reduction in emissions that can be economically achieved with currently available techniques. This sounds simple to operate but immediately generates arguments about what constitutes the "best available technique" (BAT) and in particular whether this is economically viable. In theory this is a precautionary approach since it can go beyond what is strictly necessary to protect the environment; however, in some cases even applying the best available techniques may not reduce emissions sufficiently to do this.

Environmental Controls are based on the principle of assimilative capacity and derive emission standards from the desired environmental quality. These are economically efficient and centred on environmental protection. However, to be effective they require substantial knowledge of the potential environmental consequences of the discharge.

Each of these systems has had advocates over the years but as we move forward into the 21st century, 40 years after the Stockholm Conference, there is a growing consensus that effective controls on residual effluents are likely to be a combination of all these techniques. In other words an operator will be expected to ensure that all effluent discharges are treated using BAT; the receiving water complies with all relevant environmental quality standards; and all monitoring and compliance data is in the public domain. The operator will be prohibited from using some substances and may be subject to charges per unit of discharge for some substances or wastes.

5 Improvements in the Environment Since 1970

In 1972, Barbara Ward and Rene Dubois, in their unofficial report, *Only One Earth*,[82] commissioned by the Secretary General of the UN Conference on the Human Environment in Stockholm included the following two statements in the concluding section of Part 1:

> *"With all the growth, all the enrichment, all the apparent market success of the fifties and sixties, men are left in deep unease about the current condition of the planet."*

and

> *"If all man can offer in the decades ahead is the same combination of scientific drive, economic cupidity and national arrogance then we cannot rate very highly the chances of reaching the year 2000 with our planet still functioning safely and our humanity securely preserved."*

Now in 2014 we can see that the condition of the aquatic environment in many parts of the world has indeed been transformed since the 1970s.

Although old problems still exist in some parts of the developing world and new problems are emerging we can see that their worst fears have not been realised and that the 1972 UN Conference in Stockholm[11] marked a pivotal moment in the *Care and Maintenance of a Small Planet*, the sub title of *Only One Earth*.

There is a wealth of data to demonstrate this improvement which, in the space available, this chapter cannot do justice to. The following are merely an indication of some of these improvements:

- "…efforts to regenerate waters have resulted in some impressive achievements: Atlantic salmon and brown trout have returned to rivers such as the Thames, Tyne, Wear and Mersey to breed for the first time in more than a century. After being virtually extinct in the early 1970s, the otter has made a dramatic return and is now present in every English county…"[83]
- A river monitoring and classification scheme, called the General Quality Assessment (GQA) scheme, has been used (in the UK) to assess changes in water quality over the last 20 years. In 1990, the chemical quality of 55% of monitored rivers was good or excellent. This had improved to 80% by 2009. For biological quality the improvement was from 63% to 73% over the same period.[83]
- In European rivers, the oxygen demanding substances measured as BOD and total ammonium have decreased by 55% (from 4.9 mg l^{-1} to 2.2 mg O_2 l^{-1}) and 73% (from 587 to 159 µg N l^{-1}), respectively, from 1992 to 2010.[84]
- The quality of water at designated bathing waters in Europe (coastal and inland) has improved significantly since 1990. Compliance with mandatory values in EU coastal bathing waters increased from just below 80% in 1990 to 93.1% in 2011. Compliance with guide values likewise rose from over 68% to 80.1% in 2011. Compliance with mandatory values in EU inland bathing waters increased from over 52% in 1990 to 89.9% in 2011. Similarly, the rate of compliance with guide values moved from over 36% in 1990 to 70.4% in 2011.[85]
- The concentrations of some hazardous organic substances have been monitored in a number of Swedish lakes since the 1960s. The concentration of PCBs and DDT in pike tissue has fallen since the late 1960s. In addition to this, concentrations of α-HCH and HCB have also fallen. Contrary to this, the concentrations of brominated flame retardants have been stable in Lake Bolmen after an increase during the 1970s.[86]
- The load of halogenated substances entering the River Rhine was reduced by 74% between 1986 and 1993 from 3800 tonnes AOX to 1000 tonnes AOX.[87]
- The Rhine Action Programme, initiated in 1987 had led by 2000 to a 70–100% reduction in the inputs of cadmium, chromium, copper, lead, mercury, nickel and zinc.[88]

- Riverine inputs of heavy metals show in general a significant decrease over the period 1990–2006 in the OSPAR Regions Arctic Waters (Region I), Greater North Sea (Region II) and the Celtic Seas (Region III), except for mercury in Region I and lead in Region III... Riverine inputs of cadmium, lead and mercury show in many cases statistically significant decreases. Cadmium inputs decreased in Region I (−40%), Region II (−20%) and Region III (221260%) in the period 1990–2006. Riverine inputs of lead have fallen in Region I (−85%) and Region II (−50%) and riverine inputs of mercury have significantly reduced in Region II (−75%) and Region III (−85%). For Region II, statistically significant reductions in the main catchments for example cadmium in the Elbe (−40%), mercury in the Rhine and Meuse (−70%) and lead in the Seine (−90%) confirm the overall regional trend.[89]
- Concentrations of DDE and PCBs in freshwater fish in the United States fell by 83% between 1969 and 1986.[90]
- Concentrations of DDE, PCBs, HCB, toxaphene, mirex and dieldrin in fish in the Great Lakes declined by 80–90% between 1974 and 1989.[91,92]

6 Future Prospects

In the 1970s, as we noted in the previous section, serious concerns were being raised about pollution of the aquatic environment resulting from effluent discharges. In particular, concerns were being raised about the negative trends in environmental quality that were then becoming apparent. In this chapter we have seen how these concerns were addressed and the very significant improvements to the quality of the environment that have ensued. There is now a much better appreciation by dischargers, regulators and other stakeholders about the potential environmental impacts arising from liquid waste disposal and a general consensus that waste disposal should be the last option in the general waste hierarchy.[93] However, this does not mean that all problems have now been identified and resolved. Old problems still exist and future challenges will continue to emerge:

Major improvements have indeed taken place in the developed world but more progress is needed in some newly industrialised countries such as India[94] and China[95] where untreated or partially treated effluents are still being released inappropriately into the aquatic environment. There is a need to share information and assistance on technical, managerial and regulatory best practice and, where possible to improve global harmonisation in this area. Global harmonisation is difficult to achieve and very time consuming, as demonstrated by experience with the Globally Harmonized System of Classification and Labelling of Chemicals (GHS).[96] This was initiated in 1992 at the United Nations Conference on the Environment in Rio,[97] but was only agreed in 2002 after 10 years of discussion and is not expected to be fully adopted until 2020. Nevertheless it is probable that significant further global harmonisation in regard to effluent management will be achieved in the next 30 years.

The United Nations has now started to take an active interest in this issue and following the World Summit on Sustainable Development in Johannesburg in 2002,[98] the United Nations Environment programme (UNEP) was instrumental in establishing the Strategic Approach to International Chemicals Management (SAICM) project. The key objective of SAICM is "the achievement of sound management of chemicals throughout their life-cycle so that, by 2020, chemicals are produced and used in ways that minimize significant adverse impacts on human health and the environment".[99] This project encompasses both the use of chemicals and the appropriate disposal of waste and is primarily a vehicle for the sharing of information and expertise between developed countries, newly industrialised countries and developing countries.

Future challenges arise from three sources: improvements in analytical science, improvements in toxicology and human ingenuity.

In the 1970s, analytical techniques were only able to routinely detect substances in the environment at concentrations of a few mg l^{-1} and above. Using more advanced, but non-routine techniques, some substances could also be determined at μg l^{-1}. Consequently, attention at that time was focussed on organic carbon (BOD, chemical oxygen demand or COD, and total organic carbon or TOC), nutrients, chlorinated insecticides and a small number of metallic elements, particularly cadmium, lead and mercury. Since that time analytical science has advanced dramatically with routine detection limits being reduced by at least six orders of magnitude. Thus it is now possible to routinely measure substances in the environment at ng l^{-1} and with specialised instrumentation at pg l^{-1}. As a result we can now identify many substances in the environment that would have been present in the 1970s but which, at that time, were invisible. A typical example would residues of human pharmaceuticals, such as aspirin, which are now known to be widely distributed in the aquatic environment[100,101] although their significance is open to question.[102]

Myriad additional micropollutants can be expected to emerge from the shadows over the next decade as a result of the wider application of current analytical science.[103] However, analytical science will advance and limits of detection and quantification will continue to improve. Consequently, we can expect the number of detectable "pollutants" in the environment to increase dramatically over the next few decades.

Unfortunately, our understanding of toxicology has not kept pace with the improvements in analytical techniques and so, although we can now see thousands of substances in the environment that were previously invisible, our ability to understand their potential impact, if any, on human beings and the wider environment is poor. Furthermore, the general public are convinced that "presence" is equivalent to "harmful", a proposition that has no scientific justification but which, nevertheless, remains a powerful social driver. Consequently, it is likely that much effort will be expended on popular issues rather than those that are important.

Currently there are two key areas of active research: mixtures and endocrine-disrupting substances. It has been recognised for a long time that although environmental risk assessments are substance specific, aquatic life is actually exposed to a mixture of many substances at different and variable concentrations. This is frequently referred to as the "cocktail effect".[104] Whilst this issue has been recognised for a long time, there is still no consensus on how to incorporate this into risk assessments although a lot of research is underway.

Similarly, the potential for some substances to disrupt a range of endocrine systems, potentially causing irreversible damage has also been known for a long time,[105] but discerning what this means for environmental risk assessment is proving to be difficult and highly controversial.[106–108] There have been calls, in Europe, to outlaw the use of any substance that has any effect on the endocrine system as a precautionary measure,[109] but this would have such a major impact on trade, commerce and society that politicians and regulators are seeking more scientific understanding of the issues before they take action. The science is expected to be much clearer in the next few years and this could have a major impact on emissions standards and quality objectives related to effluent discharges.

So far we have been concerned with "existing" substances, however mankind is very ingenious and new substances are being created all the time. From 1981 to 2001 a total of 2700 substances were registered as "new substances" in commercial use in the EU.[110] The majority of these substances have been of little concern, however, the discovery of the fullerenes in 1985[111] opened up the new science of nano chemistry. The development and application of nanoscience and associated nano materials has been exceptionally rapid, the "Project on Emerging Nanotechnologies"[112] estimates that over 1800 manufacturer-identified nanotech products are publicly available, with new ones hitting the market at a pace of several per week. However, our toxicological understanding of these materials and any subsequent regulatory processes has currently been left behind.

The impact of effluent discharges on the aquatic environment has been dramatically reduced in the 40 years since the Stockholm Conference. However, both incremental advances and step changes in scientific knowledge will continue to create new challenges for effluent management in the decades to come.

References

1. Anon, Holy Bible (King James Version), 1611, Deuteronomy, 23, 13.
2. Anon, *12 RIC II*, HMSO, London, 1388 , pp. C10–C13.
3. Anon, *Off. J.*, 1991, **L135**, 40.
4. S. Johnson, *The Ghost Map: The Story of London's Most Terrifying Epidemic – and How it Changed Science, Cities and the Modern World*. Riverhead Books, London, 2006, p. 307.

5. S. Halliday, *The Great Stink of London: Sir Joseph Bazalgette and the Cleansing of the Victorian Metropolis*, Sutton Publishing Ltd, Stroud, 1999, p. 224.

6. Anon, Royal Commission on Rivers Pollution: 1868-74, Final report, 1874, HMSO, London.

7. Anon, Rivers Pollution Prevention Act 39 & 40 Vict, HMSO, London, 1876, C.75.

8. USEPA, Cuyahoga River. http://www.epa.gov/greatlakes/aoc/cuyahoga/index.html.

9. Anon, Pollution in some British Estuaries and Coastal Waters, Royal Commission on Environmental Pollution 3rd Report, HMSO, London, 1972.

10. Anon, *Pollution in Four Industrialised Estuaries: Tees, Mersey, Humber and Clyde, Royal Commission on Environmental Pollution*, HMSO, London, 1973.

11. Anon, Report of the UN Conference on the human environment (Stockholm), UNEP, Nairobi, 1972.

12. Anon, River Boards Act 11 & 12 Geo. 6, HMSO, London, 1948, C.32.

13. Anon, Water Resources Act, HMSO, London, 1963, C.38.

14. Anon, *Fed. Regist.*, 1970, **35**, 15623.

15. Anon, International Commission for the Protection of the Rhine. Website: http://www.iksr.org/index.php?id=58.

16. Anon, International Commission for the Protection of the Danube. Website: http://www.icpdr.org/main/.

17. Anon, *Off. J.*, 2000, **L327**, 1.

18. Anon, *Off. J.*, 2008, **L164**, 1.

19. Anon, London Convention, Website: http://www.imo.org/About/Conventions/ListOfConventions/Pages/Convention-on-the-Prevention-of-Marine-Pollution-by-Dumping-of-Wastes-and-Other-Matter.aspx.

20. Anon, OSPAR Commission. Website: http://www.ospar.org/welcome.asp?menu=0.

21. Anon, Water Act, HMSO, London 1973, C.37.

22. Anon, Water Act Part 1, HMSO, London, 1989, C.15.

23. Anon, Environment Act, HMSO, London, 1995, C. 25.

24. Anon, *Off. J.*, 1976, **L31**, 1.

25. Anon, *Off. J.*, 1980, **L229**, 11.

26. Anon, *Off. J.*, 1975, **L194**, 26.

27. Anon, *Off. J.*, 1976, **L129**, 23.

28. Anon, *Off. J.*, 1978, **L22**, 1.

29. Anon, *Off. J.*, 1979, **L281**, 47.

30. Anon, *Off. J.*, 1980, **L20**, 43.

31. Anon, Environmental Protection Act (Part 1), HMSO, London, 1990, C43.

32. Anon, *Off. J.*, 1996, **L257**, 26.

33. Anon, *Off. J.*, 2010, **L334**, 17.

34. Anon, *Off. J.*, 1992, **C191**, 1.

35. J. Porritt, *J. R. Soc. Arts*, 1989, **137**, 488.
36. Anon, Environmental Industry News, 2007, 17 October.
37. Anon, *Off. J.*, 1999, **L182**, 1.
38. Anon, Royal Commission on Sewage Disposal – Final Report, HMSO, London, 1915.
39. (a) L. Carter, *Effluent Water Treat. J*, 1973, **13**, 647; (b) D. Taylor, Managing the impact on the aquatic environment caused by the discharge of waste from a factory making complex chemicals, in *Proceedings of the International Symposium on Management of rivers for the future*, Kuala Lumpur, 1993.
40. R. E. Lewis and A. M. Riddle, *Mar. Pollut. Bull.*, 1989, **20**, 124.
41. J. H. Clarke, *Chemistry of Waste Minimisation*, Chapman Hall, Glasgow, 1995.
42. Anon, *Off. J.*, 2006, **L64**, 37.
43. I. George, P. Crop and P. Servais, *Water Res.*, 2002, **36**, 2607.
44. WRc, Long Sea Outfalls: Proceedings of a Conference, Telford, London, 1989.
45. A. N. Jha, T. H. H. Hutchinson, J. M. Mackay, B. M. Elliott and D. R. Dixon, *Mutat. Res., Genet. Toxicol. Environ. Mutagen.*, 1997, **391**, 179.
46. Anon, *Off. J.*, 2006, **L 136**, 3.
47. UNEP, The Stockholm Convention, 2001, Web: http://chm.pops.int/default.aspx.
48. UNEP, The Minamata Convention, 2014, Web: http://www.mercuryconvention.org/.
49. S. Beder, Getting into Deep Water: Sydney's Extended Ocean Sewage Outfall', in *A Herd of White Elephants: Australia's Science & Technology Policy*, ed. P. Scott, Hale and Iremonger, Sydney, 1992, pp. 62–74.
50. Anon. The Treaty of Rome, 1957, Web: http://europa.eu/legislation_summaries/institutional_affairs/treaties/treaties_eec_en.htm.
51. Anon, *Off. J.*, 1987, **L169**, 1.
52. Anon, *Off. J.*, 1997, **C340**, 10.
53. M. Bennett and P. James, Cost of waste at Zeneca, in *Eco Management Accounting*, ed. M. Bartolomeo, Springer Science & Business Media, 1999, pp. 236–239.
54. R. Critchley and D. Marshalsay, Progress towards a shared water resources strategy in the South East of England Report Phase 2B. WRSE 2013.
55. Anon, AstraZeneca Annual Reports.
56. S. J. Wishart, S. W. Mills and J. C. Elliott, *Water Environ. J.*, 2000, **14**, 284.
57. K. J. Ormerod and C. A. Scott, *Sci. Technol. Hum.Values*, 2013, **38**, 351.
58. C. P. L. Grady Jr., G. T. Daigger, N. G. Love and C. D. M. Filipe, *Biological Wastewater Treatment*, CRC Press, Boca Raton, 3rd edn, 2011, p. 1022.
59. D. G. Rao, R. Senthilkumar, J. A. Byrne and S. Feroz, *Wastewater Treatment: Advanced Processes and Technologies*, CRC Press, Boca Raton, 2012, p. 388.

60. S. Masuda, Y. Watanabe and M. Ishiguro, *Water Sci. Technol.*, 1991, **23**, 1355.
61. J. R. Harrison and G. T. Daigger, *J. – Water Pollut. Control Fed.*, 1987, **59**, 679.
62. E. Ardern and W. T. Lockett, *J. Soc. Chem. Ind., London*, 1914, **33**, 523.
63. J. K. Baars, *Bull. W. H. O.*, 1962, **26**, 465.
64. H. Kubota, Y. Hosono and K. Fujie, *J. Chem. Eng. Jpn.*, 1978, **11**, 319.
65. M. Pribyl, F. Tucek, P. A. Wilderer and J. Wanner, *Water Sci. Technol.*, 1997, **35**, 27.
66. A. Ghorpade, C. Cabral, M. Ekenberg, T. Welander and C. Johnson, Novel Treatment for Challenging Pharmaceutical Waste, in *Proceedings of the Water Environment Federation Conference: Microconstituents and Industrial Water Quality*, 2009, pp. 251–261.
67. H. Brix, *Water Sci. Technol.*, 1994, **30**, 209.
68. O. Legrini, E. Oliveros and A. M. Braun, *Chem. Rev.*, 1993, **93**, 671.
69. C. Pétrier and A. Francony, *Ultrason. Sonochem.*, 1997, **4**, 295.
70. P. Xu, G. M. Zeng, D. L. Huang, C. L. Feng, S. Hu, M. H. Zhaoa, C. Laia, Z. Weia, C. Huanga, G. X. Xie and Z. F. Liua, *Sci. Total Environ.*, 2012, **424**, 1.
71. P. T. Anastas and J. C. Warner, *Green Chemistry: Theory and Practice*, Oxford University Press, USA, 1998.
72. S. Jobling, M. Nolan, C. R. Tyler, G. Brighty and J. P. Sumpter, *Environ. Sci. Technol.*, 1998, **9**, 2498.
73. R. Owen and S. Jobling, *Nature*, 2012, **483**, 441.
74. R. Lofstedt, *J. Risk Res.*, 2014, **17**, 543.
75. N. W. Moore, *Bird Study*, 1965, **12**, 222.
76. J. Tafuri and J. Roberts, *Ann. Emergency Med.*, 1987, **16**, 193.
77. W. A. Virtue and J. W. Clayton, *Sci. Total Environ.*, 1997, **194**, 207.
78. USEPA, Toxic Release Inventory Website http://www2.epa.gov/toxics-release-inventory-tri-program.
79. EPER, European Pollutant Release & Transfer Register Website http://www.eea.europa.eu/data-and-maps/data/member-states-reporting-art-7-under-the-european-pollutant-release-and-transfer-register-e-prtr-regulation-8.
80. Anon, Effluent charging schemes in the EU Member States. EU DG Research Working Paper, 2001, ENVI 104 EN.
81. USEPA, Acid Rain Program - 2007 Progress Report, Clean Air Markets - Air & Radiation. USEPA. January 2009.
82. B. Ward and R. Dubois, *Only One Earth: The Care and Maintenance of a Small Planet*, Andre Deutsch, 1972, p. 301.
83. Anon, Water for life and livelihoods, UK Environment Agency, Bristol, 2013, p. 50.
84. Anon, Oxygen consuming substances in rivers (CSI 019), EU Water Information System for Europe, 2012.
85. Anon, Bathing water quality (CSI 022), EU Water Information System for Europe, 2012.

86. Anon, Hazardous substances in lakes, EU Water Information System for Europe, 2003.
87. I. D. Frijters and J. Leentvaar, *Rhine Case Study*, UNESCO Series Publication, 2003, p. 33.
88. A. Schulte-Wulwer-Leidig, From an Open Sewer to a Living Rhine River, in: Proceedings 12th International Rivers Symposium, Brisbane, 2009.
89. Anon, Trends in waterborne inputs – Assessment of riverine inputs and direct discharges of nutrients and selected hazardous substances to OSPAR maritime areas in 1990 – 2006, OSPAR Commission, Monitoring and Assessment Series, 2009.
90. Anon, State of the Lakes Ecosystem Conference (SOLEC), 1995.
91. R. J. Hesselberg and J. E. Gannon, Contaminant trends in Great Lakes fish, in *Our Living Resources*, ed. E. T. LaRoe, G. S. Farris, C. E. Puckett, P. D. Doran and M. J. Mac, Department of the Interior, National Biological Service, Washington, DC, 1995, pp. 242–244.
92. J. P. Hickey, S. A. Batterman and S. M. Chernyak, *Arch. Environ. Contam. Toxicol.*, 2006, **50**, 97.
93. W. Hansen, M. Christopher and M. Verbuecheln, *EU Waste Policies and Challenges for Local and Regional Authorities*, Ecologic, Berlin, 2002.
94. D. G. L. Larsson, C. de Pedro and N. Paxeus, *J. Hazard. Mater.*, 2007, **148**, 751.
95. D. Li, M. Yang, J. Hu, L. Ren, Y. Zhang and K. Li, Antibiotic Pollution from Chinese Drug Manufacturing, *Environ. Toxicol. Chem.*, 2008, **27**, 80.
96. GHS, Globally Harmonized System of Classification and Labelling of Chemicals Website http://www.unece.org/trans/danger/publi/ghs/ghs_welcome_e.html.
97. S. P. Johnson, *The Earth Summit*, International Bar Association Series, Kluwer Law International, 2001, p. 576.
98. I. Von Frantzius, *Environ. Polit.*, 2004, **13**, 467.
99. SAICM, Strategic Approach to International Chemicals Management Website. http://www.saicm.org/index.php?option=com_content&view=article&id=72&Itemid=474.
100. D. Kolpin, E. T. Furlong, M. T. Meyer, E. M. Thurman, S. D. Zaugg, L. B. Barber and H. T. Buxton, *Environ. Sci. Technol.*, 2002, **36**, 1202.
101. B. Roig, *Pharmaceuticals in the Environment – Current Knowledge and Needs Assessment to Reduce Pressure and Impact*, IWA Publishing, London, 2010, p. 256.
102. D. Taylor and T. Senac, *Chemosphere*, 2014, **115**, 95.
103. F. Hernández, M. Ibáñez, T. Portolés, M. I. Cervera, J. V. Sancho and F. J. López, *J. Hazard. Mater.*, 2015, **282**, 86.
104. A. Kortenkamp, *Environ. Health Perspect.*, 2007, **115**, 98.
105. R. M. Sharpe and N. E. Skakkebaek, *Pure Appl. Chem.*, 2003, 75, 2023.
106. D. R. Dietrich, S. V. Aulock, H. Marquardt, B. Blaauboer, W. Dekant, J. Kehrer, J. Hengstler, A. Collier, G. B. Gori, O. Pelkonen, F. Lang,

F. A. Barile, F. P. Nijkamp, K. Stemmer, A. Li, K. Savolainen, A. W. Hayes, N. Gooderham and A. Harvey, *Chem. Biol. Interact.*, 2013, **205**, A1.

107. A. Bergman, A. M. Andersson, G. Becher, M. van den Berg, B. Blumberg, P. Bjerregaard, C. G. Bornehag, R. Bornman, I. Brandt, J. V. Brian, S. C. Casey, P. Fowler, H. Frouin, L. C. Giudice, T. Iguchi, U. Hass, S. Jobling, A. Juul, K. A. Kidd, A. Kortenkamp *et al.*, *Environ. Health*, 2013, **12**, 68.

108. P. Grandjean and D. Ozonoff, *Environ. Health*, 2013, **12**, 70.

109. A. Westlund, Report on the protection of public health from endocrine disrupters, European Parliament Committee on the Environment, Public Health and Food Safety 2012, Report 2066.

110. European Commission White Paper Strategy for a future Chemicals Policy COM(2001)88, 2001.

111. H. W. Kroto, J. R. Heath, S. C. O'Brien, R. F. Curl and R. E. Smalley, *Nature*, 1985, **318**, 162.

112. Project on Emerging Nanotechnologies Website http://www. nanotechproject.org/cpi/browse/.

Subject Index